NEW MEXICO VEGETATION

PAST PRESENT AND FUTURE

NEW PAST
MEXICO PRESENT
VEGETATION AND FUTURE

William A. Dick-Peddie

with contributions by
W. H. Moir
and
Richard Spellenberg

University of

New Mexico Press

Albuquerque

Library of Congress Cataloging in
Publication Data

Dick-Peddie, William A.
New Mexico vegetation, past,
present, and future
William A. Dick-Peddie
with contributions by W. H. Moir
and Richard Spellenberg.
—1st ed.
          p.   cm.
Includes bibliographical references
and index.
ISBN 0-8263-1361-2
1. Botany—New Mexico.
2. Plant communities—New
   Mexico.
3. Botany—New Mexico—
   Ecology.
4. Phytogeography—New Mexico.
I. Moir, William H.
II. Spellenberg, Richard.
III. Title
QK 176.D53 1992
581.9789—dc20
92-3138
CIP

Designed by Emmy Ezzell.

CONTENTS

ACKNOWLEDGMENTS xi

INTRODUCTION xiii

**1 PHYSICAL ENVIRONMENT OF
NEW MEXICO** 1
Landscape Diversity 1
Landscape Development 1
Climate 4
Soils 6

**2 HISTORY OF NEW MEXICO
VEGETATION** 9
Reconstruction of Past Vegetation 9
Tertiary-Quaternary Vegetation 10
Pleistocene-Holocene Vegetation 15
Nineteenth and Early Twentieth
    Century Vegetation 18

3 VEGETATION AND ECOLOGY 27
Vegetation Patterns 27
Vegetation Dynamics 31
Vegetation Typing 32

**4 TERRESTRIAL VEGETATION
OF NEW MEXICO** 35
Classification and Mapping
    Vegetation 35
Classification of New Mexico
    Vegetation 35
Vegetation Classification for
    New Mexico 37
Current New Mexico Vegetation 37
New Mexico Vegetation Map 38
Use of Common and Scientific
    Plant Names 39

**5 ALPINE TUNDRA AND
CONIFEROUS FOREST** 47
ALPINE TUNDRA 47
Climate 47
Vegetation and Environment 48

CONIFEROUS FOREST 50
Subalpine Coniferous Forest 51
Upper Montane Coniferous Forest 58
Lower Montane Coniferous Forest 66

ASPEN DISTURBANCE
FOREST 70

**6 WOODLAND AND SAVANNA
VEGETATION** 85
WOODLAND 85
Coniferous Woodland 85
Coniferous and Mixed Woodland 87
Pinyon-Juniper Woodland 87
Mixed Woodland 90

SAVANNA 91
Juniper Savanna 91
Disturbance and Succession 91

7 GRASSLAND VEGETATION 101
Subalpine-Montane Grassland 101
Plains-Mesa Grassland 104
Desert Grassland 106

8 SCRUBLAND VEGETATION   123
    Montane Scrub   123
    Plains-Mesa Sand Scrub   128
    Great Basin Desert Scrub   129
    Chihuahuan Desert Scrub   131

9 RIPARIAN VEGETATION   145
    Concept of Riparian Vegetation   145
    Overview of Riparian Vegetation
      in New Mexico   147
    Alpine Riparian   147
    Montane Riparian   148
    Floodplain-Plains Riparian   151
    Arroyo Riparian   152
    Closed Basin–Playa–Alkali Sink
      Riparian   153

**10 VEGETATION OF SPECIAL
     HABITATS**   165
    Aquatic Vegetation   165
    Vegetation of Rocky Unweathered
      (shallow or no soil) Surfaces   168

**11 SPECIES OF SPECIAL
     CONCERN**   179

    Nature of the Problem   179
    Why Save Species   181
    Present Action   183
    Process of Listing a Plant Species   185
    Features of New Mexico Leading to
      Endemism   187
    Endemism in New Mexico   188
    Taxonomy and Plant Species in
      New Mexico   190
    Nature of Threats to Species
      of Concern   191
    Examples of Species of Special
      Concern in New Mexico   192
    Review of Sources of Threats   193

**12 THE FUTURE OF
     NEW MEXICO VEGETATION**   225
    Need for Preservation   225
    Concept of Natural Area   226
    Natural Areas of New Mexico   228
    Projections and Proposals   228

INDEXES
    Subject Index   237
    Index of Selected Key Plants   241

FIGURES

1.1 Physiographic regions in New Mexico 2

1.2 Ancient land-water boundaries in New Mexico 3

1.3 Number of days per year without killing frost in New Mexico 4

1.4 Average annual precipitation in New Mexico 5

1.5 Soils of New Mexico 7

2.1 Primeval vegetation types in New Mexico 11

2.2 Repeat photography of areas in New Mexico 12

2.3 Tertiary geofloras and current derivatives in North America 14

2.4 New Mexico vegetation patterns 18,000 years ago 17

3.1 Typical effects of latitude, altitude, and exposure on vegetation 28

3.2 Typical soil moisture and vegetation changes following disturbance in northern New Mexico 30

5.1 Distribution of Tundra and Forest Vegetation in New Mexico 51 & 58

6.1 Distribution of Woodland and Savanna Vegetation in New Mexico 86

7.1 Distribution of Grassland Vegetation in New Mexico 102 & 103

8.1 Distribution of Scrubland Vegetation 124 & 125

8.2 Common occurrences of major shrubs of Montane Scrub Vegetation 126

9.1 Elevation zones of greatest dominance of montane riparian trees and shrubs 149

9.2 Distribution of Closed Basin Riparian Vegetation in New Mexico 153

11.1 Distribution of federal threatened and endangered plant species in New Mexico 188

TABLES

2.1 New Mexico studies containing paleo-vegetation information 21

2.2 Nineteenth-century accounts containing information on New Mexico vegetation 22

4.1 Classification of terrestrial vegetation of New Mexico 39

4.2 Synonomy of selected plants that are important members of the vegetation of New Mexico 42

5.1 Major Tundra and Forest Vegetation types in New Mexico 73

5.2 Major plants comprising Tundra Vegetation in New Mexico 75

5.3 Major plants comprising Forest Vegetation in New Mexico 76

6.1 Woodland and Savanna Vegetation types in New Mexico 93

6.2 Major plants comprising Woodland and Savanna Vegetation in New Mexico 94

7.1 Grassland Vegetation types in New Mexico 110

7.2 Major plants comprising Subalpine-Montane Grassland in New Mexico 112

7.3 Major plants comprising Plains-Mesa Grassland in New Mexico 113

7.4 Major plants comprising Desert Grassland in New Mexico 116

8.1 Scrubland Vegetation types in New Mexico 132

8.2 Major plants comprising Montane Scrub Vegetation in New Mexico 134

8.3 Major tree-shrubs and shrubs of the Montane Scrub, by elevational zone 136

8.4 Major plants comprising Plains-Mesa Sand Scrub Vegetation in New Mexico 137

8.5 Major plants comprising Great Basin Desert Scrub Vegetation 139

8.6 Major plants comprising Chihuahuan Desert Scrub Vegetation 140

9.1 Major obligate riparian and semiriparian plants in New Mexico 154

9.2 Riparian Vegetation types in New Mexico 159

10.1 Major floating and submersed plants comprising aquatic vegetation in New Mexico 170

10.2 Major attached plants comprising aquatic vegetation in New Mexico 171

10.3 Major plants comprising vegetation of the Grants and Carrizozo Malpaises in New Mexico 173

10.4 Major plants comprising early successional vegetation of exposed rock sites in New Mexico 174

10.5 Major plants of gypsum or strongly gypseous soils 175

11.1 Distribution of rare, endemic, and threatened or endangered plants in New Mexico by geographic region and county 209

11.2 Distribution of rare, endemic, and threatened or endangered plants in New Mexico by number of named entities 210

11.3 Plant species that are endemic, rare, or in decline in New Mexico 211

11.4 Nature of endemism in New Mexico plant species 222

12.1 Natural and partial natural areas of New Mexico 230

12.2 Vegetation found in some natural and partial natural areas of New Mexico 234

PLATES

Color plates appear immediately following page xv.

1   Response of vegetation to elevation.   xv
2   Response of vegetation to exposure.   xv
3   Response of vegetation to exposure.   xvi
4   Canyon Effect on vegetation.   xvi
5   Early succession.   xvii
6   Late succession.   xvii
7   Alpine Tundra vegetation above timberline.   xviii
8   Alpine Tundra fellfield community.   xviii
9   Spruce-Fir Forest.   xix
10   Subalpine-Montane Coniferous Forest.   xix
11   Douglas-fir-White Fir Forest.   xx
12.   Ponderosa Pine Forest.   xx
13   Pinyon-Juniper Woodland.   xxi
14   Juniper Savanna.   xxi
15   Subalpine Grassland.   xxii
16   Montane Grassland.   xxiii
17   Plains Grassland.   xxiii
18   Tobosa Swale.   xxiv
19   Desert Grassland.   xxiv
20   Montane Scrub.   xxv
21   Plains Sand Scrub.   xxvi
22   Mesa Sand Scrub.   xxvi
23   Great Basin Desert Scrub.   xxvii
24   Chihuahuan Desert Scrub.   xxvii
25   Montane Riparian.   xxviii
26   Floodplain Riparian.   xxix
27   Floodplain Riparian.   xxix
28   Arroyo Riparian.   xxx
29   Closed Basin Riparian.   xxx

ACKNOWLEDGMENTS

I view this book as the culmination of my interests and work over a thirty-four year period at New Mexico State University as a professor of plant ecology in the Biology Department.

My primary interests over the years were in vegetation changes (both slow, largely prehistoric and relatively rapid, historic) and possible causal factors initiating these changes. In this book I have attempted to delineate, characterize, and locate the major vegetation types of New Mexico and have proposed the relative stability of the boundaries and the species composition of these types.

Reference material for this book is included up to 1990.

I thank W. H. Moir and R. W. Spellenberg for authoring Chapters 5 and 11 respectively; G. C. Cunningham, R. D. Pieper, and G. B. Donart for their draft reviews of Chapters 7 and 8; J. P. Hubbard for supplying major material for Chapter 10; and W. Dunmire for updated material for Chapter 12. I thank G. L. Cunningham, R. A. Fletcher, W. H. Moir, E. Muldavin, R. D. Pieper, R. Spellenberg, and R. C. Szaro for use of their slides and prints for some of the colored plates in the book.

W. A. (Sandy) Dick-Peddie

INTRODUCTION

Much information is available in the United States concerning plant species in regional, state, and local manuals and floras, which often include keys and/or classifications. Many cover limited life-forms such as trees or wild flowers. Others cover species found in special habitats like aquatic plants or desert plants. The material ranges from highly technical accounts such as *Flora of Texas* (Correll and Johnston, 1970) and *Flora of New Mexico* (Martin and Hutchins, 1980–81) to a nontechnical manual like *Desert Tree Finder* (Watts and Watts, 1980).

Less information is available about groupings of different plant species occurring together in a region or on a site. These aggregations are the result of biotic events such as evolution and migration and abiotic events such as continental drift, vulcanism, mountain building, glaciation, erosion, floods, fires, and human activities. Clusters of plant species often tend to be repeated on the landscape. This makes it possible to group these combinations, give them names, and attempt to order (classify) the units in some useful way. The plant species present in an area constitute the area's vegetation.

In today's complex society, people have found it increasingly desirable to attempt to manage their environment. Also, ecologists are increas-ingly interested in attempting to understand the dynamics of aggregations of organisms sharing common geographic locations. How do these systems maintain stability? How do they respond to various disturbances? The desire to understand ecosystems and to manage the environment results in a need to know more about vegetation. Some publications on state plant communities reflect this increased interest in vegetation. Two such books are *Terrestrial Vegetation of California* (Barbour and Major, 1977) and *Natural Vegetation of Oregon and Washington* (Franklin and Dyrness, 1973).

It is hoped that this book on the vegetation of New Mexico will be both interesting and useful. Along with descriptive inventory, the dimension of time has been added to the treatment; hence the title, *Vegetation of New Mexico: Past, Present, and Future.* It is intended that the book's scope and style will appeal to a wide audience— from the scientist to the resource manager, from the environmentalist to the interested citizen. Each section in the book is followed by references, which not only include literature cited in the sections but, in addition, include references that may be of general interest because they deal primarily with vegetation features in New Mexico.

INTRODUCTION REFERENCES

Barbour, M. G. and J. Major (eds.), 1977. Terrestrial vegetation of California. John Wiley and Sons, New York.

Correll, D. S. and M. C. Johnston, 1970. Manual of vascular plants of Texas. Texas Research Foundation, Renner, Texas.

Franklin, J. F. and C. T. Dyrness, 1973. Natural vegetation of Oregon and Washington. U.S. Department of Agriculture, Forest Service, Genereal Technical Report PNW-8, Pacific Northwest Forest and Range Experiment Station, Portland, Oregon.

Martin, W. C. and C. R. Hutchins, 1980–81. A flora of New Mexico. J. Cramer, Vaduz, West Germany. 2 vols.

Watts, M. T. and T. Watts, 1980. Desert Tree Finder. Nature Study Guild, Berkeley, California.

New Mexico Counties

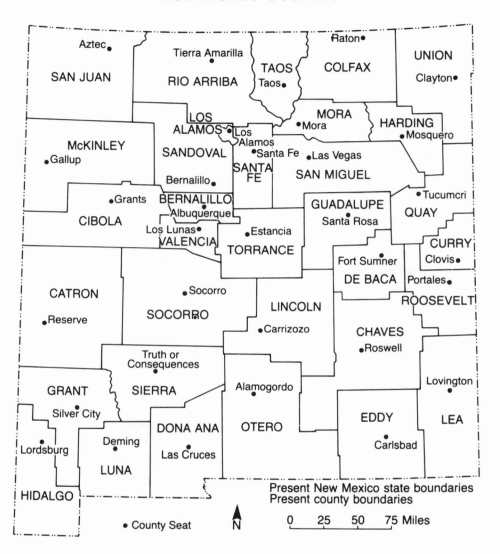

Present New Mexico state boundaries
Present county boundaries

0   25   50   75 Miles

• County Seat

N

NEW MEXICO VEGETATION

PAST PRESENT AND FUTURE

Plate 1. Response of vegetation to elevation. Lower elevation (foreground) with Ponderosa Pine. Higher elevations, (background) with Douglas-fir and White Fir. Sierra County.

Plate 2. Response of vegetation to exposure. South facing slope (left) with Pinyon-Juniper woodland and north facing slope with Douglas-fir and White Fir forest. Catron County.

Plate 3. Response of vegetation to exposure. South facing slope (left) with Chihuahuan Desert Scrub and North facing slope covered by Desert Grassland. Socorro County.

Plate 4. Canyon effect on vegetation. Ponderosa Pine forest (center) at lower elevation than surrounding Pinyon-Juniper woodland. Catron County.

Plate 5. Early
succession. Douglas-fir
growing under mature
Ponderosa Pine stand.
Lincoln County.

Plate 6. Late succession.
Spruce growing through
mature Aspen stand. Taos
County.

Plate 7. Alpine Tundra vegetation above timberline. Taos County.

Plate 8. Alpine Tundra fellfield community. Santa Fe County.

Plate 9. Spruce-Fir Forest. Rio Arriba County.

Plate 10. Subalpine-Montane Coniferous forest. Otero-Lincoln Counties.

Plate 11. Douglas-fir-White Fir Forest. Lincoln County.

Plate 12. Ponderosa Pine Forest. Catron County.

Plate 13. Pinyon-Juniper Woodland. Lincoln County.

Plate 14. Juniper Savanna. Otero County.

Plate 15. Subalpine Grassland. Thurber Fescue
Meadow. Lincoln County.

Plate 16. Montane Grassland Mountain Muhly Meadow. Lincoln County.

Plate 17. Plains Grassland. Blue Grama grassland. DeBaca County.

Plate 18. Tobosa Swale. Doña Ana County.

Plate 19. Desert Grassland. Black Grama with scattered shrubs. Otero County.

Plate 20. Montane Scrub, Grant County.

Plate 21. Plains Sand Scrub. Sand Sagebrush and Shinoak. Eddy County.

Plate 22. Mesa Sand Scrub. Dominant shurb is Indigobush. Socorro County.

Plate 23. Great Basin Desert Scrub. Shadescale and Fourwing Saltbush with Greasewood in the gullies. San Juan County.

Plate 24. Chihuahuan Desert Scrub. Creosotebush and Tarbush. Doña Ana County.

Plate 25. Montane Riparian. Arizona Alder community. Grant County.

Plate 26. (Facing page, top) Floodplain Riparian. Freemont Cottonwood gallery forest. Rio Arriba County.

Plate 27. (Facing page, bottom) Floodplain Riparian. Understory of New Mexican Olive in mature Fremont Cottonwood gallery forest. Rio Arriba County.

Plate 28. Arroyo
Riparian. Cutleaf
Bricklebush. Littleleaf
Sumac, and Desert
Willow. Doña Ana
County.

Plate 29. Closed Basin
Riparian. Dominant shrub
is Fourwing Saltbush.
Hidalgo County.

1

PHYSICAL ENVIRONMENT
OF NEW MEXICO

LANDSCAPE DIVERSITY

The New Mexico landscape is one of the most diverse in the United States. Much of it is the result of old and new volcanic activity. There are mountains with high peaks, such as Mount Taylor near Grants and Sierra Blanca near Ruidoso, which are isolated volcanic masses. Spectacular features such as Shiprock and Cabezon Peak in northwestern New Mexico are cores of ancient volcanos (volcanic necks). Capulin Mountain, east of Raton, is a recent cinder cone. Fresh black basalt lava flows cover many square miles near Grants and Carrizozo. The Jemez Mountains near Los Alamos contain a giant crater, Valle Grande, which at sixteen miles in diameter is one of the world's largest calderas. Nonvolcanic formations include the highest peaks in New Mexico, found on the fault block mountain mass Sangre de Cristo. Other fault block masses are the Manzano, Sandia, San Andres, and Sacramento mountains.

The state has numerous closed basins, such as the Tularosa Basin near Alamogordo and the Jornada near Las Cruces. There are plateaus and escarpments. The Trinidad and Canadian escarp-

ments in the northern part of the state contain extensive erosion canyons. East-central New Mexico is the western portion of the vast Llano Estacado Plateau, which contains some of the flattest surface in the world. There are three major river systems in the state containing broad floodplains. These are the San Juan, Pecos, and Rio Grande. Centuries of erosion on deep alluvial deposits has resulted in the beautiful color-banded palisades near Gallup and the Bisti Badlands near Farmington. Pleistocene glaciation has also left its mark, as on Sierra Blanca peak near Ruidoso, where a cirque attests to the peak being the southernmost glaciated peak in North America.

The complicated development of New Mexico's landscape has resulted in conflicting maps of its physiography. Figure 1.1 is a physiographic map of the regions found in New Mexico, part of a physiographic map of the conterminous United States made by Brown and Kerr (1979). This map was chosen because Brown and Kerr used vegetation as an aid in determining region boundaries.

LANDSCAPE DEVELOPMENT

The portion of the continent now called New Mexico has been largely or entirely under the sea for all but 60 million of the last 480 million years. In the late Cambrian, 480 million years before present (mybp), a small island existed in what is now north-central New Mexico. This island had sunk below the sea by late Ordovician (430 mybp),

and no land was above water for the next 70 million years, or until Devonian time. In the late Devonian (360 mybp), the same island area appeared, only to disappear under the water again by middle Mississippian times (330 mybp). During the Pennsylvanian period (310–265 mybp), there were five land areas (Fig. 1.2A), but by

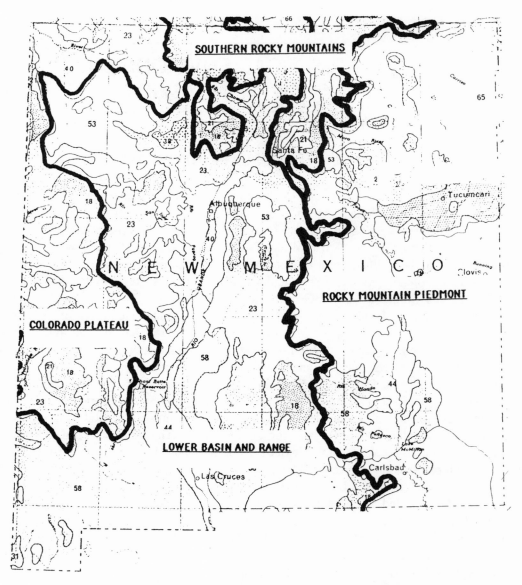

Figure 1.1 Physiographic Regions of New Mexico. Modified from *Physiographic Regions of the United States,* 1979. K. Brown and R. Kerr. U.S. Dept. of Interior, Bureau of Land Management, Albuquerque, N.M.

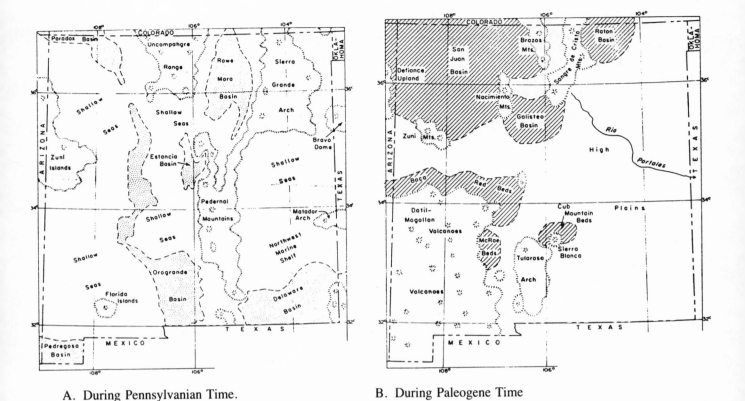

A. During Pennsylvanian Time.

B. During Paleogene Time

Figure 1.2 Ancient Land-Water Boundaries in New Mexico. Modified from *Mosaic of New Mexico's Scenery, Rocks, and History,* 1967. P. W. Christiansen and F. Kottlowski. State Bureau of Mines and Mineral Resources, Socorro, N.M.

the middle Cretaceous (100 mybp), the entire area of present New Mexico was again under the sea.

It was not until the early Cenozoic (60 mybp) that the land rose and the sea retreated for the last time. Figure 1.2B is a map depicting the New Mexico area 50 million years ago. Approximately 49 mybp, extensive geological turbulence began, including volcanic activity, mountain building, and crustal sinking (graben forming). These activities persisted for 15 to 20 million years; toward the end of this period many of the

major mountain ranges found today were formed. The last landscape features were formed less than 1 million years ago, during the Pleistocene. Taking place during this period were glacial deposits, scouring of the Sierra Blanca cirque, formation of post-Pleistocene lake beds such as the San Agustin Plains, extrusion of the Carrizozo and Grants malpais, final explosions of Valle Grande, formation of the White Sands gypsum dunes, and severe erosion of plateau areas into canyon topography.

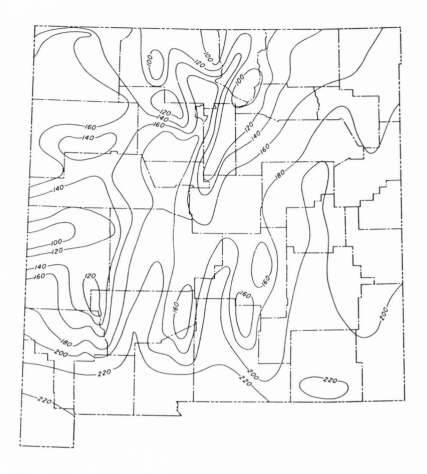

Figure 1.3 Number of Days per Year Without Killing Frost in New Mexico. Modified from *Historical Atlas of New Mexico,* 1976. W. Beck and Y. Haase. University of Oklahoma Press. Norman.

CLIMATE

New Mexico spans approximately five degrees of longitude and three degrees of latitude. This considerable geographic spread gives the state many different microclimates. As a result, New Mexico has many and varied vegetation types. Temperature regimes are very different around the state. The growing season (frost-free period) ranges from 148 to 220 days (Fig. 1.3). The maximum average number of days without killing frost is found in the Deming-Lordsburg area of southwestern New Mexico, along the lower end of the Rio Grande Valley, and in a small area in the Pecos River valley south of Carlsbad. The

minimum average number of days without killing frost is 100, around Taos, Tres Piedras, and Gavilan. Climate means convey little information in New Mexico. For example, the mean annual temperature of the state is about 11.5°C (53°F), yet it varies as much as 14.4°C (26°F) at different stations. The variations are caused more by altitude than latitude. The temperature variation northward is from .83°to 1.4°C (1.5° to 2.5°F) for every degree of latitude. However, for every 300 m (1,000 ft.) of elevation, the temperature gradient is 2.8°C (5°F). This effect of altitude upon temperature is readily noted by comparing

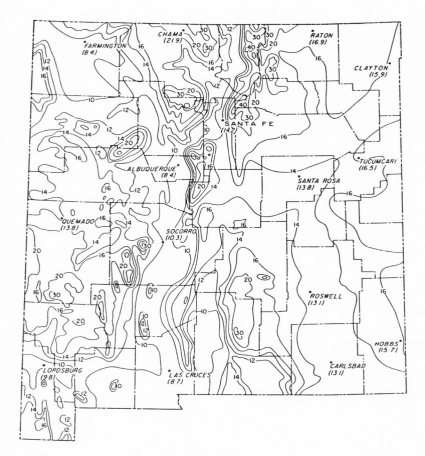

Figure 1.4 Average Annual Precipitation in New Mexico. Modified from *Historical Atlas of New Mexico,* 1976. W. Beck and Y. Haase. University of Oklahoma Press, Norman.

the July average of 20.7°C (69.2°F) in Santa Fe (elevation 2,030 m or 6,696 ft.) with that of Santa Rosa (elevation 1,400 m or 4,616 ft.), which has a July average of 25.7°C (77.5°F). Deming (elevation 1,443 m or 4,331 ft.) records a July average of 26.8°C (80.2°F), whereas Silver City (elevation 1,867 m or 5,595 ft.), only a short distance away, has an average of 23.8°C (74.9°F) for the same month.

The average annual rainfall in New Mexico is 380 mm (15 in.). Normally, precipitation increases with elevation, particularly along the Rocky Mountains. A high of 1,000 mm (40 in.) of rainfall is reached in the upper parts of the Jemez and Sangre de Cristo mountains. The drier por-

tions of New Mexico, which receive only 200 to 250 mm (8 to 10 in.) of rainfall a year, are the basin areas of the San Juan and Chaco rivers, the Rio Grande Valley from Española south, the southern part of the Tularosa Valley, and the area known as the Jornada del Muerto. At any given elevation, precipitation is generally greatest in the eastern third of the state and least in the western portion (Fig. 1.4), since the source of most New Mexico summer moisture is the Gulf of Mexico; the farther one goes from that body, the less precipitation.

Most of the precipitation in the state comes in the form of local high-intensity summer storms of short duration. Such storms occur when warm, moist Gulf air moves inland and becomes unstable over the hot terrain, causing thundershowers. These usually occur in late afternoon and are especially common on the southern slopes of the Rocky Mountains. Precipitation can be quite effective in much of the state because it occurs during the growing season (April to October). In the drier areas of the state, precipitation is erratic. Season, quantity, frequency, and intensity of precipitation typically vary greatly from place to place and from year to year. The natural vegetation has, of course, evolved under these conditions, and the erratic precipitation pattern is the predictable norm in these areas of New Mexico.

SOILS

New Mexico's turbulent physiographic history has resulted in surface outcrops of many kinds. Mountain building, graben formation, vulcanism, and erosion have placed varied rocks and minerals at the surface, so that subsequent weathering of these surface materials has resulted in soils of many kinds. These soils can have a direct effect on vegetation, such as the presence of toxic or beneficial chemicals, and indirect effects, such as modification of the amount of water available to plants. There are assemblages of plants peculiar to gypsum soils, limestone surfaces, soils with high selenium, and saline soils. Soils may be considerably more varied than the parent materials themselves for reasons well detailed in the following excerpt from *Soils of New Mexico* (Maker, et al., 1978).

The kind of soil that develops in any area is the result of the interaction of five soil-forming factors: climate, vegetation, parent material, topography, and time. The first two are called "active" factors because they act on the soil parent material as conditioned by topography over varying periods of time.
Climate and vegetation frequently are considered together because climate is the major determinant of vegetation. Soils of the high mountains of northern New Mexico are commonly leached, well-developed, and acid because precipitation is relatively high, temperatures are low and the dominant vegetation is coniferous trees which are best suited to these climatic and soil conditions. On the other hand, grasses and desert shrubs are dominant in the hot dry desertic region. Here, the soils are not leached, are less developed, and are neutral or alkaline. Whether the soil determines the kind of vegetation or whether vegetation determines the kinds of soil is a much-debated point, but climate certainly is the deciding factor for both.

Three general categories of soil based upon color are found in New Mexico (Maker et al., 1978). These are (1) dark and moderately dark-colored soils, (2) moderately dark-colored soils, and (3) light-colored soils. Each of these soil color areas is subdivided into two geographic regions: High Plains Region and Mountainous Region (soil color category 1); Central Plains Region and Plateau Region (soil color 2); and Cool Desertic Region and Warm Desertic Region (soil color 3). Figure 1.5 is a map of the soil regions found in New Mexico including the size (acreage) of each region and the number of soil associations (from official soil taxonomy adopted by the U.S. Department of Agriculture in 1965) found in each region. The 110 soil associations found in New Mexico attest to the diversity of the state's geology, topography, climate, and vegetation.

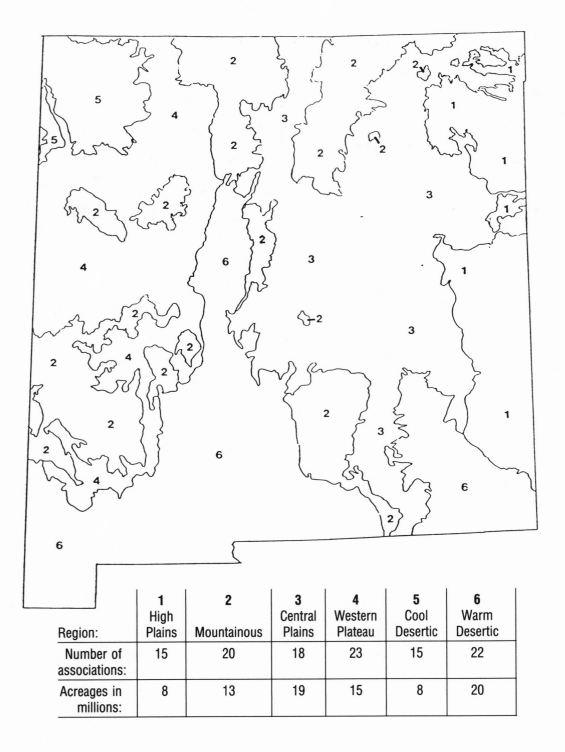

Region:	**1** High Plains	**2** Mountainous	**3** Central Plains	**4** Western Plateau	**5** Cool Desertic	**6** Warm Desertic
Number of associations:	15	20	18	23	15	22
Acreages in millions:	8	13	19	15	8	20

Figure 1.5 Soils of New Mexico. Modified from *Soils of New Mexico*, 1974. H. Maker, H. Dregne, V. Link, and J. Anderson. New Mexico State University, Las Cruces.

CHAPTER 1 REFERENCES

PHYSICAL ENVIRONMENT OF
NEW MEXICO

Beck, A. and Y. D. Haase, 1976. Historical altas of New Mexico. University of Oklahoma Press, Norman, Oklahoma.

Brown, K. F. and R. M. Kerr, 1979. Physiographic regions of the United States (map). U.S. Department of Interior Bureau of Land Management, Albuquerque, New Mexico.

Christiansen, P. W. and F. E. Kottlowski, 1967. Mosaic of New Mexico's scenery, rocks, and history. Bureau of Mines and Mineral Resources, New Mexico Institute of Mining and Technology, Socorro.

Maker, H. J., H. E. Dregne, V. G. Link, and J. U. Anderson, 1978. Soils of New Mexico. Agricultural Experiment Station, Research Paper 285, New Mexico State University, Las Cruces.

2

HISTORY OF
NEW MEXICO VEGETATION

RECONSTRUCTION OF PAST VEGETATION

Research into the composition, extent, and dynamics of vegetation that existed millions (or even thousands) of years ago draws upon evidence from various fields. For many years, fossils found in rocks were the only form of evidence thought to be of value. These records of bits and pieces of plant material (macrofossils) could be assigned an age through the use of various dating methods. If enough fossils having the same age were found from a given region, a picture of the vegetation of the time could be obtained. For plants or plant parts to become fossilized, though, they had to have been trapped in a condition where decay was prevented or greatly delayed. The chance that a plant or plant part would end up in such a situation is limited; consequently, the number of such fossils is relatively small and, of those fossilized, only a fraction are ever likely to be found.

One portion of plants commonly fossilized is pollen. Plant pollen (microfossils) is now considered to be more useful for early (1 mybp) vegetation reconstruction than macrofossils. Pollen grains have an almost airtight, and tough outer shell, are produced in great quantities, and are relatively light. As a consequence, pollen grains can easily be transported by wind or water and are more likely to end up in situations leading to fossilization than are other plant parts. Pollen grains have surprisingly varied morphologies unique to specific taxa and can be used as a reliable plant identification feature. The large quantities of fossil pollen also allow estimations of past relative densities or importance of the species of a past vegetation. Carefully extracted layers of pollen in peat bogs, sedimentary lake or playa beds, and even early human sites yield sequences of older (deeper) to younger (shallower) fossil pollen. From this record, sequences of vegetation in an area can be inferred. Some of the oldest continuous fossil pollen records known are found in New Mexico on the San Agustin Plains (Clisby and Sears, 1956; Clisby, Foreman, and Sears, 1957; Potter and Rowley, 1960). Other evidence of past vegetation patterns may also be obtained from archaeological sites. For example, the presence of bones from grassland animals, rather than from forest or desert animals, implies the existence of grassland vegetation in a region (Vankat, 1979).

An interesting recent avenue of research in reconstruction of past vegetation has been the careful excavation and dating of fossilized animal nest material (midden). Some species of animals in arid and semiarid regions use the same nest sites for thousands of years. The slow decomposition inherent under dry (xeric) conditions plus the preservative nature of crystallized urine (amberat), results in almost perfectly preserved plant material thousands of years old. The middens most commonly used for study in New Mexico and vicinity are those of the wood or packrat (*Neotoma*). The most useful middens are located in miniature caves or crevices in escarpments or other rocky outcrops. It can be assumed that macrofossil midden plant material is from plants growing near the midden site. Through carbon dating, some packrat midden plant material has been found to be more that 40,000 years old (Van Devender, 1980). An improved dating technique called the tandem accelerator mass spectrometer method (TAMS) has both increased the precision of dating and increased the kinds of plant and animal material that can be effectively dated (Van Devender et al., 1985).

Table 2.1 lists major studies that provide information on vegetation patterns in New Mexico from 25,000 to 400 years ago. Some of these studies obtained pollen profiles from early human sites, while others constructed profiles from midden material.

Diaries from early explorations, military missions, and surveys for proposed railroads can be useful in vegetation reconstruction of the nineteenth century in New Mexico. Some of the early accounts that contain information about New Mexico vegetation are given in Table 2.2, which is modified from Leopold (1951). Paul B. Sears (1921) found that century-old vegetation patterns in Ohio could be reconstructed through the analysis of data from the field records of the United States Territorial Survey. This technique has subsequently been used to reconstruct presettlement vegetation patterns for much of the eastern United States, as well as for most of the Midwest. Territorial survey records have been used to recon-struct vegetation patterns in New Mexico by Buffington and Herbel (1965), York and Dick-Peddie (1969), and Gross and Dick-Peddie (1979). Figure 2.1 is a map of primeval vegetation in New Mexico (Gross and Dick-Peddie, 1979). The word *primeval* was chosen because in New Mexico, unlike farther east, the survey was not conducted prior to settlement, nor was it all-inclusive as it had been in the East and Midwest, because land grants and tribal lands were not surveyed. These lands were extensive and, as a consequence, it was necessary to interpolate vegetation boundaries through them.

A technique used to chart changes in vegetation over recent periods is repeat photography. If an old photograph can be found that was taken of, or at least includes, some vegetation, and the location and date are known, it is often possible to take a photograph from the exact spot and compare the vegetation included in the two pictures. Sometimes files of federal or state agencies will contain photographs taken of similar areas at different times (say 30 to 100 years apart) and can be used for this purpose. An entire book, *The Changing Mile* by Hastings and Turner (1965), was written using repeat photography on the Sonoran Desert vegetation in Arizona. Sallach (1986) used this technique for vegetation in New Mexico (Fig. 2.2A and B).

TERTIARY-QUATERNARY VEGETATION

Toward the end of Cretaceous time (70 mybp), the continental sea that had covered most of New Mexico had receded. Also, the newly evolved (100 mybp) angiosperms were rapidly replacing gymnosperms, and vegetation began to reoccupy areas covered by the sea. The history of New Mexico vegetation for a period of approximately 65 million years, from the beginning of the Tertiary to this century, shows that vegetation patterns have been the result of three factors: relief, climate, and human impact. Early (65 million to 30 thousand ybp) vegetation patterns in New Mexico can be understood more easily within the context of continental and regional events. The following continental overview is based on an excellent book by John L. Vankat, *The Natural Vegetation of North America* (1979).

At the beginning of the Tertiary, the North American continent was greatly different from what it is today. It was relatively flat, and the

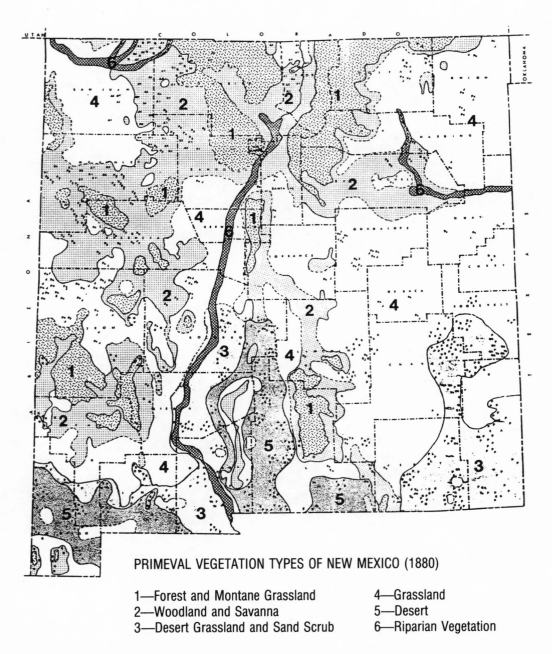

PRIMEVAL VEGETATION TYPES OF NEW MEXICO (1880)

1—Forest and Montane Grassland   4—Grassland
2—Woodland and Savanna   5—Desert
3—Desert Grassland and Sand Scrub 6—Riparian Vegetation

Figure 2.1 Primeval Vegetation Types in New Mexico. Modified from A Map of Primeval Vegetation in New Mexico, 1979. F. A. Gross and W. A. Dick-Peddie. The Southwestern Naturalist, Austin, Texas.

1946 photo

1916 photo

1986 photo

1986 photo

A. Area west of Jicarilla Reservation in northwestern New Mexico. Top photo taken in 1946, by R. King. Note grassland on the plain with some big sagebrush at far left. Bottom photo taken 40 years later, by B. Sallach. Sagebrush covers entire plain.

B. Area in north-central New Mexico. Top photo taken in 1916 at Pecos National Monument. Note scattered juniper forming a savanna. Bottom photo, taken in 1986 by B. Sallach, shows a closed stand of juniper constituting juniper scrubland.

Figure 2.2 Repeat Photography of Areas in New Mexico. Negatives courtesy of New Mexico Museum of Natural History, Albuquerque.

Appalachian and the newly forming Rocky mountains were low in elevation. The Sierra Nevada, Cascade, and Coast ranges had not yet appeared. The climate has been characterized as equible, that is, warm with ample precipitation, over the entire continent. Even though data are limited, it appears that the relatively uniform environment resulted in less diverse vegetation. Three geofloras have been described for this period: Arcto-Tertiary, Neotropical-Tertiary, and Madro-Tertiary (Fig. 2.3A).

The Arcto-Tertiary geoflora covered the northern half of the continent. It was a forest rich in species, a mixture of gymnosperms and deciduous angiosperms. This geoflora had three major elements: (1) species that today are characteristic of eastern North America, including hickories, beech, elms, maples, and ashes; (2) species more common today in western North America, such as firs, cedars, spruces, sequoias, and pines; (3) east Asian species such as tree of heaven, ginkgo, and dawn redwood. The Neotropical-Tertiary geoflora covered most of the southern half of North America. This geoflora was composed of rich forests similar to the tropical and subtropical forests of southern Mexico and Panama today. The Madro-Tertiary geoflora, which had evolved from the other two geofloras, was relatively young and restricted to dry areas in north-central Mexico, southwestern New Mexico, and southeastern Arizona. This geoflors consisted of a variety of types, including woodland and scrubland.

At the beginning of the Tertiary Period, a flat North America had a mild climate that supported three general types of vegetation. Western North American environments became more varied between early and middle Tertiary, due to the increased relief caused by the continued rise of the Rocky Mountains. Cooler temperatures were accompanied by drier conditions in the rain shadow east of the Rocky Mountains. These environmental changes produced changes in the vegetation. The Arcto-Tertiary geoflora extended southward as the climate cooled. Many deciduous angiosperms were lost due to the drier winters. The Neotropical-Tertiary geoflora could survive only by moving southward and coastward. The Madro-Tertiary, on the other hand, made great gains during these times, expanding into much of the drier region. A new type of vegetation, the grasslands, developed in the Rocky Mountains rain shadow and spread east, northeast, and southeast. Tundra vegetation occupied the far north and the highest mountains.

During late Tertiary times, the uplift of the Sierra Nevada, Cascades, and Coast ranges proceeded at a geologically rapid pace. The entire continent continued to become cooler and drier. There were now two tall western mountain chains producing rain shadows to their east. Desert vegetation made its appearance during this time and occupied the rain-shadow lowlands east of the Sierra Nevadas and dry areas in eastern Mexico. Figure 2.3B shows the location of derivatives of the Tertiary geofloras in North America. Note that New Mexico has one of the most complicated ancestral vegetation patterns in North America.

Supplementing this continental view is the following regional picture from Tidwell (1972).

According to Axelrod (1950) the upper Cretaceous floras of western North America provide evidence of four broad climatic provinces, none of which can be said to indicate low rainfall. In the central Rocky Mountain region and extending westward into the area of the present Great Basin were forests requiring a warm-temperate to subtropical climate. Northward climates were more temperate while southwardly the climate was warmer with some evidence of a tropical savanna climate in the southern Great Basin. The proximity of most of these floras to the Cretaceous seas doubtlessly minimized seasonal temperature variations.

The climatic history of the Rocky Mountains during this period and extending into the Quaternary was dominated by three trends: (1) a general, often interrupted cooling throughout the Tertiary and Quaternary; (2) a progressive drying

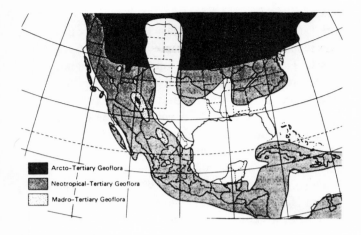

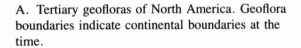

A. Tertiary geofloras of North America. Geoflora boundaries indicate continental boundaries at the time.

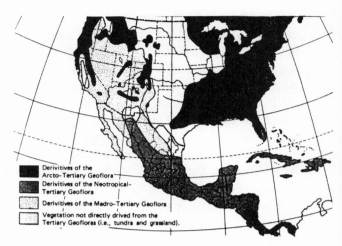

B. Current distribution of derivatives of Tertiary geofloras.

Figure 2.3 Tertiary Geofloras and Current Derivatives in North America. Modified from The Natural Vegetation of North America, 1979. J. L. Vankat. John Wiley and Sons, New York.

from the early Miocene into the Pleistocene because of the increasing elevation of the Sierra Nevada and the Cascades; and (3) Pleistocene fluctuations between glacial-pluvial and warm-dry ages (Antevs, 1952).

In conjunction with these trends, the vegetational patterns of the Tertiary were marked by extensive migrations of plant groups in response to these climatic changes. These migrations were basically a withdrawal of tropical and semi-tropical vegetation (Neotropical Geoflora) toward the Equator as the climate of the continents grew cooler and drier in response to the rising mountain ranges. Most genera of angiosperms (at least the woody ones) apparently evolved during the Cretaceous, with evolutionary processes during the Tertiary concentrating on a proliferation of taxa at the species level.

In the early Tertiary, during the Paleocene epoch (65–55 mybp), there were climatic zones in the western United States apparently based upon latitude. According to Dix (1962) "In the western United States, Mexico and Central America, the luxuriant Neotropical-Tertiary flora, containing such plants as *Ficus* and *Persea,* occupied an area as far north as latitude 55 in western North America."

Temperate genera such as ginkgo, birch, hazel, and maple were found no farther south than the northern tier of western states. Southern Wyoming had breadfruit and cinnamon trees. Palms occurred as far north as Montana on the Yellowstone River. During the Paleocene (65–25 mybp), three low basins existed in northwestern and north-central New Mexico. These basins have yielded the following information about the vegetation during that period. Silicified wood of pine, oak, and poplar is abundant in the Galisteo Formation, with logs up to 6 feet in diameter and 135 feet long. Swamps of the Raton Basin, which had a climate similar to that of Georgia, have yielded fossils of tall reeds, water lilies, fig trees, and magnolias (Christiansen and Kottlowski, 1967).

By the Eocene epoch (50–35 mybp), portions of the Madro-Tertiary geoflora had reached the central Rocky Mountain region. This flora included hackberry, nut pine (pinyon), sycamore, and squawbush. During the next 20 million years, from late Eocene to Oligocene, the climate in the Rocky Mountain West was becoming cooler and drier. In response to these conditions, coniferous forests moved south and the dry-adapted Madro-Tertiary forms continued their northeasterly march. Axelrod (1966) proposed that by this time the Madro-Tertiary geoflora had divided into Woodland Savanna, Chaparral, and Thorn Scrub vegetation units. Fossil evidence of these units has been found in Oligocene beds in the Colorado Springs region.

Continued climatic change and mountain building in the Miocene (25–12 mybp) initiated new vegetation migrations. Dix (1962) indicates how grasslands now came into the picture. The increasing aridity of the Great Plains and the Central and Rocky Mountain lowlands during the late Oligocene and Miocene forced the Arcto- and Neotropical-Tertiary floras to migrate north and south, respectively, leaving only the more xeric-adapted taxa (grasses and forbs) to colonize the plains. The major grass units came from both the Arcto- and Neotropical-Tertiary geofloras: wheatgrass (*Agropyron*) and bluegrass (*Poa*) from the Arcto-Tertiary, and grama grass (*Bouteloua*), hilaria (*Hilaria*), needlegrass (*Stipa*), and dropseed (*Sporobolus*) from the Neotropical-Tertiary. By the end of the Miocene (12 mybp), the Cascade and northern Sierra Nevada ranges were beginning to rise and initiate their rain-shadow effect. The first sagebrush pollen found was from this period. Brown (1982) gives the following overview of vegetation activity during the last half of the Tertiary in the southwestern United States.

Subsequent to the Eocene epoch, dry climates expanded throughout the second half of the Tertiary, culminating in their greatest geographic area and severity during Mio-Pliocene time. By the close of the Tertiary, more than 75 million years after the period began, the dry-adapted Madro-Tertiary Geoflora had spread over southwestern North America. This was accomplished with concomitant retreats, under expanding aridity, of the northern temperate Arcto-Tertiary Geoflora and the southern mesophytic Neotropical-Tertiary Geoflora from which Madro-Tertiary species and communities of essentially modern aspect were in place at the end of the Tertiary—at the beginning of the Quaternary and Pleistocene events (2 million years before present).

PLEISTOCENE-HOLOCENE VEGETATION

The Pleistocene epoch began 2.5 or 3 mybp and lasted until approximately 11,000 years ago. This period was characterized by glacial advances and retreats (pluvial periods). Vegetation exhibited opposite movements, by retreating south ahead of the advancing ice and then moving north again to reoccupy the barren landscape left by the melting ice. Spaulding et al. (1984) use the following dates for late Pleistocene time: 24,000 ybp, considered middle to late Wisconsin; 18,000 ybp, considered late Wisconsin maximum; 11,000 ybp, considered end of late Wisconsin (end of Pleistocene). Barry (1984) indicates that the snow line in the Sangre de Cristo Range today is 900 m (2,970 ft.) higher than during late Pleistocene. Antevs (1955) and Leopold (1951a) suggest that a decrease in temperature of 5.6°C (10°F) and an increase of 23 cm (9 in.) in annual precipitation would now be necessary to maintain the levels of such New Mexico pluvial lakes as Lake Estancia. In other words, the climate of New Mexico's lowlands 18,000 years ago must have been similar to that found in southern Oregon today.

The following summary of late Pleistocene

vegetation features in New Mexico comes from material by Wells (1970); Van Devender and Spaulding (1979); and Spaulding, Leopold, and Van Devender (1984). Even though species in the Southwest moved independently, not as groups in discrete communities, toward the late Pleistocene or early Holocene (18,000 ybp), vegetation zones in New Mexico and in the West in general were displaced 900 to 1,200 m (3,000 to 4,000 ft.) lower than the elevations at which these zones are found today. Tundra and subalpine coniferous forest were found on all the major mountain ranges of New Mexico as long as 28,000 years ago. During Wisconsin glaciation, the upper tree line in northwestern New Mexico dropped 900 m (2,970 ft.) to an elevation of 2,500 m (8,250 ft.). Spruce was found as low as 2,065 m (6,800 ft.) around the San Agustin Plains and at 2,780 m (9,200 ft.) in the Chuska Mountains of northwestern New Mexico 27,000 years ago. There was a telescopic effect, so that Ponderosa Pine was only depressed 400 m (1,300 ft.).

Most of the state was covered by montane coniferous forest, composed of Douglas Fir (*Pseudotsuga menziesii*), Southwestern White Pine (*Pinus strobiformis*), and White Fir (*Abies concolor*). Douglas Fir, Southwestern White Pine, and Dwarf Juniper (*Juniperus communis*) were in the Guadalupe Mountains of southeastern New Mexico at 2,000 m (6,600 ft.) 12,000–13,000 years ago. Today the nearest Dwarf Juniper is 400 km (248 m.) north. There was a lower woodland zone of pygmy conifers and xerophytic shrubs and succulents, restricted to elevations about 1,800 m (5,940 ft.) and latitudes south of about 37 degrees north.

On the Llano Estacado, 12,000–17,000 years ago, forest dominated by Spruce changed to woodland dominated by Pinyon Pine in a belt that stretched from southern Missouri to north Texas. Woodland vegetation covered areas in southern New Mexico 24,000 years ago that are presently covered by desert grassland or Chihuahuan desert scrub. This woodland occupied the basin floors and adjacent slopes to elevations of 1,750 m (5,800 ft.). The pinyon in question appears to have been largely Texas Pinyon (*Pinus remota*) and not Colorado Pinyon (*Pinus edulis*).

A general view of New Mexico vegetation patterns during this period is presented in Figure 2.4. Notice that grassland and desert vegetation units were absent from the state 18,000 years ago, although there was some Great Basin desert scrubland in northern New Mexico by this time.

A number of major climatic changes evidently took place between 17,000 and 600 years ago. In New Mexico there was a change to a warmer and drier mode approximately 12,000 ybp; about 8,000 ybp there was a marked reduction in winter precipitation. Another xeric (drier) shift took place between 5,000 and 4,000 ybp. There was a general cooling up to 2,500 ybp, then an increase in temperature to 800 ybp. The climate in New Mexico seems to have become relatively stable about 600 years ago (Hall, 1977). Van Devender, Betancourt, and Wimberly (1984) present an interesting sequence of vegetation changes on a site in southern New Mexico at an elevation of 1,600 m (5,300 ft.): 17,000 ybp, pinyon-juniper woodland; 10,000 ybp, juniper-oak woodland; 6,000 ybp, grassland; 4,000 ybp, desert scrub.

The movement of vegetation down and/or south during cool, wet times (Pleistocene) and then up and/or north during hot, dry times has some validity as a generalization, but the type that returned was usually quite different in species composition. In addition, vegetation zones are rarely synchronous in their movements. Van Devender, Betancourt, and Wimberly (1984) were of the opinion that the pre-Pleistocene deserts did not just retreat south into Mexico during the Pleistocene to languish in refugia until conditions were right for a return to their present ranges, but rather thought that many of the desert species were components of the Woodland-Savanna vegetation complex during the Pleistocene. Just as the Arcto- and Neotropical-Tertiary grasses that stayed behind to occupy the Rocky Mountain rain

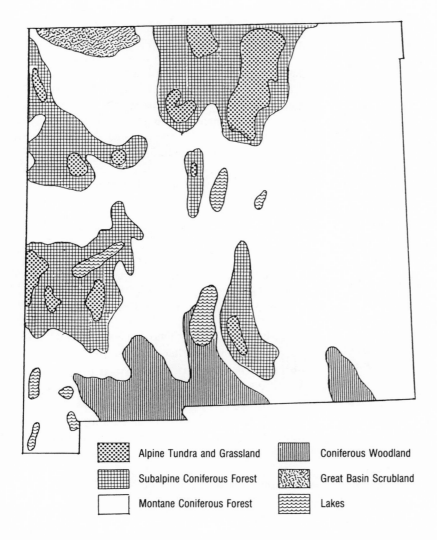

⬚ Alpine Tundra and Grassland	⬚ Coniferous Woodland
⬚ Subalpine Coniferous Forest	⬚ Great Basin Scrubland
⬚ Montane Coniferous Forest	⬚ Lakes

Figure 2.4 New Mexico Vegetation Patterns 18,000 Years Ago. Modified from History of New Mexico Vegetation, 1986. B. Sallach. Review Paper, New Mexico State University, Las Cruces.

shadow became the Grassland formation of North America, the xeric-adapted shrubs of the woodlands remained in the lower-southern portions of the Southwest to form the Desert Scrubland formation as the woodlands were forced to retreat upward and northward. It is not yet clear where the now ubiquitous creosotebush (*Larrea tridentata*) spent the cool, wet period, but it had definitely arrived in North America from South America long before the post-Pleistocene dry period.

NINETEENTH AND EARLY TWENTIETH CENTURY VEGETATION

Vegetation at the time Europeans became established in New Mexico was essentially similar to that shown on the vegetation map insert accompanying this book. The considerable changes in vegetation that have taken place during the past 150 years have been more a change in pattern than of additions or losses of major types. Some vegetation types have moved onto ground previously covered by another type, in some instances representing a permanent occupation, in other cases a persistent successional stage, which may eventually be succeeded by another type. The types whose ranges have diminished the most are also those having the greatest commercial value, i.e., Forest and Grassland. During the settlement period, forests were cut and the wood used to build towns and railroads by pioneers moving West. Later, continued reduction resulted from timber export. However, Cooper (1960) found that stands of ponderosa pine are denser now than before European settlement; he attributes the change not to climatic change, but rather to a combination of fire suppression and livestock grazing. There had already been great local reductions in grass in the vicinity of Spanish settlements by the time people from the eastern plains came to New Mexico, as documented by Leopold in *Vegetation of Southwestern Watersheds in the Nineteenth Century* (1951b).

A number of statements were written concerning the area around Santa Fe, mostly by persons attached to the Army of the West, which took over the village in the summer of 1846. Johnston wrote:
 "The grass was well eaten out before (meaning
 "previously") about camp and the country
 around Santa Fe, and today is thinly covered
 with grama grass and occasional cedar shrubs,
 betokening the greatest sterility."

Abert left Santa Fe on October 9, 1846, to make his examination of New Mexico. He noted that "the Jemez valley is very sandy; the bed of the stream

three-quarters of a mile in width, contains, in many places, no water, and when it is found, it is of a dark red color." At Santa Ana Pueblo "we had much trouble to get wood for our fires and fodder for our mules; there was no grass to be seen any where in the vicinity." The irrigated land of this pueblo was washed out during Spanish times, and the government gave the pueblo a grant of land to replace it, which was called "Los Ranchitos de Santa Ana" grant.

Cozzens crossed the Puerco on the same road in 1858.
 "This valley is quite extensive and very flat, and
 is covered with a species of coarse grass,
 valuable for sheep and goats, thousands of which
 were seen grazing on every side . . ."

Reduction in grassland in New Mexico has been so pronounced that some people today question if it was ever very extensive. However, there actually is ample evidence of the spectacular extent and quality of grassland in New Mexico, such as the following from Capt. Johnston's (1848) journal of the Cooke detachment (quoted in Leopold, 1951b):

The country from the Rio Grande to Tucson is covered with grama grass, on which animals, moderately worked, will fatten in winter . . . Southeast there was a vast plain of diluvion covered with grama grass. This plain connects with that of the Del Norte (Rio Grande), so that one can ride south of the Sierra Del Buro from the Del Norte to the Gila without crossing a single mountain.

Grasslands were still extensive and of good quality in the southern portion of the state during the last quarter of the nineteenth century and into the early years of the twentieth century. In 1883, W. G. Ritch wrote the following in *Illustrated New Mexico* (emphasis added):

The vast plains and extensive mountain ranges of Dona Ana county are covered with a species of "gramma," which grows in bunches, more or less thick, according to the locality, but it is always found sufficiently abundant to furnish stock with the most nutritious food at all seasons of the year. It does not flourish on damp or clay soil, and hence it is not found in the river bottoms. It thrives best in sand and gravel and is found in perfection on the dry sandy plains and rocky hill slopes. Horses, cattle and sheep live and thrive upon this excellent grass without other feed; flowerless and seedless, it covers broad plains and clothes the mountain sides with withered looking bunches that seem to combine the qualities of grain and the best of hay in the greatest perfection.

Cavalry officers, freighters and stock-raisers give it the very first rank among all sorts of hay, and assert that it is superior as hay, to best clover or timothy, and this opinion is shared by all who have had experience in its use. *Thousands of tons of this valuable hay can at any time be had for the cutting and baling in close proximity to a railroad track for over one hundred and fifty miles in this county.* Good gramma hay can be cut any day in the year. The best season for cutting, however, is in the months of September, October and November, or at any time after the summer rains are over and before the first frost. With thousands of square miles covered with such grasses, with a climate that permits stock to run at large unsheltered every day in the year, Dona Ana county necessarily counts stock-raising among the most important and most lucrative industries. Scarcity of water on the plains is a drawback, but one that can easily be overcome. The railroad company who have laid over two hundred and fifty miles of track in Dona Ana county have never failed to find water on the plains wherever they have bored or dug for it. Persons intending to take up stock ranches will have the benefit of this experience. *Intelligent stock men assert that the profits on cattle and sheep raising will average fifty per cent annually on the amount invested, and that the average loss will not reach two per cent. No kind of stock is ever required to be winter-fed or sheltered* [emphasis in original].

Even after the turn of the century, New Mexico grassland must still have been remarkable, con-
sidering the amount of livestock in the state. The following inventory of livestock in the Territory in 1906 (Frost and Walter, 1906) compared with an inventory for 1979 (from New Mexico Department of Agriculture and USDA Livestock Reporting Service, Las Cruces, New Mexico) attests to native forage potentials far in excess of those found today.

	1906	1979
Cattle	1,050,000	1,500,000
Sheep	5,875,000	604,000
Goats	225,000	not available
Horses	100,000	97,000
Total	7,250,000	2,201,000

And from Standley (1915), referring to the Chama Valley: "The valley lies mostly in the lower part of the Transition Zone [Merriam's life zones]. It has been overgrazed by sheep with the result that much of the grass and other natural vegetation is replaced by weeds, such as that omnipresent southwestern plant, *Ximenesia exauriculata* [current name, *Verbesina enceliodes*] which in spite of its abundance apparently has never received a common name."

Grassland reduction during the last 150 years has resulted primarily in an increase of Desert Scrubland, including both Great Basin Desert Scrub and Chihuahuan Desert Scrub (see Fig. 2.1 and vegetation map insert). In addition, the transition (ecotone) Desert Grassland type now occupies areas previously covered by Mesa-Plains Grassland (Figs. 2.1 and 7.1C). Occasionally, Desert Grassland appears to be a successional stage on the way from Grassland to Desert Scrubland. There has been an extensive advance of Juniper Savanna, also at the expense of Grassland. In northern New Mexico, the juniper advance downslope has been accompanied by a sagebrush (*Artemisia tridentata*) advance upslope, resulting in an open juniper-sagebrush woodland rather than the juniper-grass savanna found in most of the state. There has been very little change in the Pinyon-Juniper Woodland type. Sallach (1986) found that much of the recent Pinyon (*Pinus edulis*) expansion has been into sites that were actually Pinyon sites at an earlier time but that had been cleared, possibly to increase grass cover.

Recent changes in New Mexico vegetation can be attributed to human activities. The relationships of Forest to Woodland, Forest and Woodland to Grassland, and Grassland to Scrubland have been altered through logging, grazing, farming, and modifications of fire frequencies. Riparian vegetation patterns have been modified by these same factors but, in addition, the construction of dams with their reservoirs has had a marked effect upon this vegetation, both in New Mexico and in the Southwest in general (Harris, 1966).

Given the great span of time species have had to adapt to the current climate, it is understandable that native species are relatively well adapted to their primeval areas. It is unlikely that a species adapted to a similar but different habitat could out-compete a species on its home ground. In the absence of a substantial climatic shift, introduced (alien) species that are increasing, apparently at the expense of native species, are probably doing so because human activities have modified the habitat of the native species. This appears to be the case for invaders such as Russian Thistle (*Salsola* spp.), Saltcedar (*Tamarix* spp.), and Russian Olive (*Elaeagnus angustifolia*). These human activities may also allow occupation by native species formerly restricted to other areas. This is true for juniper (*Juniperus* spp.), mesquite (*Prosopis* spp.), cholla (*Opuntia* spp.), and Creosotebush (*Larrea tridentata*) where they have expanded into areas previously under grass. It is important for research efforts that we remind ourselves that these changing vegetation patterns are primarily the result of habitat modification prior to a so called "invasion" by the "villainous" species.

Table 2.1. New Mexico Studies Containing Paleo-Vegetation Information.

Site	Location	Source(s)
EASTERN		
San Jon	Quay Co.	Hafsten, 1961
Blackwater Draw	Roosevelt Co.	Hafsten, 1961
		Van Devender and Spalding, 1979
Arch Lake	Roosevelt Co.	Hafsten, 1961
Wolf Ranch Canyon	Chavez Co.	Hafsten, 1961
Dry Cave, McKittrick Hill	Eddy Co.	Harris, 1970
		Harris, 1980
		Van Devender, Moodie, and Harris, 1976
Rocky Arroyo and Last Chance Canyon	Eddy Co.	Van Devender, 1980
Loving Marsh and Salt Lake	Eddy Co.	Hafsten, 1961
CENTRAL		
Isleta Caves	Bernalillo Co.	Harris and Findley, 1964
Estancia Valley	Torrance Co.	Hafsten, 1961
Rhodes Canyon	Sierra Co.	Van Devender and Toolin, 1983
Big Boy Canyon	Otero Co.	Lanner and Van Devender, 1981
Gardner Springs	Doña Ana Co.	Freeman, 1972
Robledo Cave	Doña Ana Co.	Van Devender, Moodie, and Harris, 1976
Bishop's Cap	Doña Ana Co.	Harris and Crews, 1983
		Lanner and Van Devender, 1981
		Thompson, Van Devender, Martin, Foppe, and Long, 1980
		Van Devender, Moodie, and Harris, 1976
		Van Devender and Everitt, 1977
		Van Devender and Spaulding, 1979
Anthony Cave	Doña Ana Co.	Harris, 1979
WESTERN		
Chuska Mountains	San Juan Co.	Bent and Wright, 1963
		Betancourt, Martin, and Van Devender, 1983
		Wright, Bent, Hansen, and Maher, 1973
Chaco Canyon	San Juan Co.	Betancourt, Martin, and Van Devender, 1983
		Hall, 1977
Plains of San Agustin	Catron Co.	Clisby and Sears, 1956
		Foreman, Clisby, and Sears, 1959
		Potter and Rowley, 1960
Howell's Ridge Cave	Hidalgo Co.	Van Devender and Wiseman, 1977

Table 2.2. Nineteenth-Century Accounts Containing Information on New Mexico Vegetation.

Author	Date	Upper Rio Grande	Lower Rio Grande	Raton–Las Vegas	Galisteo	Rio Puerco	Chaco
Gregg	1843	X				X	
Abert	1846	X	X		X	X	
Cooke	1846		X		X		
Emory	1846		X	X	X		
Johnston	1846	X	X				
Ruxton	1846	X	X		X		
Hughes	1846	X			X	X	
Wislizenus	1846		X	X	X		
Marcy	1849	X		X	X		
Simpson	1849					X	X
Sitgreaves	1852					X	
Woodhouse	1852					X	
Beale	1857					X	
Campbell	1857		X				
Cozzens	1858					X	
Macomb	1859	X				X	
Beadle	1869					X	
Ruffner	1874	X				X	X
Jackson	1877					X	X

CHAPTER 2 REFERENCES

HISTORY OF NEW MEXICO VEGETATION

Abert, J. W., 1848. Examination of New Mexico, in the years 1846–47. pp. 456–460. In: 30th Congress, 1st Session, House Executive Document No. 41, Washington, D.C.

Antevs, E., 1952. Arroyo cutting and filling. Journal of Geology, 60: 182–191.

———, 1954. Climate of New Mexico during the last glacial-pluvial. Journal of Geology, 62.

———, 1955. Geologic-climatic dating in the west. American Antiquity, 20: 317–335.

Axelrod, D. I., 1950. Evolution of desert vegetation in western North America. pp. 215–306. In: D. I. Axelrod. Studies in late Tertiary paleobotany. Carnegie Institution of Washington. Publication 590.

———, 1958. Evolution of the Madro-Tertiary geoflora. Botanical Review, 24: 433–509.

———, 1966. The early Pleistocene Soboba flora of Southern California, University of California Publications in Geological Science, 60: 1–109.

Bachhuber, F. W. and W. A. McClellan, 1977. Paleoecology of marine formations in pluvial Estancia Valley, central New Mexico. Quaternary Research, 7: 254–267.

Baker, R. G., 1984. Holocene vegetational history of western United States. pp. 109–127. In: H. E. Wright, Jr. (ed.). Late Quaternary environments of the United States. University of Minnesota Press, Minneapolis.

Barry, R. G., 1984. Late-Pleistocene climatology. pp. 390–407. In: S. E. Porter (ed.). Late

Quaternary environments in the United States.

Bartlett, J. R., 1954. (Personal narrative of explorations and incidents in Texas, New Mexico, California, Sonora, and Chihuahua. Volumes I and II). The Botanical Review, 24: 193–252.

Bent, A. M. and H. E. Wright, Jr., 1963. Pollen analysis of surface materials and lake sediments from the Chuska Mountains. New Mexico Geological Society of America Bulletin, 74: 481–500.

Brown, D. E., 1982. Biotic communities of the American Southwest—United States and Mexico. 4: 1–4. University of Arizona, Tucson.

Betancourt, J. L. and T. R. Van Devender, 1981. Holocene vegetation in Chaco Canyon, New Mexico. Science 214: 656–658.

Bryant, V. M., Jr., 1977. Late Quaternary pollen records from the east-central periphery of the Chihuahuan desert. pp. 3–22. In: R. H. Waur and D. H. Riskind (eds.). Transactions of the symposium of the biological resources of the Chihuahuan desert region, United States and Mexico. U.S. Dept. of the Interior, National Park Service Transactions and Proceedings Series No. 3, U.S. Government Printing Office, Washington, D.C.

Buffington, L. C. and C. H. Herbel, 1965. Vegetational changes on a semi-desert grassland range from 1858–1963. Ecological Monographs, 35: 139–164.

Christiansen, P. W. and F. E. Kottlowski, 1964. Mosaic of New Mexico's scenery, rocks, and history. State Bureau of Mines and Mineral Resources, New Mexico Institute of Mining and Technology, Socorro.

Clisby, K. H., F. Foreman, and P. B. Sears, 1957. Pleistocene climatic changes in New Mexico, U.S.A. Transactions of the Fourth International Session of the Botanical Quarternary, 1957. Geobot. Inst. Rubel, 34: 21–26.

Clisby, K. H. and P. B. Sears, 1956. San Augustin Plains—Pleistocene climatic changes. Science, 124: 537–539.

Cooke, P. G., 1938. (Journal of the march of the Morman Battalion, 1846–1847). The Botanical Review, 24: 193–252.

Cooper, C. F., 1960. Changes in vegetation, structure, and growth of southwestern pine forests since White settlement. Ecological Monographs, 30: 129–164.

Cozzens, S. W., 1865. The marvelous country. London.

Darrow, R. A., 1961. Origin and development of the vegetational communities of the southwest. pp. 30–47. In: L. M. Shields and J. L. Gardner (eds.). Bioecology of the arid and semiarid lands of the southwest. New Mexico Highlands University Bulletin 212, Las Vegas.

Delcourt, H. R., P. A. Delcourt, and W. Thompson III. Dynamic plant ecology: the spectrum of vegetational change in space and time. Quarternary Science Reviews, 1: 153–175.

Dick-Peddie, W. A., 1965. Changing patterns on southern New Mexico. pp. 234–235. In: Guidebook of southwestern New Mexico. 2. New Mexico Geological Society 16th Field Conference. Bureau of Mines and Mineral Resources, Socorro, New Mexico.

Dix, R. L., 1962. A history of biotic and climatic changes within the North American grassland. Contribution No. 331, Department of Plant Ecology, University of Saskatchewan, Saskatoon.

Foreman, F., K. H. Clisby, and P. B. Sears, 1959. Plio-Pleistocene sediments and climates of the San Augustin Plains, New Mexico. pp. 117–120. In: New Mexico Geological Society Guidebook, 10th Field Conference. Bureau of Mines and Mineral Resources, Socorro, New Mexico.

Fountain, A. J., 1885. Dona Ana county, her people and resources. Bureau of Immigration of the Territory of New Mexico, Rio Grande Republican Print., Las Cruces, New Mexico.

Freeman, C. E., 1972. Pollen study of some Holocene alluvial deposits in Dona Ana county, southern New Mexico. Texas Journal of Science, 24: 203–220.

Frost, M. and P. A. F. Walter (eds.), 1906. The land of sunshine. New Mexico Bureau of Immigration of the Territory of New Mexico. New Mexico Printing Company, Santa Fe.

Frye J. E., A. B. Leonard, and H. O. Class, 1978.

Late Cenozoic sediments, molluscan faunas, and clay minerals in northeastern New Mexico. New Mexico Bureau of Mines and Mineral Resources, Circular 160. Bureau of Mines and Mineral Resources, Socorro, New Mexico.

Gross, F. A. and W. A. Dick-Peddie, 1979. A map of primeval vegetation in New Mexico. The Southwestern Naturalist, 24: 115–122.

Hafsten, U., 1961. Pleistocene development of vegetation and climate in the southern High Plains as evidenced by pollen analysis. pp. 59–91. In: F. Wendorf (ed.). Paleoecology of the Llano Estacado. Museum of New Mexico Press, Fort Burgwin Research Center Publication No. 1 Museum of New Mexico Press, Albuquerque.

Hall, S. A., 1977. Late Quaternary sedimentation and paleoecologic history of Chaco Canyon, New Mexico. Geological Society of America Bulletin, 88: 1593–1618.

Hansen, B. S. and E. J. Cushing, 1973. Identification of pine pollen of Late Quaternary age from the Chuska Mountains, New Mexico. Geological Society of America Bulletin 84: 1181–1200.

Harris, A. H., 1970. The Dry Cave mammalian fauna and Late Pluvial conditions in southeastern New Mexico. Texas Journal of Science, 22: 3–27.

———, 1977. Wisconsin age environments in the northern Chihuahuan desert: evidence from the higher vertebrates. pp. 23–52. In: R. H. Wauer and D. H. Riskind (ed.). Transactions of the symposium on the biological resources of the Chihuahuan desert region, United States and Mexico. U.S. Department of the Interior. National Park Service Transactions and Proceedings Series No. 3. U.S. Government Printing Office, Washington, D.C.

———, 1979. A record of spruce (Pinaceae: *Picea*) from the Pleistocene of south-central New Mexico. The Southwestern Naturalist, 24: 710.

———, 1980. The paleoecology of Dry Cave, New Mexico. National Geographic Society Research Reports 12: 331–338.

Harris, A. H. and J. S. Findley, 1964. Pleistocene-Recent fauna of the Isleta Caves, Bernalillo County, New Mexico. American Journal of Science, 262: 114–120.

Harris, D. R., 1966. Recent plant invasions in the arid and semi-arid southwest of the United States. Annals of the Association of American Geographers, 56: 408–422.

Hastings, J. R. and R. M. Turner, 1965. The Changing Mile. University of Arizona Press, Tucson.

Johnston, A. R., 1848. Journal of Captain A. R. Johnston, First Dragoons. pp. 565–614. In: 30th Congress, 2nd Session, Senate Document 608, Washington, D.C.

Lanner, R. M. and T. R. Van Devender, 1981. Late Pleistocene pinyon pines in the Chihuahuan desert. Quaternary Research, 15: 278–290.

Leopold, L. B., 1951a. Pleistocene climates in New Mexico. American Journal of Science, 249: 152–168.

Leopold, L. B., 1951b. Vegetation of the southwestern watersheds in the nineteenth century. Geographical Review, 41: 295–316.

Litle, E. L., Jr. and R. S. Campbell, 1943. Flora of Jornada Experimental Range, New Mexico. American Midland Naturalist, 30: 626–670.

Marcy, R. B., 1850. Report on a route from Fort Smith to Santa Fe. pp. 197–198. In: 31st Congress of the United States Senate Executive Document 64, Washington, D.C.

Martin, P. S., 1964. Pollen analysis and the full-glacial landscape. pp. 66–75. In: J. J. Hester and J. Schoenwetter (eds.). The reconstruction of past environments. Fort Burgwin Research Center Publication No. 3. University of New Mexico Press, Albuquerque.

Meinzer, O. E., 1922. Map of the Pleistocene lakes of the Basin and Range province and its significance. Geological Society of America Bulletin, 33: 541–552.

Oliver, C. D., 1981. Forest development in North America following disturbances. Forest Ecology and Management, 3: 153–168.

Potter, L. D. and J. Rowley, 1960. Pollen, rain and vegetation, San Augustin Plains, New Mexico. Botanical Gazette, 122: 1–25.

Richmond, G. M., 1964. Glacial deposits on Sierra Blanca Peak, New Mexico. pp. 79–81. In:

Guidebook of the Ruidoso country, 5th Field Conference, New Mexico Geological Society, Bureau of Mines and Mineral Resources, Socorro.

Ritch, W. G., 1883. Illustrated New Mexico. Bureau of Immigration of the New Mexico Territory (3rd edition), New Mexico Printing Company, Santa Fe.

Sallach, B. K., 1986. Vegetation changes in New Mexico documented by repeat photography. Thesis, New Mexico State University, Las Cruces.

Schoenwetter, J. and F. W. Eddy, 1964. Alluvial and palynological reconstruction of environments, Navajo Reservoir District. Museum of New Mexico Papers in Anthropology No. 13. University of New Mexico Press, Albuquerque.

Sears, P. B., 1921. Vegetation mapping. Science, 53: 325–327.

Sears, P. B., 1950. Pollen analysis in Old and New Mexico. Geological Society of America Bulletin, 61: 1171.

Smith, G. I. and A. Street-Perrott, 1983. Pluvial lakes of the western United States. In: S. C. Porter (ed.). Late-Quaternary environments of the United States. Volume 1. The late Pleistocene. University of Minnesota Press, Minneapolis.

Spaulding, W. G., E. B. Leopold, and T. R. Van Devender, 1984. Late Wisconsin paleoecology of the American southwest. pp. 259–293. In: H. E. Wright (ed.). Late-Quaternary Environments of the United States, Volume 2.

Standley, P. C., 1915. Vegetation of the Brazos Canyon, New Mexico. Plant World, 18: 179–191.

Thompson, R. S., T. R. Van Devender, P. S. Martin, T. Foppe, and A. Long, 1980. Shasta ground sloth (*Nothrotheriops shastense*) at Shelter Cave, New Mexico: environment, diet, and extinction. Quarterly Research 14: 360–376.

Tidwell, W. D., 1972. Evolution of the floras in the intermountain region. pp. 190–204. In: A. Cronquist, A. H. Holmgren, N. H. Holmgren, and J. L. Revean. Intermountain Flora, Volume 1. Hafner Publishing Company, Inc., New York.

Vale, T. R., 1975. Presettlement vegetation in the intermountain west. Journal of Range Management, 28: 32–36.

———, 1982. Plants and people—vegetation change in North America. Resource Publications in Geography, Association of American Geographers, Washington, D.C.

Van Cleave, M., 1936. Vegetation changes in the Middle Rio Grande Conservancy District. Thesis, University of New Mexico, Albuquerque.

Van Devender, T. R., 1977. Holocene woodlands in the Southwestern deserts. Science 198: 189–192.

———, 1980. Holocene plant remains from Rocky Arroyo and Last Chance Canyon, Eddy County, New Mexico. The Southwestern Naturalist, 25: 361–372.

Van Devender, T. R., J. L. Betancourt, and M. Wimberly, 1984. Biogeographical implications of a packrat midden sequence from the Sacramento Mountains, south-central New Mexico. Quaternary Research, 22: 344–360.

Van Devender, T. R. and B. L. Everitt, 1977. The latest Pleistocene and Recent vegetation of the Bishop's Cap, south-central New Mexico. The Southwestern Naturalist, 22: 337–352.

Van Devender, T. R., P. S. Martin, A. M. Phillips III, and W. G. Spaulding, 1977. Late Pleistocene biotic communities from the Guadalupe Mountains, Culberson County, Texas. pp. 107–113. In: R. H. Wauer and H. Riskind (eds.). Transactions of the symposium on the biological resources of the Chihuahuan desert region, United States and Mexico. U.S. Department of the Interior. National Park Service Series of Transactions and Proceedings Series No. 3. U.S. Government Printing Office, Washington, D.C.

Van Devender, T. R., P. S. Martin, R. S. Thompson, K. L. Cole, A. J. Timothy Jull, A. Long, L. J. Toolin, and D. J. Donahue, 1985. Fossil packrat middens and the tandem accelerator mass spectrometer. Nature, 317: No. 6038. pp. 610–613.

Van Devender, T. R., K. B. Moodie, and A. H. Harris, 1976. The desert tortoise (*Gopherus agassizi*) in the Pleistocene of the northern Chihuahuan desert. Herpetologica, 32: 298–304.

Van Devender, T. R. and D. H. Riskind, 1979. Late Pleistocene and Early Holocene plant remains from Hueco Tanks State Historical Park: the development of a refugium. The Southwestern Naturalist, 24: 127–140.

Van Devender, T. R. and W. G. Spaulding, 1979. Development of vegetation and climate in the southwestern United States. Science, 204: 701–710.

Van Devender, T. R., W. G. Spaulding, and A. M. Phillips III, 1979. Late Pleistocene plant communities in the Guadalupe Mountains, Culberson County, Texas. pp. 13–30. In: H. H. Genoways and R. J. Baker (eds.). Biological Investigation in the Guadalupe Mountains. Proceedings of a Symposium held at Texas Tech University, Lubbock, Texas, U.S. Department of the Interior. National Park Service, Washington, D.C.

Van Devender, T. R. and L. J. Toolin, 1983. Late Quaternary vegetation of the San Andres Mountains, Sierra County, New Mexico. pp. 33–54. In: P. L. Eidenbach (ed.). The prehistory of Rhodes Canyon. 6585th Test Group, Holloman Air Force Base, Holloman Air Force Base, Holloman, New Mexico.

Van Devender, T. R. and F. M. Wiseman, 1977. A preliminary chronology of bioenvironmental changes during the Paleoindian Period in the monsoonal Southwest. pp. 13–27. In: E. Johnson (ed.). Paleoindian Lifeways. West Texas Museum Association, The Museum Journal XVII. West Texas Museum Association, Dallas.

Vankat, J. L., 1979. The natural vegetation of North America. John Wiley and Sons, New York.

Wells, P. V., 1966. Late Pleistocene vegetation and degree of pluvial climatic change in the Chihuahuan desert. Science, 153: 971–975.

———, 1970. Vegetational history of the Great Plains. Science, 167: 1574–1582.

———, 1977. Postglacial origin of the present Chihuahuan desert less than 11,500 years ago. pp. 67–84. In: R. H. Wauer and H. Riskind (eds.). Transactions of the symposium on the biological resources of the Chihuahuan desert region, United States and Mexico. U.S. Department of the Interior. National Park Service Transactions and Proceedings Series No. 3. U.S. Government Printing Office, Washington, D.C.

———, 1979. An equable glaciopluvial in the west: Pleniglacial evidence of increased precipitation on a gradient from the Great Basin to the Sonoran and Chihuahuan deserts. Quaternary Research, 12: 311–325.

———, 1983. Late Quaternary vegetation of the Great Plains. Transactions of the Nebraska Academy of Science, XI (Special Issue): pp. 83–89.

Wislizenus, F. A., 1912. A journey to the Rocky Mountains in the year 1839. Missouri Historical Society, St. Louis.

Wright, H. E., 1976. The dynamic nature of Holocene vegetation. Quaternary Research, 6: 581–596.

———, 1981. Vegetation east of the Rocky Mountains 18,000 years ago. Quaternary Research, 15: 113–125.

Wright, H. E., Jr., A. M. Bent, B. S. Hansen, and L. H. Maher, Jr., 1973. Present and past vegetation of the Chuska Mountains, northwestern New Mexico. Geological Society of America Bulletin, 84: 1155–1180.

York, J. C. and W. A. Dick-Peddie, 1969. Vegetation changes in southern New Mexico during the past hundred years. pp. 157–166. In: W. G. McGinnies and B. J. Goldman (eds.). Arid lands in perspective. University of Arizona Press, Tucson.

3

VEGETATION AND ECOLOGY

VEGETATION PATTERNS

While people are vaguely aware of vegetation, few actually see it. Seeing vegetation involves the recognition of differences in its structure (physiognomy) and in its composition, such as the number and kind of dominant species. In addition, with some experience, one can notice physical features associated with particular types of vegetation.

Gross differences in vegetation structure, such as forest versus grassland, are closely correlated with differences in climatic features, such as mean maximum and minimum temperatures and mean annual precipitation. In New Mexico, available moisture alone can be used to segregate most vegetation types. With the addition of frost information, most vegetation throughout the world falls into some major category. In New Mexico, generally the vegetation changes on an available-moisture gradient (lower to higher) as follows: scrubland to grassland to woodland to forest. *Moisture availability* is considered here to be moisture available to plants for their establishment and maintenance. *Available moisture* is considered here as the moisture which is free to enter the roots. Moisture availability as the primary factor influencing vegetation patterns in New Mexico was recognized by Watson in 1912, when he wrote: "The chief factor determining this change is moisture, the supply of which is largely determined by precipitation, ability to hold it, and protection from drying winds and sun. . . ." In 1942, Daubenmire wrote that species comprising vegetation zones in the Rocky Mountains were restricted at their lower elevations by reduced available moisture and at their higher margins by

low temperatures. Temperature and available moisture together can be considered primary factors producing vegetation patterns. Features that singly or in combination affect these two factors can be thought of as secondary. Examples of secondary factors affecting moisture availability are latitude, altitude, annual precipitation, evaporation rates, soil texture and structure, slope, exposure (aspect), salinity, and season.

Vegetation patterns tend to change with altitude on a mountain and also with latitude staying at the same elevation. It is interesting that in the United States and elsewhere in the Northern Hemisphere, vegetation changes from lower to higher elevations are similar to those from south to north. In other words, altitude simulates latitude when considering these vegetation changes, according to changes in temperature and moisture regimes. Plate 1 illustrates vegetation response to elevation.

A given vegetation type on the south side of a mountain will occur at higher elevations than on the north side. (Fig. 3.1) The angle at which the sun's rays strike the earth on the south-facing exposure results in higher daytime temperatures than on the northern exposure. At any given atmospheric humidity, the higher temperatures on the south side will result in greater evaporative water loss from plants and soil than on the north side at the same elevation. Therefore, for a vegetation type on the south side to have the same amount of available moisture as on the north side, the vegetation would have to occur higher so that cooler temperatures would result in a water loss similar to that on the north side at a lower ele-

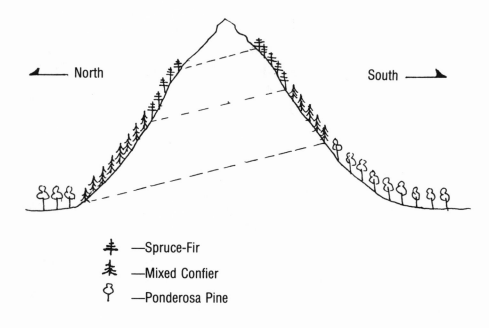

‡ —Spruce-Fir

⁂ —Mixed Confier

☖ —Ponderosa Pine

A. Typical zonation of vegetation on a high mountain
in northern New Mexico

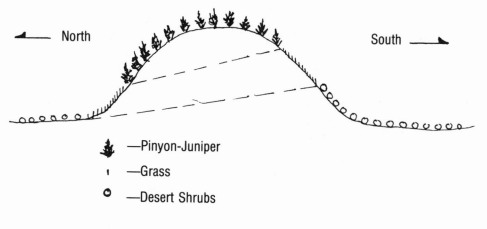

♣ —Pinyon-Juniper

ı —Grass

○ —Desert Shrubs

B. Typical zonation of vegetation on a low mountain
in southern new Mexico

Figure 3.1 Typical Effects of Latitude, Altitude, and Exposure on Vegetation.

vation. In New Mexico, these exposure effects and the resultant elevational displacements are striking. (Pl. 2) Exposure and soil type as secondary factors often responsible for reduced amounts of available moisture for vegetation have been well documented by Aide and Van Auken (1985) for vegetation in an area of west Texas. Hasse (1970) emphasized the effects of exposure on moisture availability in Arizona. In his study of Cochiti and Bland canyons of the Jemez Mountains, Robertson (1968) stated that exposures may be more important than altitude, because opposing exposures of equal elevation usually support floras of different life zones. The exposure phenomenon can also be observed by looking at the vegetation on either side of an east-west running canyon or gully. (Pl. 3) Assuming that vegetation is an indirect indicator of available moisture, it is likely that the conditions on the south-facing slope are too dry to support the type occupying on the north-facing slope, and the more moist (mesic) conditions on the north-facing slope allow this vegetation type to out-compete the type dominating the south-facing slope.

It is not uncommon to find trees growing here and there in grassland or desert scrub vegetation. Upon close inspection, it is often apparent that the trees are associated with rock outcrops or young erosion patterns. When rain falls on level or gently rolling landscapes, it tends to enter the soil where it falls. If the same amount of precipitation falls on boulders and rocks, it will run off and accumulate in pockets between them. In these catchments there is sufficient moisture for junipers, or in some cases even ponderosa pines, to grow. This *water catchment* phenomenon is common on grasslands and scrublands in New Mexico. Recent changes in vegetation patterns in portions of the state appear to be the result of this differential accumulation of available moisture. In north-central New Mexico, near the Rio Grande at Taos, the *bajadas* (piedmont or erosion slopes) are commonly covered by savannas of juniper and the desert shrub, Big Sagebrush (*Ar-temisia tridentata*). Territorial survey records document that most of these bajadas were covered by grassland less than 120 years ago. Remembering that the expected sequence of vegetation on an available moisture gradient is desert scrub to grassland to woodland, we have what would appear to be a paradox. We see a savanna of trees and desert shrubs without grassland, which requires the intermediate amount of moisture. Figure 3.2 diagrams a likely explanation for the paradox. Under livestock grazing, the grass cover was reduced. The exposed bare ground allowed rain water to flow rather than infiltrate where it fell. The water then accumulated in small depressions that provided more water than normal under grass cover so that junipers could become established. Slightly raised areas were left with less soil moisture than when the sites were under grass, and as a consequence the sites can now only support desert shrubs. Under these conditions erosion increases, accelerating the process. Farther south in the state, the desert shrubs mixed with junipers are usually snakeweed (*Gutierrezia* spp.) and rabbit brush (*Chrysothamnus* spp). This situation is common in New Mexico, and such changes are still taking place.

There has been a large-scale expansion of junipers onto past grassland sites in the state during the past 80 to 100 years, due to the foregoing process. Other explanations have been proposed, such as climate change and fire suppression, but in areas which have never been grazed, the juniper-grassland line is the same as it was at the time of the territorial survey, 120 to 140 years ago (York and Dick-Peddie, 1969). Cottam and Stewart (1940) found that in Utah, juniper moves both up and down when grass competition is reduced. They also found a complete change of vegetation after 1862 because of overgrazing followed by erosion, and that juniper had started to move by 1880. They suggested, however, that fire suppression may have been important in later moves by juniper.

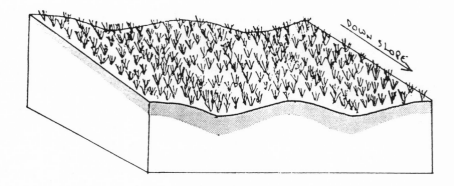

A. Predisturbance

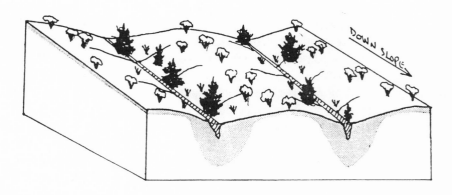

B. Postdisturbance

Juniper—

Sagebrush—

Grass—

Soil moisture
distribution
following a rain—

Figure 3.2 Typical Soil Moisture and Vegetation Changes Following
Disturbance in Northern New Mexico

Another example of vegetation patterns reflecting local differences in water availability is the very common canyon effect. When banks or canyon walls are steep enough to substantially reduce the amount of direct sunlight striking the canyon floor, vegetation patterns will differ from those found on surrounding open sites. Most of the time, these conditions are associated with drainage systems, such as rivers, streams, and arroyos. The sheltered portions are cooler, so that evaporation rates from the soil and vegetation are lower than from the soil and vegetation in the surrounding open areas. Therefore, there will be more available moisture for the canyon vegetation even independent of moisture associated with the drainage.

In 1912, Watson recognized this canyon effect on New Mexico vegetation when he wrote that, "The tendency of the higher zones to creep down the canyons and of the lower zones to creep up the ridges receives a much more plausible explanation in connection with the supply of moisture in the two situations, than through the cooling effects of descending currents and the warming effects of ascending ones." In the mountains of New Mexico, the next higher-elevation vegetation type is typically found in these situations. (Pl. 4) Good examples of canyon habitats similar to habitats of higher open slopes were recorded by Wagner (1977) in the Animas Mountains and by Robertson (1968) in the Jemez Mountains.

Often vegetation found along drainage systems is dominated by plants peculiar to such habitats and not normally found on open slopes or mesas at any elevation. This type of vegetation is covered in Chapter 9, Riparian Vegetation.

VEGETATION DYNAMICS

The vegetation of an area usually forms layers. In a forest, the crowns of the tallest trees constitute the highest, or top canopy, layer. There may be a second layer under the top canopy, composed of species that remain in the understory, possibly accompanied by some immature individuals belonging to the top canopy species. Next in this vertical stratification is usually a shrub layer, which may also include young individuals belonging to species of the higher layers. Finally there is a ground layer, usually dominated by herbaceous species composed of grasses, forbs, sedges, ferns, and so forth. There are also likely to be tree and shrub seedlings in this ground layer. For purposes of vegetation classification in New Mexico, only the tree (top canopy), shrub, and ground-cover layers are used.

Interesting information about vegetation can be obtained by looking for this vertical stratification and learning what species dominate at each level. For example, a stand with a top canopy composed of Ponderosa Pine and lower layers composed of Douglas-fir and White Fir have few or no seedlings or immature Ponderosa Pine, then it is eventually going to be a Douglas-fir and White Fir stand. Such a stand is said to be a successional, or seral stage because the Ponderosa Pine, which currently dominates, is in the process of becoming much less important, while the Douglas-fir and White Fir trees will form the top canopy in the future. Such a successional stand usually occupies a site where Douglas-fir and White Fir have been removed by fire or logging. After clearing, the site is drier, warmer, and brighter, conditions that are too severe for Douglas-fir and White Fir seedling establishment, but optimal for Ponderosa Pine seedlings. As Ponderosa Pine trees mature, they create shade that cools the site. The cooler site, in turn, results in increased soil moisture. As the process continues, the site becomes suited for the establishment and maintenance of Douglas-fir and White Fir seedlings, but is no longer suitable for ponderosa seedling establishment. But this example simplifies stand dynamics; moisture availability may not be the only factor responsible for promoting Douglas-fir and White Fir establishment. Other subdominant and understory species would also be undergoing changes in their relative positions of importance in the stand. Plate 5 illustrates early succession and Plate 6 late succession in coniferous forest.

It can be assumed that succession is taking place when the vegetation contains insufficient numbers of immature plants to replace mature members of the species dominating the site. Succession can be found in most types of vegetation. In New Mexico, much of the forest, woodland, and grassland areas are currently in some stage of succession. This correctly suggests that much of the state's vegetation has been subjected to some kind of disturbance during the past century. Changes in vegetation patterns due to changes in climate result from combinations of evolution, extinctions, and migrations. However, the concept of plant succession is usually reserved for vegetation change within a climatic regime (approximately a 500-year period) and may be initiated by natural disturbances, such as fire, flood, vulcanism, and avalanche, and/or human disturbances, such as fire, flood, logging, grazing, plowing, and fire suppression (Barbour, Burk, and Pitts, 1980). Because of its impact upon veg-

etation types economically important to man, fire has been the focus of a great deal of research, but results have been inconclusive. The relationship between fire and vegetation can be very complicated; broad generalizations, such as fire favoring grasses over shrubs, are often invalid (Kozlowski and Ahlgren, 1974). Kittams (1972) and Ahlstrand (1981) found that most desert shrubs are marginally affected by fire and that many of the shrubs of Montane Scrub crown sprout after fire. Cable (1967 and 1972) found that fire as a management tool in grama grassland is only partially effective, with a required fire frequency far greater than ever occurred naturally. Cable also found that grama grass, particularly Black Grama (*Bouteloua eriopoda*) is more negatively affected by fire than most shrub species. According to Cornelius (1988), there is a lower postfire recovery response in perennial grasses than in either shrubs or forbs. He concluded that not only was fire not important in the maintenance of Black Grama Desert Grassland, but that with extensive fires there would have been a decrease in grass abundance and an increase in that of shrubs and forbs. Buffington and Herbel (1965) found that fire frequency in southern New Mexico was so low that a thorough review of early accounts revealed no reports of fire at all. Fire suppression may have caused and may still be causing some changes in the patterns of forests, woodlands, and mountain meadows, but is doubtful that fires or their suppression have had much effect upon nonmontane grasslands or scrublands in the state.

VEGETATION TYPING

It is apparent that stands of vegetation undergoing succession could be difficult to classify or map for future reference; it would be best to find nonsuccessional, stable stands for these purposes. Such stands can be recognized by the presence in the understory of sufficient numbers of seed-

ling and immature plants for the replacement of mature individuals comprising the various layers of the stand. Even though there may be variations in the number of individuals through time, these species are said to be in dynamic equilibrium with the environment under the existing climate. Such stable stands are sometimes said to represent the termination of succession and are considered to be the mature or climax vegetation of an area.

Commonly, vegetation will contain a number of stands similar in species composition and density that can be grouped into a category or type. While such grouping is an abstraction and somewhat arbitrary, typing makes it easier to study and understand vegetation by concentrating on relatively homogeneous portions of the vegetation for the purpose of more detailed analyses and/or comparisons with other types. Just as similar stands may be grouped into a common type, similar types may be grouped into a more inclusive category. Such groupings constitute a vegetation classification. The degree of similarity necessary for grouping at any level in the hierarchy is subjective and open to debate. Choices

of stand characteristics that contribute to similarity are also open to disagreement. Nevertheless, there has been some concensus over the years as to the best stand characteristics to use for similarity comparisons and also as regards the degree of similarity necessary to form groups. Statistical techniques are constantly being developed in an attempt to standardize these subjective decisions, so that types can be consistently recognized and so that categories of vegetation will be comparable in classifications. As our understanding of vegetation has increased, these proposed category designations have had varied acceptance. Some of the more commonly used category names are Biome, Formation, Series, Association, Community Type, and Society. Association and Community Type are used as synonyms in this book.

CHAPTER 3 REFERENCES

VEGETATION AND ECOLOGY

Ahlstrand, G. M., 1981. Ecology of fire in the Guadalupe Mountains and adjacent Chihuahuan desert. Carlsbad Caverns and Guadalupe National Parks, Carlsbad, New Mexico.

Aide, M. and O. W. Van Auken, 1985. Chihuahuan desert vegetation of limestone and basalt slopes in west Texas. The Southwestern Naturalist, 30: 533–542.

Barbour, M. G., J. H. Burk, and W. D. Pitts, 1980. Terrestrial Plant Ecology. Benjamin/ Cummings Publishing Company, Inc., Menlo Park, California.

Buffington, L. C. and C. H. Herbel, 1965. Vegetation changes on a semi-desert grassland range from 1858–1963. Ecological Monographs, 35: 139–164.

Cable, D. R., 1967. Fire effects on semidesert grasses and shrubs. Journal of Range Management, 20: 170–176.

———, 1972. Fire effects in southwestern semi-desert grass-shrub communities. pp. 109–127. In: Proceedings, Annual Tall Timbers Fire Ecology Conference, No. 12. Lubbock, Texas.

Cornelius, J. M., 1988. Fire effects on vegetation of a northern Chihuahuan desert grassland. Dissertation, New Mexico State University, Las Cruces.

Cottam, W. P. and G. Stewart, 1940. Plant succession as a result of grazing and of meadow desiccation by erosion since settlement in 1862. Journal of Forestry, 38: 613–626.

Daubenmire, R. F., 1942. Soil temperature versus drought as a factor determining the lower altitudinal limits of trees in the Rocky Mountains. Botanical Gazette, 105: 1–13.

Hasse, E. F., 1970. Environmental fluctuations on south-facing slopes in the Santa Catalina Mountains of Arizona. Ecology, 51: 959–974.

Kittams, W. A., 1972. Effect of fire on vegetation of the Chihuahuan Desert Region. pp. 427–444. In: Proceedings, Annual Tall Timbers Fire

Ecology Conference, No. 12. Lubbock, Texas.

Kozlowski, T. T. and C. E. Ahlgren, 1974. Fire and ecosystems. Academic Press, New York.

Moir, W. H., 1966. Influence of ponderosa pine on herbaceous vegetation. Ecology, 47: 1045–1048.

Mueller-Dombois, D. and H. Ellenberg, 1974. Aims and Methods of Vegetation Ecology. John Wiley and Sons, New York.

Pearson, G. A., 1914. The role of aspen in the reforestation of mountain burns in Arizona and New Mexico. Plant World, 17: 249–260.

Robertson, C. W., 1968. A study of the flora of Cochiti and Bland Canyons of the Jemez Mountains. Thesis, University of New Mexico, Albuquerque.

Wagner, W. L., 1977. Floristic affinities of the Animas Mountains, southwestern New Mexico. Thesis, University of New Mexico, Albuquerque.

Watson, J. R., 1912. Plant geography of north central New Mexico. Botanical Gazette, 54: 194–217.

York, J. C. and W. A. Dick-Peddie, 1969. Vegetation changes in southern New Mexico during the past hundred years. pp. 157–166. In: W. G. McGinnies and B. J. Goldman (eds). Arid lands in perspective. University of Arizona Press, Tucson.

4

TERRESTRIAL VEGETATION OF NEW MEXICO

CLASSIFICATION AND MAPPING VEGETATION

Typing vegetation is based on patterns of dominant species, but subjective decisions must ultimately be made as to the discreteness necessary to create types. Grouping types into a system of classification involves similar judgments, as does mapping vegetation. Types often show gradual transitions, but for mapping purposes these ecotones cannot be effectively displayed; it is usually thought best to use single lines in depicting type boundaries. When an ecotone is extensive, it may be considered and mapped as a type. For example, pinyon-juniper vegetation is sometimes considered an ecotone between Forest and Grassland formations, but because in the Southwest this kind of vegetation is as extensive as forest or grassland, it is commonly typed as a Woodland.

Most vegetation classifications include types down to and including the level of Association (Community Type). However, due to their scale, including these smaller units would diminish the utility of most maps. Consequently, some categories in a classification hierarchy must be excluded from a map. A number of vegetation maps have been made for or include New Mexico, with varying classifications and scales. Both the vegetation classification and vegetation map presented in this book differ in varying degrees from existing ones, in response to the material covered.

CLASSIFICATION OF NEW MEXICO VEGETATION

This classification is built upon vegetation units recognized at the level of Association (Community Type). Recently, a considerable amount of quantitative and qualitative research has been sponsored by the United States Forest Service for typing forest vegetation. Vegetation units recognized as a result of this research are referred to as Habitat Types, roughly equivalent to the more subjectively created units referred to as associations or community types in most classifications (USDA, 1986, 1987a, and 1987b). This habitat type system is a modification of one proposed by Daubenmire (1952), who also made a case for the validity of the concept of the association in biogeographical classifications. A community type or association can become a habitat type if the vegetation structure and composition and the physical environment are consistent enough to be found in a number of mature stands. Habitat typing is currently being carried out in woodland and grassland vegetation by various workers. Associations in the present classification have been based upon habitat types where possible; eventually, a classification may be based entirely upon habitat type units created using relatively objective techniques.

The units comprising a vegetation classification should be named in a brief but informative way. Since vegetation usually exhibits some vertical layering, it is helpful for a classification to indicate the presence of vertical stratification in the type name, along with the name(s) of dominant species. In New Mexico it is desirable to recognize three basic strata: tree layer, shrub layer, and forb-grass (ground-cover) layer. For example, one type in the Ponderosa Pine Series is named *Pinus ponderosa*/MS/*Bouteloua gracilis*. This name indicates that there is a tree layer dominated by Ponderosa, a shrub layer comprised of mixed species (MS), and a ground cover dominated by Blue Grama grass. If a vegetation type has more than one dominant species in a layer, the species will be listed and connected by a dash. For example, *Abies lasiocarpa–Picea engelmannii*/*Juniperus communis*/S. In this example, Subalpine Fir and Engelmann Spruce are codominants in the tree layer, the shrub layer is dominated by Dwarf Juniper, and the forb-grass layer has sparse (S) vegetation. In the present system, each potential layer always contains species names or symbols in the association community type or habitat type name. If there are no consistent, diagnostic dominant species in a layer, the following symbols are used: MS, mixed shrub species; MG, mixed grass species; MF, mixed forb species; MG-F, mixed grass and forb species; S, sparse, few plants in the layer.

The classification and type naming used in this book depart in minor ways from the Forest Service Habitat Type system. In the Forest Service system, only one dominant species is listed in a Series name, even though there may actually be two or more equally dominant species in a layer. In the present system, codominants may constitute a Series name. In most cases, Series names are the common names of the upper canopy dominant(s). In some cases, a generic common name may be used, such as Oak Series. Occasionally, a collective name is used, such as Bunchgrass Series or Encinal Series.

D. E. Brown and C. H. Lowe (1980) collaborated in the construction of a vegetation classification for the southwestern United States and northwestern Mexico. This classification includes a digital system for all units of classification, allowing the classification to be easily expanded to accommodate other regions, countries, and continents, which in turn could lead to a degree of standardization in world vegetation classification. For a number of reasons explained later, the Brown-Lowe system was not ideally suited for New Mexico vegetation as presented here.

Another departure from most classifications is the inclusion here of a Successional-Disturbance category. Most of New Mexico's vegetation is in some stage of succession, as a result of natural or human-initiated disturbances, such as fire, logging, and grazing. Although this vegetation is sometimes slow to change and may appear to be stable, it is helpful to recognize the type if it is extensive and also to indicate that the structure and/or composition may change. The vegetation type eventually likely to follow succession, providing there is no more disturbance, follows the association name. Some vegetation in New Mexico resulting from human disturbance may actually be relatively stable. An example of this kind of vegetation is the commonly found dense stands of the alien Saltcedar (*Tamarix* spp.) that have resulted from human disruption of stream flow, livestock grazing, and the clearing of native riparian trees. The altered hydrology is probably permanent and, therefore, so are the salt cedar communities. This type of vegetation is also included in the Successional-Disturbance category.

The emphasis of this book is on terrestrial vegetation; submergent and emergent vegetation, including the vegetation of marshes, sloughs, cienegas, and bogs are considered in this book to be portions of wetlands vegetation. This topic is covered in Chapter 10, Vegetation of Special Habitats.

VEGETATION CLASSIFICATION FOR NEW MEXICO

A classification of terrestrial vegetation through the Series category is presented in Table 4.1. The number in parentheses following each series indicates the number of different plant Associations found in the Series. Detailed classifications at the Association level are given in Tables 5.1, 6.1, 7.1, 8.1, and 9.1, which are introduced in the discussions of each major type (Tundra and Forest, Woodland–Savanna, Grassland, Scrubland, and Riparian). This classification should be considered open ended; subsequent research will probably result in additions to, deletions from, and some lumping or splitting of some of the Series and/or Associations. Vegetation classifications should never be considered definitive or natural, although they should be based upon objective observations and attempts should be made to group plant aggregations in relatively discrete units, recognizable in the field. Such classifications are of greater use to resource managers, researchers, students, and the public in general.

CURRENT NEW MEXICO VEGETATION

The major types of New Mexico vegetation are Tundra, Forest, Woodland, Grassland, Scrubland, and Riparian. The highest mountain ridges and peaks in the state support small amounts of Alpine Tundra vegetation, found above timberline. Alpine Tundra consists of grasses, sedges, and dwarf cushion plants, some of which are woody. Forest here includes the Spruce-fir, Mixed Conifer, and Ponderosa Formations of some authors. Scrubland includes the Chaparral and Desert Formations of some authors.

Extensive areas in New Mexico are occupied by a mix of the major types. Such mixes can be considered transitional or ecotonal areas. Here the ecotone between Woodland and Grassland is referred to as Savanna and treated as a type. In New Mexico, Savanna consists of juniper scattered in a matrix of grama grass. Juniper savanna is extensive enough to be classified (Tables 4.1 and 6.1), mapped (Fig. 6.B and vegetation map insert), discussed as a type. Another extensive ecotone is a mosaic of grassland and desert scrubland, called Desert Grassland; it is mapped (Fig. 7.1C and vegetation map insert), classified (Tables 4.1 and 7.1), and discussed as a Grassland type. This vegetation was first called Desert Grassland by Shantz (1911). In 1929, Weaver and Clements called it Desert Plains Grassland, and Shreve in 1942 called this type Semidesert Grassland. Shreve (1942) recognized the ecotonal nature of Desert Grassland, calling it a "transitional region in which the conditions are intermediate between the optimum ones for grassland and the desert." He further wrote that "true grassland was in favorable valleys, higher mountains, and mesas with deep soils at elevations of 1500-1800 meters." Both Juniper Savanna and Desert Grassland have been expanding in New Mexico, at the expense of Plains-Mesa Grassland.

Forest in New Mexico is made up of Subalpine Coniferous Forest and Montane Coniferous Forest types. These forest types and Alpine Tundra are mapped (Fig. 5.1A-C and vegetation map insert) and described together in the text. Woodland in New Mexico is primarily the Coniferous Woodland type, composed of Pinyon Pine and various species of juniper. There are small patches of Mixed Woodland (pine-oak), covered in the text but not mapped separately from Coniferous Woodland.

Grassland areas in the state are of two principal kinds: Plains-Mesa Grassland and Montane (mountain meadow) Grassland. Plains-Mesa

Grassland includes a species combination called Great Basin Grassland by Donart, Sylvester, and Hickey (1978a). Weaver and Albertson (1956) included the Great Basin Grassland of Donart et al. as part of their Mixed Prairie type, called Plains-Mesa Grassland in this book. The Great Basin phase is discussed in the text but is not classified or mapped as a distinct type, because it differs from Plains-Mesa Grassland only by the addition of a few cool season grasses to the dominant Blue Grama grass. This variation is not extensive in New Mexico.

New Mexico Scrubland has been subdivided into Great Basin Desert, Chihuahuan Desert, Montane, Plains-Mesa Sand, and Closed Basin Scrubland types. These are classified (Tables 4.1 and 8.1), mapped (Fig. 8.1A-D and vegetation map insert), and discussed as separate types. There are extensive areas in southeastern New Mexico dominated by a scrub oak (*Quercus havardii*). Vegetation dominated by this species has traditionally been called shinnery. Shinnery vegetation is found on deep sand that tends toward dunes. Grass species always associated with sand

are also found here, as is Sand Sagebrush (*Artemisia filifolia*). There are a number of deep sand areas in the state dominated by various sand-adapted species, the total area of which is sufficient to create a Plains-Mesa Sand Scrubland category in the classification and on the vegetation maps. Another scrubland type, Closed Basin Scrubland, is dominated by Four-Wing Saltbush (*Atriplex canescens*). The sites are typically dry and somewhat salty closed basins, which are broad, gently sloping flats where water spreads and moves slowly over the surface or may even stand for short periods of time. This type is mapped as a Scrubland type and is the only riparian vegetation extensive enough to be mapped at the scale used here.

Riparian vegetation is treated in this book as a major type. Because riparian vegetation is relatively independent of surrounding upland vegetation, separate treatment should be useful. Even though the riparian vegetation type is not formally considered a Formation, it is treated as one here.

NEW MEXICO VEGETATION MAP

Vegetation maps of a region differ considerably, for a number of reasons. Vegetation patterns may have changed between mapping dates; the scales used for maps may not be the same; the people making the maps may have been using different philosophies of vegetation classification; and there may be disagreement or error in determining type boundaries. All of these reasons explain the differences between our map and others. Earlier vegetation maps made for or including New Mexico are:

1913. Life Zones of New Mexico, Bailey.
1924. Natural Vegetation. Atlas of American Agriculture, Shantz and Zon.

1956. Distribution of Vegetation Zones in New Mexico, Castetter.
1957. Vegetation Type Map of New Mexico, Department of Animal Husbandry–Range Management, New Mexico A&M.
1964. Potential Natural Vegetation of Conterminous United States, Kuchler.
1977. Vegetation and Land Use in New Mexico, Morain, Budge, and White.
1978. Potential Natural Vegetation: New Mexico, Donart, Sylvester, and Hickey (1987b).
1980. Biotic Communities of the Southwest, Brown and Lowe.

The vegetation map accompanying this book depicts the following units of vegetation: Tundra,

Subalpine Coniferous Forest, Montane Coniferous Forest, Woodland, Savanna, Plains-Mesa Grassland, Montane Grassland, Desert Grassland, Montane Scrub, Plains-Mesa Sand Scrub, Closed Basin Scrub, Great Basin Desert Scrub, and Chihuahuan Desert Scrub. Color choices for types and units follow convention in part and in part were chosen to reflect transition zones. Woodland is dark green, Grassland is yellow, and Savanna (an ecotone) is yellow-green. Chihu-ahuan Desert and Great Basin Desert Scrublands are shades of red, and Desert Grassland, the ecotone between Desert Scrubland and Plains-Mesa Grassland, is a peach color. Other colors were chosen for contrast, to make the map easier to use. Blank white areas on the map depict urban areas and farmlands, where the native vegetation has been removed. Some white areas contain the symbols D = dunes, L = lava beds (malpais), W = lakes and reservoirs.

USE OF COMMON AND SCIENTIFIC PLANT NAMES

Sources for plant ranges and scientific and common names used in this book were *Flora of New Mexico* (Martin and Hutchins, 2 vol., 1980, 1981), *Flora of New Mexico* (Wooton and Standley, 1915), *Manual of the Vascular Plants of Texas* (Correll and Johnston, 1970), *Arizona Flora* (Kearney and Peebles, 1969), *Manual of the Plants of Colorado* (Harrington, 1954), *Trees, Shrubs, and Woody Vines of the Southwest* (Vines, 1960), *Recommended Plant Names* (Beetle, 1970), and *Trees of North America* (Elias, 1980). Listed in Table 4.2 are names of important plants of the vegetation that have more than one scientific name commonly used in the literature. These names are included and the names used in this book are indicated. In this book, specific epithets are never capitalized and common names for genera are not capitalized, but plant species common names are capitalized. In species common names, words are combined and hyphens deleted, wherever possible.

Table 4.1. Classification of Terrestrial Vegetation of New Mexico. Number in parentheses ( ) indicates number of Associations or Habitat Types found in Series.

TUNDRA VEGETATION
  ALPINE TUNDRA
    Fellfield Series (1)
    Sedge Series (2)
    Alpine Avens Series (2)
    Willow Series (1)

FOREST VEGETATION
  SUBALPINE CONIFEROUS FOREST
    Bristlecone Pine Series (1)
    Engelmann Spruce–Bristlecone Pine Series (2)
    Corkbark Fir–Engelmann Spruce Series (5)

    Corkbark Fir–Engelmann Spruce–White Fir
      Series (3)
    Engelmann Spruce–Douglas-fir Series (5)
    Engelmann Spruce–Limber Pine Series (1)

  UPPER MONTANE CONIFEROUS (MIXED CONIFER)
  FOREST
    Blue Spruce Series (6)
    White Fir–Douglas-fir Series (5)
    Douglasfir–Southwestern White Pine Series (2)
    White Fir–Douglas-fir–Ponderosa Pine Series (7)
    Douglas-fir–Limber Pine (Bristlecone Pine)
      Series (2)

Table 4.1. (continued).
    Douglas-fir–Gambel Oak Series (5)
    Douglas-fir/Silverleaf Oak Series (1)

  LOWER MONTANE CONIFEROUS FOREST
    Ponderosa Pine/Gambel Oak Series (5)
    Ponderosa Pine–Silverleaf Oak Series (2)
    Ponderosa Pine–Pinyon Pine–Gambel Oak Series (4)
    Ponderosa Pine–Pinyon Pine–Gray Oak Series (2)
    Chihuahuan Pine Series (1)

  SUCCESSIONAL-DISTURBANCE FOREST
    Aspen Successional Series (1)

WOODLAND AND SAVANNA VEGETATION
  CONIFEROUS WOODLAND
    Bristlecone Pine Series (3)
    Colorado Pinyon–One-seed Juniper Series (10)
    Colorado Pinyon–Alligator Juniper Series (4)
    Colorado Pinyon–Utah Juniper Series (1)
    Colorado Pinyon–Rock Mountain Juniper Series (1)
    Colorado Pinyon–Mixed Juniper Series (3)
    Border Pinyon-Mixed Juniper Series (1)

  MIXED WOODLAND
    Colorado Pinyon–Oak–Juniper Series (2)
    Encinal Series (3)

  SAVANNA (ECOTONE)
    Arizona White Oak Series (1)
    One-seed Juniper Series (8)
    One-seed Juniper–Rocky Mountain Juniper Series (1)
    Utah Juniper Series (1)

GRASSLAND VEGETATION
  SUBALPINE–MONTANE GRASSLAND
    Mixed Sedge Series (3)
    Fescue Series (3)
    Pine Dropseed–Mountain Muhly Series (1)
    Kentucky Bluegrass Series (1)

  PLAINS-MESA GRASSLAND
    Bluestem–Sideoats Series (1)
    Grama–Buffalograss Series (1)
    Grama–Feathergrass Series (2)
    Grama–Galleta Series (3)
    Grama–Threeawn Series (2)
    Grama–Western Wheatgrass Series (2)
    Grama–Indian Ricegrass Series (1)
    Galleta–Indian Ricegrass–Needlegrass Series (1)
    Indian Ricegrass–Dropseed Series (1)
    Grama Grass Series (2)
    Grama–Dropseed Series (1)
    Tobosa Series (3)
    Sacaton Series (3)

  DESERT GRASSLAND (ECOTONE)
    Shrub–Alkali Sacaton Series (3)
    Shrub–Western Wheatgrass Series (1)
    Shrub–Indian Ricegrass Series (3)
    Shrub–Galleta Series (3)
    Shrub–Blue Grama Series (1)
    Shrub–Black Grama Series (4)
    Shrub–Mixed Grass Series (16)

SCRUBLAND VEGETATION
  MONTANE SCRUB
    Mountain Mahogany–Mixed Shrub Series (6)
    Mixed Evergreen Series (3)

  PLAINS-MESA SAND SCRUB
    Shinoak Series (5)
    Sand Sagebrush Series (4)
    Mixed Shrub Series (4)

  GREAT BASIN DESERT SCRUB
    Sagebrush Series (3)
    Saltbush Series (4)

  CHIHUAHUAN DESERT SCRUB
    Creosotebush Series (1)
    Creosotebush–Mixed Shrub Series (3)

  SUCCESSIONAL-DISTURBANCE MONTANE SCRUB
    Oak Successional Series (1)
    Oak–Locust Successional Series (1)

RIPARIAN VEGETATION
  ALPINE RIPARIAN
    Sedge Series (3)

  MONTANE RIPARIAN
    Willow Series (1)
    Willow–Alderbuck Series (1)

Table 4.1. (continued).

Willow–Dogwood Series (1)
Blue Spruce Series (2)
Aspen Series (1)
Aspen–Maple Series (1)
Boxelder Series (2)
Boxelder–Alder Series (1)
Alder Series (1)
Narrowleaf Cottonwood Series (2)
Narrowleaf Cottonwood–Mixed Deciduous
   Series (5)
Broadleaf Cottonwood Series (3)
Broadleaf Cottonwood–Mixed Deciduous
   Series (7)
Sycamore Series (1)
Arizona Walnut Series (2)
Ash Series (1)
Ash-Arizona Walnut Series (1)
Hackberry Series (1)

FLOODPLAIN-PLAINS RIPARIAN
   Cottonwood Series (5)

Cottonwood-Willow Series (3)
Mesquite Series (2)
Little Walnut Series (1)
Little Walnut–Soapberry Series (1)
Soapberry Series (1)

ARROYO RIPARIAN
   Greasewood–Desert Shrub Series (4)
   Saltbush Series (2)
   Rabbitbrush Series (1)
   Apache Plume Series (2)
   Burrobrush Series (1)
   Brickelbush Series (1)

CLOSED BASIN–PLAYA–ALKALI SINK RIPARIAN
   Iodineweed Series (2)
   Fourwing Saltbush Series (4)
   Mesquite Series (1)

SUCCESSIONAL-DISTURBANCE
   Russian Olive Successional Series (1)
   Saltcedar Successional Series (1)

Table 4.2. Synonomy of Selected Plants That Are Important Members of the Vegetation of New Mexico.

Common Name (used here)	Scientific Name (used here)	Scientific Name (also currently in use)
TREE		
Chihuahua Pine	*Pinus leiophylla*	*Pinus chihuahuana*
Fremont Cottonwood	*Populus fremontii*	*Populus wislizenii*
Border Pinyon Pine	*Pinus discolor*	*Pinus cembroides*
Velvet Ash	*Fraxinus pennsylvanica*	*Fraxinus velutina*
TREE-SHRUB		
Black Cherry	*Prunus serotina*	*Prunus serotina* subsp. *virens*
Soapberry	*Sapindus drummundi*	*Sapindus sponaria* var. *drummundi*
Texas Madrone	*Arbutus xalapensis*	*Arbutus texana*
SHRUB		
Big Sagebrush	*Artemisia tridentata*	*Artemisia tridentata* subsp. *tridentata*
Black Sagebrush	*Artemisia arbuscula*	*Artemisia arbuscula* subsp. *nova*
Bluets	*Hedyotis intrica*	*Houstonia fasciculata*
Broom Indigobush	*Psorothamnus scoparius*	*Dalea scoparia*
Broom Snakeweed	*Gutierrezia sarothrae*	*Xanthocephalum sarothrae*
Creosotebush	*Larrea tridentata*	*Larrea divaricata*
Graythorn	*Ziziphus obtusifolia*	*Condalia lyciodes*
Honey Mesquite	*Prosopis glandulosa*	*Prosopis juliflora*
Jimmyweed	*Isocoma heterophyllus*	*Haplopappus heterophyllus*
Mountain Mahogany	*Cercocarpus montanus*	*Cercocarpus breviflorus*
Range Ratany	*Krameria glandulosa*	*Krameria parvifolia*
Skunkbush	*Rhus trilobata*	*Rhus aromatica* var. *flabelliformis*
Spicebush	*Aloysia wrightii*	*Lippia wrightii*
Turpentinebush	*Ericameria laricifolia*	*Haplopappus laricifolius*
Winterfat	*Ceratoides lanata*	*Eurotia lanata*
GRASS		
Fluffgrass	*Erioneuron pulchellum*	*Tridens pulchellus*
Little Bluestem	*Schizachyrium scoparium*	*Andropogon scoparius*
Saltgrass	*Distichlis stricta*	*Distichlis spicata* var. *stricta*
FORB		
Bursage	*Ambrosia acanthicarpa*	*Franseria acanthicarpa*
Pale Trumpets	*Ipomopsis longiflora*	*Gilia longiflora*
James Rushpea	*Caesalpinia jamesii*	*Hoffmanseggia jamesii*
Skyrocket	*Ipomopsis aggregata*	*Gilia aggregata*
	Nerisyrenia camporum	*Greggia camporum*

CHAPTER 4 REFERENCES

TERRESTRIAL VEGETATION OF NEW MEXICO

Animal Husbandry–Range Management Department, 1957. Vegetation type map of New Mexico. New Mexico College of Agriculture and Mechanic Arts, Las Cruces, New Mexico.

Bailey, R. G., 1980. Description of the ecoregions of the United States. U.S. Department of Agriculture. Forest Service Miscellaneous Publication No. 1391. Washington, D.C.

Bailey, V., 1913. Life zones and crop zones of New Mexico. U.S. Department of Agriculture, North American Fauna, 35. U.S. Department of Agriculture, Washington, D.C.

Baker, W. L., 1983. Alpine vegetation of Wheeler Peak, New Mexico, USA; gradient analysis, classification, and biogeography. Arctic and Alpine Research, 15: 223–240.

———, 1984. A preliminary classification of the natural vegetation of Colorado. Great Basin Naturalist, 44: 647–676.

Barkman, J. J., J. Moravec, and S. Rauschert, 1976. Code of phytosociological nomenclature. Vegetatio, 32: 131–185.

Barnes, F. J., 1986. Carbon gain and water relations in pinyon-juniper habitat types. Dissertation, New Mexico State University, Las Cruces.

Bedker, E. J., 1966. A study of the flora of the Manzano Mountains. Thesis, University of New Mexico, Albuquerque.

Brown, D. E. and C. H. Lowe, 1980. Biotic communities of the southwest, (map). Rocky Mountain Forest and Range Experiment Station, U.S. Department of Agriculture. Forest Service. Tempe, Arizona.

Campbell, R. S. and I. F. Campbell, 1938. Vegetation of gypsum soils of the Jornada plain, New Mexico. Ecology, 19: 572–577.

Castetter, E. F., 1956. The vegetation of New Mexico. New Mexico Quarterly, 26: 256–288.

Correll, D. S. and H. B. Correll, 1972. Aquatic and wetland plants of southwestern United States. Water Pollution Control Research Series, 16030 DNL, Environmental Protection Agency, Washington, D.C.

Correll, D. S. and M. C. Johnston, 1970. Manual of the vascular plants of Texas. Texas Research Foundation, Renner, Texas.

Daubenmire, R., 1943. Vegetation zonation in the Rocky Mountains. Botanical Review, 9: 325–393.

———, 1952. Forest vegetation of northern Idaho and adjacent Washington, and its bearing on concepts of vegetation classification. Ecological Monographs, 22: 301–330.

Dick-Peddie, W. A. and W. H. Moir, 1970. Vegetation of the Organ Mountains, New Mexico. Range Science Department Science series No. 4. Colorado State University, Ft. Collins.

Ditmer, H. J., 1951. Vegetation of the southwest—past and present. Texas Journal of Science, 3: 350–355.

Donart, G. B., D. Sylvester, and W. Hickey, 1978a. A vegetation classification system for New Mexico, USA. pp. 488–490. In: D. N. Hyder (ed). Proceedings of 1st International Rangeland Congress, Denver, Colorado.

Donart, G. B., D. Sylvester, and W. Hickey, 1978b. Potential Natural Vegetation—New Mexico. New Mexico Interagency Range Committee Report II, M7-PO-23846. USDA Soil Conservation Service. (map, scale 1:1,000,000).

Elenowitz, A. S., 1983. Habitat use and population dynamics of transplanted desert bighorn sheep in the Peloncillo Mountains, New Mexico. Thesis, New Mexico State University, Las Cruces.

Elias, T. S., 1980. Trees of North America. VanNostrand Company, Inc., New York.

Fletcher, R. A., 1978. A floristic assessment of the Datil Mountains. Thesis, University of New Mexico, Albuquerque.

Gehlbach, F. R., 1967. Vegetation of the

Guadalupe Escarpment, New Mexico-Texas. Ecology, 48: 404–419.

Gleason, H. A., 1939. The individualistic concept of the plant association. American Midland Nauralist, 21: 91–110.

Gross, F. A., 1975. A computerized approach for standardizing the mapping of current and presettlement vegetation. Thesis, New Mexico State Univeristy, Las Cruces.

Gross, F. A. and W. A. Dick-Peddie, 1979. A map of primeval vegetation in New Mexico. The Southwestern Naturalist, 24: 115–122.

Harrington, H. D., 1954. Manual of the plants of Colorado. Sage Books, Denver, Colorado.

Hayes, J. A., 1982. Distribution and habitat selection of Rocky Mountain bighorn sheep (*Ovis canadensis*) along the San Francisco River canyon in New Mexico. Thesis, New Mexico State University, Las Cruces.

Huschle, G. and M. Hironaka, 1980. Classification and ordination of seral plant communities. Journal of Range Management, 33: 179–183.

Kearney, T. H. and R. H. Peebles, 1969. Arizona Flora. University of California Press, Berkeley and Los Angeles.

Kelley, N. E., 1973. Ecology of the Arroyo Hondo Pueblo site. Thesis, University of New Mexico, Albuquerque.

Kuchler, A. W., 1964. The potential natural vegetation of the conterminous United States (manual and map). American Geographical Society Special Publication: 361. New York.

Lebgue, T., 1982. Flora of the Fort Stanton Experimental Ranch, Lincoln County, New Mexico. Thesis, New Mexico State University, Las Cruces.

MacKay, H. A., 1970. A comparative floristic study of the Rio Hondo Canyon–Lake Fork–Wheeler Peak locale, New Mexico and the Huerfano River–Blanca Peak locale, Colorado. Thesis, University of New Mexico, Albuquerque.

Martin, S. C., 1972. Semidesert ecosystems. Journal of Range Management, 25: 317–319.

Martin, W. C. and C. R. Hutchins, 1980–81. A Flora of New Mexico, Vol. I and II. J. Cramer, Vaduz, West Germany, 2,591 pp.

Merriam, C. H., 1898. Life-zones and crop-zones of the United States. U.S. Department of Agriculture Division of Biological Survey, Bulletin 10. Washington, D.C.

Moir, W. H., 1963. Vegetational analyses of three southern New Mexico mountain ranges. Thesis, New Mexico State University, Las Cruces.

Morain, S. A., T. K. Budge, and M. E. White, 1977. Vegetation and land use in New Mexico (map). Albuquerque, New Mexico.

Naylor, J. N., 1964. Plant distributions of the Sandia Mountains area, New Mexico. Thesis, University of New Mexico, Albuquerque.

New Mexico Interagency Range Committee, 1978. Potential natural vegetation of New Mexico (map). Interagency Range Committee Report 2. U.S. Department of Agriculture. Soil Conservation Service. Portland, Oregon.

Osborn, N. C., 1962. The flora of Mount Taylor. Thesis, University of New Mexico, Albuquerque.

————, 1966. A comparative floristic study of Mount Taylor and Redondo Peak, New Mexico. Dissertation, University of New Mexico, Albuquerque.

Perez-Garcia, A., 1971. A vegetational survey of the Teseachic ranch of the University of Chihuahua, Chihuahua, Mexico. Thesis, New Mexico State University, Las Cruces.

Pfister, R. D. and S. F. Arno, 1980. Classifying forest types based on potential climax vegetation. Forest Science, 26: 52–70.

Riffle, N. L., 1973. The flora of Mount Sedgwick and vicinity. Thesis, University of New Mexico, Albuquerque.

Robertson, C. W., 1968. A study of the flora of the Cochiti and Bland Canyons of the Jemez Mountains. Thesis, University of New Mexico, Albuquerque.

Robinson, J. L., 1969. Forest survey of the Guadalupe Mountains, Texas. Thesis, University of New Mexico, Albuquerque.

Shantz, H. L., 1911. Natural vegetation as an indicator of the capabilities of land for crop production in the Great Plains area. U.S. Department of Agriculture, Bureau of Plant

Industry, Bulletin 201. Washington, D.C.

Shantz, H. L. and R. Zon, 1924. Natural vegetation. Atlas of American Agriculture. Part 1, Section E (map). U.S. Department of Agriculture, Washington, D.C.

Shimwell, D. W., 1971. Description and classification of vegetation. University of Washington Press, Seattle.

Shreve, F., 1942. Grassland and related vegetation in northern Mexico. Madrono, 6: 190–198.

Standley, P. C., 1915. Vegetation of the Brazos Canyon, New Mexico. Plant World, 18: 179–191.

Tatschl, A. K., 1966. A floristic study of the San Pedro Parks Wild Area, Rio Arriba County. Thesis, University of New Mexico, Albuquerque.

Thomas, M. G., 1977. A floristic analysis of the Sevilleta Wildlife Refuge and Ladron Mountains. Thesis, University of New Mexico, Albuquerque.

Townsend, C. H. T., 1893. On life zones of the Organ Mountains and adjacent region in southern New Mexico, with notes on the fauna of the range. Science, 22: 313–315.

UNESCO, 1973. International classification and mapping of vegetation, Series 6, ecology and conservation. United Nations Educational, Scientific, and Cultural Organization, Paris.

USDA Forest Service, 1986. Forest and Woodland habitat types (plant associations) of southern New Mexico and central Arizona (north of Mogollon Rim), 2nd ed., USDA Forest Service Southern Region, Albuquerque, New Mexico.

———, 1987a. Forest and woodland habitat types (plant associations) of northern New Mexico and northern Arizona, 2nd ed., ibid.

———, 1987b. Forest and woodland habitat types (plant associations) of southern Arizona (south of Mogollon Rim) and southwestern New Mexico, 2nd ed., ibid.

Vines, R. A., 1960. Trees, shrubs, and woody vines of the southwest. University of Texas Press, Austin.

Von Loh, J. D., 1977. A flora of the San Andres National Wildlife Refuge, Dona Ana County, New Mexico. Thesis, University of New Mexico, Albuquerque.

Wagner, W. L., 1977. Floristic affinities of the Animas Mountains, southwestern New Mexico. Thesis, University of New Mexico, Albuquerque.

Watson, J. R., 1912. Plant geography of northcentral New Mexico. Botanical Gazette, 54: 194–217.

Weaver, J. E. and F. W. Albertson, 1956. Grassland of the Great Plains. Johnson Publishing Company, Lincoln, Nebraska.

Weaver, J. E. and F. Clements, 1929. Plant Ecology. McGraw-Hill Book Company, New York.

Wooton, E. O. and P. C. Standley, 1915. Flora of New Mexico. Contributions from the United States National Herbarium No. 19, Washington, D.C.

Wright, M. E., 1965. Habitat analysis of the genus *Neotoma* in the southern Rio Grande Valley, New Mexico. Thesis, New Mexico State University, Las Cruces.

5

ALPINE TUNDRA AND CONIFEROUS FOREST

W. H. MOIR

ALPINE TUNDRA

Alpine tundra is a mountaintop vegetation at elevations mostly above 11,500 feet (3,500 m). Because so little such topography exists in the state, alpine tundra is the least extensive of the major vegetation forms (see Fig. 5.1A). There are also relatively few studies, and therefore our classification (Table 5.1) is incomplete. New Mexico's alpine tundra is extremely scenic and heavily used for recreation, although a very fragile vegetation, susceptible to deterioration by trampling from humans and concentrations of domestic livestock. When the fabric of tundra vegetation is broken, soil erosion quickly takes place and is difficult to arrest because of severe climate and freeze-thaw soil action.

Tundra may be abruptly separated from timberline forests by sharp contrast between tall trees of the essentially closed subalpine forests and the diminutive stature of tundra plants. There may also be a transition zone between forest and tundra where gradually smaller trees are broken into patches by interspersed tundra (Pl. 7). The upper elevational limit of this forest-alpine tundra ecotone features tree species dwarfed and shaped by wind, severe evapotranspiration, and snow or ice abrasion (Wardle, 1968). The low, shrubby patches of Engelmann Spruce, Subalpine Fir, and sometimes Bristlecone Pine and Willows thus shaped by climatic adversity are often referred to

as krummholz (literally, "twisted wood"). The vegetative growth and downwind development of krummholz islands have been described in Colorado (Marr, 1977). In New Mexico, as in Colorado, the thermal cover and food resources of krummholz makes this a preferred habitat for Ptarmigan (especially Willow krummholz), Bighorn Sheep, and Snowshoe Hares at their upper elevational limits. In addition, high-country backpackers also find the shelter of krummholz alluring. People insensitive to the slow growth of krummholz can rapidly deplete the wood resource by campfires. Knowledgeable high country users protect the fragile "wind timber" by using primus stoves for cooking and foregoing campfires. The wind-sculpted resource is thereby shared and sustained for both wildlife and other human users.

CLIMATE

The climatic demarcation between forest at timberline and tundra has sometimes been simplified to the mean maximum July isotherm of 10°C (50°F). The tundra growing season in New Mexico averages less than 90 days: approximately June through August. The soil tempera-

47

ture regime is pergelic,[1] although freeze-thaw phenomena as expressed by sorted polygons and other soil patterns are not as well expressed as in tundras to the north. Generally strong winds, high intensity solar insolation, and severe evapotranspiration during the growing season are other characteristics of alpine tundra climates. Mean annual precipitation probably exceeds 35 in. (89 cm), mosty as snow. Whether snow or rain, most precipitation is redistributed by wind so that patterns of soil moisture are extremely variable, from snowfields lasting into summer to windswept, dry ridgetops.

VEGETATION AND ENVIRONMENTS

Alpine tundra vegetation is patterned on a fine-grained scale and closely reflects adaptations to length of growing season, high evapotranspiration rates, and soil conditions (especially drainage, freeze-thaw processes, and duration of water during the growing season). Published studies of alpine plant communities suggest a strong floristic and structural resemblence of New Mexico and Colorado tundra communities (Baker, 1983; Moir and Smith, 1970). Some of the common plants comprising tundra vegetation in New Mexico are listed in Table 5.2.

Krummholz communities are found at the lower tundra edge. The wind-sheared vegetation is dominated by Engelmann Spruce and Subalpine Fir usually less than 1 m tall. Other plants in these low thickets include Common Juniper (*Juniperus communis*), Mountain Currant (*Ribes montigenum*), Grouse Whortleberry (*Vaccinium scoparium*), Shrubby Cinquefoil (*Potentilla fruticosa*), and Smoothleaf Willow (*Salix glauca*). These woody thickets are usually found in patches and "waves" conforming to microtopography and wind patterns, with other tundra communities interrupting. The lee side of krummholz often creates a snow fence effect (Berg, 1988). Patterns of krummholz growth and deformation by wind have been described by Marr (1977) and Hansen-Bristow (1986).

Fellfield, or rock field, communities feature the cushion-plant life-form, as exemplified by compact perennials hugging the ground at rock level that have well-developed below-ground food storage organs. Some common cushion plants are Alpine Forget-me-not (*Eritrichium nanum*), alpine Clovers (*Trifolium dasyphyllum, T. parryi,* and *T. nanum*), Moss-pink (*Silene acaulis*), and Alpine Sandwort (*Minuartia obtusiloba*) (see Table 5.2). A fellfield community dominated by Sierra Blanca Cinquefoil (*Potentilla sierraeblancae*) and Ellis Primula (*Primula ellisiae*) is the southernmost tundra in the United States, in vicinity of Sierra Blanca Peak in the Sacramento Mountains (Moir and Smith, 1970). Fellfield environments are considered to be both severely cold and dry, relative to environments in other tundra communities. Excessive drainage, windy, mostly snowfree relief, and high evaporation are major factors of plant adaptation. Soils are usually extremely cobbly as well as highly erodible, especially if the tread-sensitive cushion plants are killed by hikers or prolonged livestock or bighorn sheep usage (Pl. 8).

A Kobresia turf association is well developed in the Sangre de Cristo Range of northern New Mexico (cf. plot 11 of Andrews, 1983). Baker

[1]In this chapter soil temperature terms follow generalized definitions of the Soil Survey Staff (1990): **pergelic**—mean annual soil temperature at 50 cm depth (MAST) is less than 0°C; **cryic**—MAST is between 0–8°C; and MSMW (see below) is less than 5°C; **frigid**—MAST is between 0–8°C and the difference between mean winter (MW) and mean summer (MS) soil temperature (at 50 cm) is greater than 5°C; **mesic**—MAST is between 8–15°C.

(1983) describes stands in this association on sites with relatively low effective insolation and intermediate snow depths. Soils are both Lithic and Pergelic Cryumbrepts (USFS 1987c), with a turf bound tightly by dense fibrous roots. The principal species are Kobresia (*Kobresia bellardii*), which gives a rich coppery color to the autumn tundra, and Alpine Avens (*Geum rossii*), which adds a brilliant reddish hue to the tundra in September.

Perhaps the most common of alpine tundra communities is what Baker (1983) referred to as a *Carex rupestris*/cushion type. The environment is intermediate to the fellfield and Kobresia communities. Cushion plants mentioned above are frequent in this community, but Kobresia is minor or absent. Instead, a common sedge is *Carex rupestris,* a rhizomatous species that helps form a sod. Other common plants of this community include Alpine Avens, Bistort (*Polygonum bistortoides*), Alpine Sage (*Artemisia scopulorum*), Cushion Yellow Aster (*Tonestus pygmaeus*), Alpine Fescue (*Festuca brachyphylla*), and a tundra race of the Common Sandwort (*Arenaria fendleri*).

The wettest and more limited of New Mexico Tundras are snowbank and rivulet communities. Melting snowbanks often feature dwarf Willows (*Salix arctica* and *S. reticulata*) in matted patches less than 5 cm above the ground (cf. plot 21 in Andrews, 1983). An assortment of lichens and vascular plants are associated with the Willows, depending on the length of snow persistence and nature of the run-off waters (whether on the soil surface or at various depths below the surface). Taller Willow scrub (*Salix planifolia, S. glauca*) occurs on wet soils of less climatically severe microsites. These thickets may be up to 1 m tall and intergrade to krummholz at lower elevations near timberline.

Rivulet communities border small streams. They are spectacular with the flowering colors of plants such as Groundsels (*Senecio* spp.), Kings Crown (*Rhodiola integrifolia*), Rose Crown (*Clementsia rhodantha*), and Parry Primrose (*Primula parryi*).

Two other wet alpine communities are Drummond Rush and Tufted Hairgrass, both named after the dominant plant. The former is found near melting snowbanks, where ample soil water occurs most of the growing season. Two very common plants are the Rush (*Juncus drummondii*) and Sibbaldia (*Sibbaldia procumbens*). The latter, according to Baker (1983), are typically associated with late snowbanks on sites with fairly high insolation. Common vascular plants include Tufted Hairgrass (*Deschampsia caespitosa*), Clover (*Trifolium brandegei*), Sibbaldia, Cinquefoil (*Potentilla diversifolia*), and Coulter Daisy (*Erigeron coulteri*).

A variety of talus, stone-stripe, and rock-detritus communities are found in the high alpine. These have been lumped into rock outcrop or rubbleland (USFS, 1987c), but include some very specialized plant communities. Gravelly scree slopes provide habitat for such showy Groundsels as *Senecio atratus* and *Ligularia soldanella*. Two kinds of yellow-flowered Saxifrages (*Saxifraga chrysantha* and *S. flagellaris*) favor excessively drained gravelly soils. The most widespread alpine rubbleland vegetation, however, consists of crustose and foliose lichens. The density and coverage of these lichens indicates whether the rock substrate is geomorphically active or inactive. Sparse lichen coverage indicates accretion of rocky debris, rock turbulence from solifluction, or prolonged burial under snow.

The New Mexico alpine remains to be studied. Plant communities and their environmental relationships are not known over a broad geographic basis in the southern Rocky Mountains. Because of high recreation pressures on this resource and possible effects of warming climates and acid deposition, the study of tundra is timely, if not urgent.

CONIFEROUS FOREST

An astonishing diversity of coniferous forests exists in New Mexico (see Table 5.1). These forests have been intensively classified and studied by the U.S. Forest Service since 1972; numerous descriptions at the series and plant association levels have been published (see references). Some of the common trees, shrubs, and herbaceous plants in New Mexico forests are listed in Table 5.3 (a complete list of plants is given in appendix B of Muldavin et al., 1990). In recent years, the U.S. Forest Service has also developed a comprehensive correlation of soils and forest vegetation along major climatic gradients in the Southwest. This section summarizes our knowledge of coniferous forests from these sources.

Forests in New Mexico exist generally in early and middle successional stages. The most extensive cause of disturbance has been logging activities since settlement by Europeans. The era of railroad logging in New Mexico extended from the early 1900s to about 1945 (Glover, 1984; Glover and Hereford, 1986). Timber sales on lands designated as suitable for commercial harvest continue to the present. Prior to about 1900, wildfires were also important and widespread in forests at all elevations (Swetnam, 1990). In the pine and lower mixed-conifer regions, naturally occurring fires were very frequent, mostly low-intensity ground fires that helped maintain stands of open, parklike structure (Moir and Dieterich, 1988). At higher elevations, naturally occurring wildfires were less frequent and more disposed to high intensities (stand-replacing fires). A good example is the fire of 1891, which created thousands of acres of Aspen at Aspen Basin in the Santa Fe National Forest. Since about 1900, extensive wildfire suppression in the Southwest also had a major impact on forest succession, often resulting in very dense stands and high fuel accumulations. To offset these conditions, forest management agencies employ a variety of thinning, fuel reduction, and silvicultural techniques (including prescribed fires and precommercial thinnings with hand and power tools). The results are managed forests in a wide variety of successional and structural conditions.

A third major cause of forest disturbance and succession is insects and disease. Western spruce budworm is a principal defoliating insect in mid-elevation mixed-conifer and spruce-fir forests. Since the 1920s, five major outbreaks occurred in New Mexico, the most recent of which extended over approximately 700,000 acres during its maximum (Fellin et al., 1990). Other extensive insect and disease phenomena causing succession and changes in forest structure include dwarf mistletoe, root rots (especially in the ponderosa pine region), and mountain pine beetles (USFS, 1988).

Logging activities, fires and fire suppression, and insect damages and disease mentioned above are all disturbances on widespread or regional scales. Air pollution might also be included, but little is known at present about its effects on New Mexico forests. Local disturbances at the scale of individual stands, plots, or even single trees are also important in contributing to succession. In addition to the local effects of people (including recreational impacts), there are many examples of succession initiated by local weather, fire, insects, disease, snow avalanches, and other mass movements (for example, landslides), and impacts of grazing or browsing by wildlife and domestic livestock.

Coniferous forests are therefore found in many stages of succession, but most commonly in earlier stages developed from disturbances (including natural fire suppression) after about 1880, when the influence of Europeans became paramount. In the following sections, the major coniferous forest types and some of their known successional stages are described. The order in which the coniferous forests are discussed follows a generalized climatic gradient from cold,

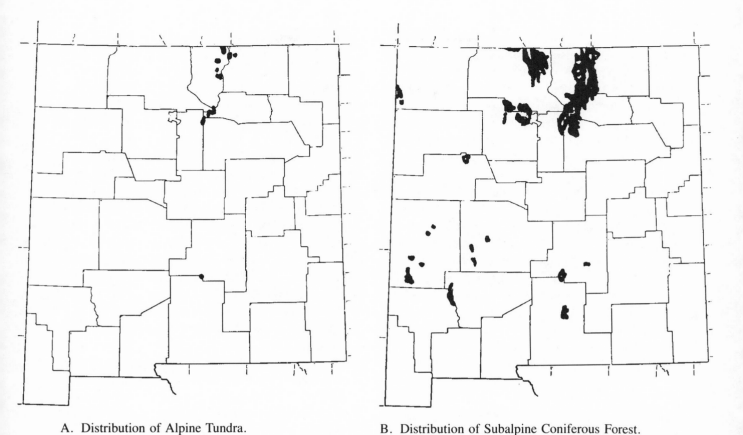

A. Distribution of Alpine Tundra.

B. Distribution of Subalpine Coniferous Forest.

Figure 5.1 Distribution of Tundra and Forest Vegetation in New Mexico. See New Mexico County Map on page xiv for reference, if needed, on county names.

wet environments at timberline to warm, dry environments at lowest elevations.

SUBALPINE CONIFEROUS FOREST

Subalpine coniferous forests (see Fig. 5.1B) occur generally from about 9,500 ft. (3,000 m) elevation to timberline, approximately 12,000 ft. (3,650 m). These forests have short growing seasons, heavy snow accumulation, and soils in the cryic temperature regime (see footnote 1, page 48). Diagnostic trees are Engelmann Spruce (*Picea engelmannii*) and Corkbark Fir (*Abies lasiocarpa* var. *arizonica*) (see Fig. 5.3A). Corkbark Fir is a southern Rocky Mountain endemic whose genetics are best expressed in southern New Mexico and northern Arizona (San Francisco Peaks); it grades into Subalpine Fir (*Abies lasiocarpa* var. *lasiocarpa*) in northern New Mexico and southern Colorado (Pl. 9).

These mountaintop forests have strong ecological and floristic affinities to high-latitude cold forests throughout the northern hemisphere (in Eurasia such forests are called taiga). During Ice Ages these forests ranged far south along the Rocky Mountain cordillera. They also existed at low elevations and onto plains, which today are grasslands or woodlands. As climates warmed in the Holocene the forests retreated to their present subalpine locations, becoming fragmented and isolated in various of the more elevated mountain islands. It is not surprising that subalpine forests in New Mexico exhibit considerable biological endemism of both plants and animals. These forests are also very important as watershed, more so than any forests at lower elevations, because of water storage and discharge from deep snowpack. Subalpine forests are very important to recreation and tourism as well. Major users are the ski industry, hikers, backpackers, and packsaddle parties of high-mountain trails and wildernesses, campers, hunters, and the scenic touring industry.

Bristlecone Pine Series

Bristlecone Pine forests (woodland would be more descriptive of the short tree stature) are found here and there on gravelly or rocky outcrops near upper timberline. The pine (*Pinus aristata*) is the major tree but Engelmann Spruce is sometimes present in minor numbers. Sites with scree or cobbly soils along ridges generally between 10,500 and 11,800 ft. (3,200–3,600 m) elevation have been described as Bristlecone Pine/ mountain currant habitat type[2] (USFS, 1987a). Old Bristlecone Pine are twisted and gnarled.

[2]In this chapter we refer to a habitat type as all land that can support the same plant association. A plant association is a vegetation classification referring to the assemblage of dominant plants in each of the tree, shrub, and herbaceous layers when succession has advanced to climax. In practice, the climax or potential vegetation is estimated from existing vegetation in late seral or near-climax stages (Moir and Ludwig, 1983).

These trees have many unique features of longevity, tree-ring sensitivity to climatic patterns and atmospheric carbon dioxide, and very high aesthetic appeal (LaMarche et al., 1984; Ferguson, 1968). Some beautiful stands in the Pecos Wilderness are remote enough for protection from vandalism.

Several Bristlecone Pine habitat types also occur at lower elevations and on deep meadow soils (Argic or Argic Pachic Cryoborolls), often near ridgetops or on south-facing slopes that adjoin grassy meadows. Stands are usually open and savanna-like with grasses and forbs resembling Thurber fescue (*Festuca thurberi*) meadows. Older trees are not twisted or gnarled, nor much over 300 years of age. These Bristlecone Pine forests (or, if you prefer, woodlands) have been described as Bristlecone Pine/Thurber Fescue association found locally in the Sangre de Cristo Mountains of northern New Mexico (DeVelice et al., 1986). Another Bristlecone Pine forest of lower elevations is more appropriately assigned to the Douglas-fir–Limber Pine–Bristlecone Pine series, discussed within the mixed-conifer region.

Engelmann Spruce–Bristlecone Pine Series

This is a widespread high-elevation, cold forest, often found at timberline or just below a krummholz zone. Engelmann Spruce is the only major tree, for the most part. Locally there may be a codominance of Bristlecone Pine, and sometimes one can find an occasional Corkbark Fir at upper timberline.

These high-elevation forests have visual, recreational, and wildlife values. Mature forests also have watershed value as snow catchment and water supply for municipal, agricultural, and other uses throughout the state.

The climate is at the cold extreme for forests. Mean annual air temperature is around 34°–36°F (1°–2°C), and at timberline the mean July air temperature is about 54°F (12°C). The soil temperature regime is near the coldest extreme of cryic, around 33°–34°F (1°C). Indeed, in high-elevation fire-created openings, mean annual soil temperatures at 50 cm hover very near 0°C, suggesting a tundra-like environment that retards reforestation (Carleton et al., 1974). At these high elevations the favorable growing season for forest plants is less than 110 days.

High winds and a high radiation regime create strong plant evaporative stresses at these elevations. Transpiration stress can be especially severe at cold soil temperatures. Mean annual precipitation is around 34 in./yr. (86 cm/yr.). Much of this falls as snow which can be redistributed by wind to sheltered areas or lower elevations of the subalpine region. Although snow can locally be very deep, the mean depths are probably not so great as in the Corkbark Fir-Engelmann Spruce series at lower elevations. Because of lower air temperatures, however, the snowpack within these forests is the last to disappear in late spring.

The forest understory is characterized by a mixture of plants of the lower subalpine forests and plants of tundra affinities. Common shrubs are Myrtle Huckleberry, Currants (*Ribes montigenum* and *R. wolfii*), and Bearberry Honeysuckle. Common grasslike plants are Fringed Brome, Mountain Trisetum, Bluegrasses (*Poa glauca* var. *rupicola, Poa reflexa*), and Nodding Woodrush (*Luzula parviflora*). Other common herbs are Wood Nymph (*Monesis uniflora*), Showy Fleabane (*Erigeron peregrinus*), White-flowered Lousewort (*Pedicularis racemosa*), Timberline Groundsel (*Senecio amplectens*), Jacob's Ladder (*Polemonium pulcherrimum* ssp. *delicatum*), Strawberry (*Fragaria ovalis*), Alpine Clover (*Trifolium dasyphyllum*), and Alpine Avens (*Geum rossii*).

The severe environment strongly curtails forest productivity. Many stands require hundreds of years to attain full stocking and high wood volumes. The slowness of succession is shown at the highest elevations above Aspen Basin (outside Santa Fe), where infilling by Engelmann Spruce of meadows created by the fire of 1891 is still taking place. That fire created a tundra-like opening where tree reestablishment is a very slow and long process.

On sites where timber harvesting has occurred, clear-cutting has been practiced with predictably poor results, since forest recovery is slow. Indeed, many of these forests are noncommercial, except on sheltered sites where special silvicultural techniques are needed to assure that not too much of the stand is harvested and that existing regeneration is protected (USFS, 1987a).

Several habitat types have been described from the southern Sangre de Cristo Range. The most extensive is Engelmann Spruce/Huckleberry–Jacob's Ladder, generally from 11,200 to 11,800 ft. (3,400–3,600 m) elevation. Below about 11,200 Corkbark Fir can become codominant with Engelmann Spruce (the Corkbark Fir phase of this habitat type). However, the presence of high elevation plants and other environmental features that indicate a short growing season help identify this habitat type (DeVelice et al., 1986). One other timberline forest is noteworthy. The Engelmann Spruce/Bluebells habitat type is found on seep soils (Argiaquic Cryoborolls) of cirque basins. A typical location is the Rincon Bonito in the Pecos Wilderness.

Corkbark Fir–Engelmann Spruce Series

Middle elevations of the subalpine forest region have forests of mixed Corkbark Fir and Engelmann Spruce. An occasional Douglas-fir or Limber Pine may be found, but more typically conifers of lower elevations are absent. Aspen may or may not be an important seral tree after fires, depending mostly on the number of years since the previous fire.

Mean annual precipitation associated with this series is about 32 in./yr. (80 cm/yr.), much of this occurring as snowfall. Snowpack from two stations in the Mogollon Mountains has been measured over 17 years at average depths of 37–51 in. (94–130 cm) by around February 1, increasing to 47–72 in. (119–183 cm) around April 1 (Jones, 1981). The water content of this snow is about 12 in. (30 cm) in February and about 20 in. (50 cm) in April. These data represent measurements in openings rather than under trees. The deep snowpack helps identify this series with one of the most important water supply areas in New Mexico (Leaf, 1975)

Mean annual air temperatures are around 35°–37°F (1.7°–2.8°C), and the soil temperature regime is cryic (see footnote 1, pg. 48). Summers are cool, and the season of favorable plant growth is short (around 110 days).

Forest vegetation associated with the Engelmann Spruce and Corkbark Fir is usually a luxuriant mixture of shrubs and herbs. Some of the more widespread shrubs are Myrtle Huckleberry (*Vaccinium myrtillus*), Bearberry Honeysuckle (*Lonicera involucrata*), currants (*Ribes montigenum* and *R. wolfii*), and Mountain Lover. Some common and widespread herbs are Forest Fleabane, One-sided Wintergreen (*Ramischia secunda*), Wood Nymph, White-flowered Lousewort (*Pedicularis racemosa*), Fringed Brome, Parry Goldenweed (*Oreochrysum parryi*), Spike Trisetum (*Trisetum spicatum* ssp. *montanum*), Fireweed (*Epilobium angustifolium*), and Osha (*Ligusticum porteri*). There are some endemic plants of these high-elevation, mountain island forests, including Burnet Groundsel (*Senecio sanguisorboides,* Sacramento Mountains and southern part of the Sangre de Cristos), Narrowleaf Lousewort (*Pedicularis angustissima,* Mogollon Mountains), and Sacramento Mountain Lupine (*Lupinus sierrae-blancae*) (Pl. 10).

The most widespread forest, with affinity to similar forests along the entire Rocky Mountain Cordillera (into Canada), has been classified as the Corkbark Fir/Myrtle Huckleberry habitat type (Fitzhugh et al., 1987; DeVelice et al., 1986). The Corkbark Fir/Burnet Groundsel habitat type is found primarily in the Sacramento Mountains, with outliers at the southern extreme of the Sangre de Cristo Range (Dye and Moir, 1979; Alexander et al., 1984). This habitat type lacks Myrtle Huckleberry and contains a robust, long-lived strain of Corkbark Fir found nowhere else in the Southwest.

Forests of special environments include Corkbark Fir/Bluebells (on soils saturated during the growing season by snowmelt), Corkbark Fir/Moss (summits and ridgetops with reduced snowpack from blowoff), and Corkbark Fir/Arizona Peavine (the exact environmental characteristics are unresolved).

Succession in this spruce-fir series is poorly known. Major stand replacing disturbances include fire and clear-cutting. Among the disturbances that affect some but not all trees are windthrow, damage by Spruce Beetles (*Dendroctonus rufipennis*), and partial cutting. Seral trees include Aspen, Bristlecone Pine, Engelmann Spruce, and Corkbark Fir. Aspen is a widespread sprouter after fire, but if the fire interval is longer than about 150 years of closed-canopy coniferous dominance, then local clones of Aspen may die out of a stand. In such cases, only spruce and fir remain as the major seral trees (Limber or Bristlecone Pine may be minor).

Large openings created by fire or logging can be difficult to regenerate if shrubs or herbs become dense. Turfs of sedges and grasses are particularly adverse microenvironments for survival of young seedling spruce or Corkbark Firs. Intense solar radiation also kills seedling conifers (Ronco, 1974); established seedlings or small saplings benefit from the shade and soil microenvironment of down logs within openings (Maser et al., 1978).

Corkbark Fir–Engelmann Spruce– White Fir Series

At the lowest elevations within the subalpine forest region are forests with mixtures of Engelmann Spruce, Douglas-fir, White Fir, Corkbark Fir, Southwestern White Pine, Aspen, and sometimes Blue Spruce. Mean annual precipitation is about 29 in./yr. (74 cm/yr.). Snow measurements

(Jones, 1981) indicate mean depths from each of February and April sample dates to be about 21 in. (53 cm), an equivalent of about 6 in. (15 cm) of water (snow was measured in openings). The mean annual air temperature is about 36°–39°F (2°–4°C), which indicates a relatively warm climate within the subalpine. Similarly, soil temperatures average toward the warm end of the cryic range (3°–6°C, or 37°–43°F, Carleton et al., 1974, Carleton and Brown, 1983).

When present in stands, White Fir, Southwestern White Pine, Blue Spruce, and Aspen are early seral trees (able to reproduce and attain maturity only in an early stage of succession, although older trees persist into later stages). Douglas-fir is typically a late seral tree (able to reproduce and attain maturity well into the middle stages of succession, but seedling establishment finally fails in late stages). Of course, both Engelmann Spruce and Corkbark Fir are regarded as climax trees (reproducing best and surviving to maturity mostly in the middle and latest stages of succession, including old growth).

There are many associated understory species in these lower-elevation subalpine forests. Indeed, in most habitat types botanical diversity is high. Some of the more common shrubs include Huckleberry (especially *Vaccinium myrtillus*), Thimbleberry, Twinflower, Canada Buffaloberry (*Shepherdia canadensis*), Bearberry Honeysuckle (*Lonicera involucrata*), Rocky Mountain Maple, Common Juniper, and Mountain Lover. Numerous herb species include various grasses and sedges, especially Mountain Trisetum (*Trisetum spicatum* ssp. *montanum*), Fringed Brome, Forest Ricegrass, Foeny Sedge (*Carex foenea*), and Forest Fescue (*Festuca sororia*). Common forbs are Canada Violet, Sweet Cicely (*Osmorhiza depauperata*), Forest Fleabane, Columbines (*Aquilegia elegantula* and *A. triternata*), False Solomon Seals (*Smilacina* spp.), Pale Geranium, Wintergreens (especially *Ramischia secunda* and species of *Pyrola*), and Baneberry (*Actaea rubra*). A conspicuous endemic is Cardamine Groundsel (*Senecio cardamine*), known only from the Mogollon Mountains in southwestern New Mexico and adjoining White Mountains in Greenlee County, Arizona. In many of the shrubby habitat types there is often a dense and conspicuous carpet of mosses and lichens. Lichens may also be draped on trunks, branches, and limbs of conifers.

Some shrubby habitat types in this series are Corkbark Fir/Huckleberry-Twinflower, Corkbark Fir/Huckleberry-Thimbleberry, Corkbark Fir/Rocky Mountain Maple, and Corkbark Fir/Burnet Groundsel (Rocky Mountain Maple phase), and Corkbark Fir/Thimbleberry. By contrast the "herbaceous" habitat types, Corkbark Fir/Forest Fleabane and Corkbark Fir/Arizona Peavine, have luxuriant herbaceous understories, but usually a weak or poorly represented shrub complement. The Corkbark Fir/Moss habitat type (Douglas-fir phase) is found on sites where winds or exposures result in snow blowoff and severe temperature extremes at the soil surface. A poorly represented shrub layer and sparse herb cover is typical, so cryptogams (mosses and lichens) dominate on soil, rock, and downed logs of the forest floor, where litter is not deep. Most of these habitat types are widespread in New Mexico, and both keys and descriptions are provided in one or more of the referenced habitat type publications.

Major disturbances in the lower subalpine forest region include wildfires, western spruce budworm defoliation, and timber harvest activities.

Fire history has not been studied in this series. Inferences from succession suggest that fires were infrequent (on the order of decades). When fires did occur, they burned erratically, missing entire areas, burning along the ground as "cool" fires in other places, and spreading as "hot" crown fires here and there (Jones, 1974; Dieterich, 1983). During rare climatic episodes (on the order of centuries) a fire holocaust might consume thousands of acres (Swetnam, 1990). With fuels currently at very high levels in many contemporary stands, hot, stand-replacing fires may become more frequent and more extensive.

The widespread existence of Aspen, White Fir, Douglas-fir, Blue Spruce, and Southwestern White Pine in the lower subalpine is primarily the result of pre-1900 fires. Cessation of fires due to suppression programs since about 1900 has resulted in an increased proportion of Engelmann Spruce and Corkbark Fir, particularly as young and advanced regeneration. In addition, many of the Aspen stands which originated 80 to 150 years ago are being replaced by mid-succession conifers, and new Aspen stands are not being produced at rates comparable to their decline from forest succession.

Mid- and late-successional stands in the lower subalpine region are susceptible to defoliations by western spruce budworm. Preferred trees for larval feeding are White Fir and Douglas-fir, but spruces are also attacked. Several episodes of defoliation have occurred since the 1920s, particularly in northern New Mexico. Preliminary evidence suggests that this lower subalpine environment is somewhat too cold for sustained, severe budworm outbreaks (USFS, 1987a), but this hypothesis is currently being examined more carefully. The most severe defoliations seem to be in uneven-aged stands resembling mixed conifer forests of lower elevations. As succession advances to later stages, there is a decline in the content of White Fir and Douglas-fir and an increase in proportions of Engelmann Spruce and Corkbark Fir. Correspondingly, the forest microclimate shifts to a colder condition, which seems to be less favorable for sustaining budworm infestations.

Timber harvest activities have been extensive in this series. Preferred stands were accessible, high-volume mature or old growth, particularly where large Douglas-firs or Engelmann Spruce grew. Early logging was often a "high grade" operation that concentrated on commercially preferred trees and was not concerned with multiple use forestry or silviculture. High-grading was also accompanied by rigorous fire suppression programs. Most stands harvested before about 1930 are now dominated by shade-tolerant Engelmann Spruce and Corkbark Firs in dense, closed stands (Swetnam and Lynch, 1989).

Today's forest management philosophy sharply departs from the "pick and pluck" strategy. Instead, there is an effort to recreate mosaics of contrasting forest structures and successional stages in the local landscape. Existing old-growth stands are no longer simply perceived as high-volume timber, but rather as important elements of landscape diversity (Harris, 1983). Tree harvests are not necessarily or even primarily for a timber objective, but also serve other objectives, such as forage enhancement, disease management, improvement of water yields, or maintenance of Aspen (Franklin et al., 1986).

Engelmann Spruce–Douglas-fir Series

These are rather localized forests at the lowest elevations of the subalpine forest region. Engelmann Spruce is the only climax tree (that is, it can reproduce and survive to maturity in the latest stages of succession), while Douglas-fir is the most important seral tree (reproducing and surviving to maturity in early to late stages, although mature and old trees persist in latest stages after reproductive success has ceased). Southwestern White Pine is second in importance to Douglas-fir as a seral tree. The climate is evidently too cold for White Fir, which is usually absent. Aspen, when present, is important only in the early years of stand history and declines rapidly (in terms of overstory cover or stem density) in stands 50 to 100 years old (as measured from tree rings obtained at breast height). An occasional Ponderosa Pine may also be found, but generally this environment is too cold for ponderosa other than as a very minor or accidental presence.

Forests in this series exist within nearly the same regional climate as the Corkbark Fir–Engelmann Spruce–White Fir series described above. However, the absence of Corkbark Fir is noteworthy. Some ecologists maintain that this is simply an accident of plant distribution. In small, isolated mountains, such as the San Mateos, or in remote, refugial canyons (where subalpine forests are relicts from a more extensive distribution during the Pleistocene) the Corkbark Fir may simply never have had time or opportunity to migrate and colonize. In still other locations, such as the Capitan Mountains, Corkbark Fir occurs as only a minor tree at the warmest or driest limit of its physiological tolerance. Finally, we run out of explanations for the absence of Corkbark Fir in portions of the Jemez Mountains, for example, where nearby low-elevation subalpine forests do contain this tree.

Most of the habitat types are local to one or a few mountain ranges. The Engelmann Spruce/ Myrtle Huckleberry forest occurs in the San Mateo and Jemez Mountains. The understory is well represented by Myrtle Huckleberry (*Vaccinium myrtillus*), commonly interpreted as a shrub of deep snowpack. The Engelmann Spruce/Moss habitat type occupies dry, northerly or westerly summits and ridges in the San Mateo, Magdalena, and Mount Taylor areas. The understory of mature and old growth stands is sparse in cover and depauperate in the number of species present. The Engelmann Spruce/Beardless Wildrye habitat type is found on cobbly, alaskite parent materials around 10,000 ft. (3,050 m) in the Capitan Mountains. Some impressive old growth stands have large Douglas-fir and ancient Southwestern White Pine, with younger trees mostly of Engelmann Spruce. The Engelmann Spruce/Forest Fleabane habitat type occurs in the Jemez Mountains and locally at lowest elevations (for subalpine forest) in the Black Range and Mogollon Mountains. The soils are deep, well watered, and support a luxuriant herbaceous flora and sometimes tall shrubs such as Rocky Mountain Maple and Forest Willow (*Salix scouleriana*). Finally,

one of the most interesting habitat types is Engelmann Spruce/Rocky Mountain Maple occurring around 8,800–9,000 feet (2,680–2,740 m), in cold air drainages of the Hubbell and Sacramento Canyons in the Sacramento Mountains. These are the southernmost outliers of Engelmann Spruce in the United States (a Mexican species of spruce occurs on high summits of Mexico's Sierra Madre ranges). All of these habitat types have been described in the references cited. There are no studies on their ecology or succession, but general responses to regeneration cutting methods are given in U.S. Forest Service manuals (USFS, 1986, 1987a).

Engelmann Spruce–Limber Pine Series

This is a cold, ridgetop forest at elevations around 9,850 ft. (3,000 m) in northern New Mexico and Colorado. The environment has high winds, intense insolation, and low soil water. Exposed sites favor snow blowoff rather than accumulation. Climatic data from a ridgetop located in the Colorado Front Range have been published by Marr (1961).

A single habitat type has been described as Limber Pine/Kinnikinnick by DeVelice at al. (1986). Engelmann Spruce, Limber Pine, and Douglas-fir are important trees at all stages of succession. Aspen, when present, has poor form and low stature. The growth of trees is extremely slow, as evidenced by consistently narrow tree rings. Tree rings examined from the Cimarron Range, in northern New Mexico, reveal that mature forests take centuries to develop after fire or stand blowdown. Seedlings of Engelmann Spruce doubtless suffer enormous mortality from drought and intense solar radiation in openings (Ronco, 1970).

A shrubby understory features Kinnikinnick (*Arctostaphylos uva-ursi*), Common Juniper, and other shrubs. Herbaceous species have negligible cover, but include some southern Rocky Mountain endemic plants, such as *Erigeron vetensis*.

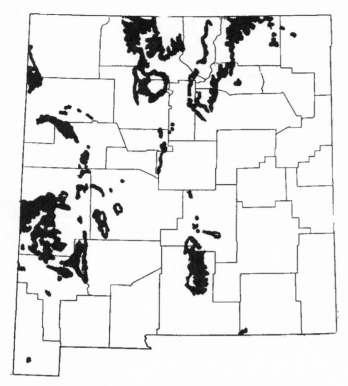

C. Distribution of Montane Coniferous Forest.

Figure 5.1 Distribution of Tundra and Forest Vegetation in New Mexico. See New Mexico County Map on page xiv.

UPPER MONTANE CONIFEROUS FOREST

Middle elevations (roughly 8,000–10,000 ft. or 2,400–3,100 m)[3] in the mountains of New Mexico (see Fig. 5.1C) feature forests of Douglas-fir (*Pseudotsuga menziesii* var. *glauca*), White Fir (*Abies concolor*), several tall Pine species,

---

[3]Exact elevations will vary depending upon the massiveness and location of mountain ranges and upon site factors such as slope, aspect, soils, and landform.

Blue Spruce (*Picea pungens*), and Aspen (*Populus tremuloides*). These forests, commonly known as mixed conifer, are the most productive in the state. High production is usually attributed to the very favorable forest climate at these elevations. Precipitation as both rainfall and snow is ample, and the soils are well watered for most or all of the growing season (the soil moisture regime is udic). There is a long growing season of favorable temperatures for tree growth. Therefore, some of the most impressive, high-biomass forests are found here, especially on the deeper and more fertile soils.

Blue Spruce Series

Frost pockets and cold air drainages often contain Blue Spruce (*Picea pungens*) as a major tree. Some of the most scenic forests in New Mexico belong to this series, whose stands are often found along streamsides and on lower valley slopes where cold night air occurs. Snow accumulates during winter, and the snowpack is late to disappear in spring.

Climatic data from weather stations are unreliable, unless instruments are located within the cold air convection layer. In addition, there is probably additional soil water from both precipitation increases (especially as snow that has been redistributed by wind during storms) and as subsurface flow coming within the rooting zone along lower slopes. Data from the Terrestrial Ecosystems Survey (USFS, 1989) indicate that Blue Spruce forests are found mostly within a regional climate of 23–28 in./yr. (60–70 cm/yr.) mean annual precipitation where soils are well watered throughout the year (udic soil moisture regime). The soil temperature regime is at the cold extreme of frigid and in some forests is cryic (see footnote 1, page 48).

Contrasting Blue Spruce forests occur in the state. Forests of mixed Blue Spruce, Douglas-fir, Ponderosa Pine, and local groves of Aspen can be found on southerly or westerly lower slopes. Understories have a grassy appearance, because

species such as Arizona Fescue, Mountain Muhly, Parry Oatgrass (*Danthonia parryi*), Ross and Foeny Sedges, Kentucky Bluegrass, and Mutton Grass are usually abundant. Other common herbs include Yarrow, Meadow Fleabane (*Erigeron formosissimus*), Strawberries (*Fragaria ovalis* and *F. americana*), Geraniums (*Geranium caespitosum* and *G. richardsonii*), and both Arizona and Grassleaf Peavines (*Lathyrus arizonica* and *L. graminifolius*). Two closely related habitat types (see source note, p. 11) are Blue Spruce/Arizona Fescue and Blue Spruce/Foeny Sedge, both widespread in the larger mountain ranges of the state (USFS, 1986, 1987a). Although within a cold air inversion layer, these two types are considered to be "warm" Blue Spruce forests because of their southerly and westerly aspects.

Different Blue Spruce forests are found on northerly and easterly lower slopes of valleys and along high streamside terraces. Here temperatures are too low for Ponderosa Pine, but Douglas-fir, White Fir, Southwestern White Pine, and occasionally Engelmann Spruce and Corkbark Fir can be found as associates of Blue Spruce. Aspen is very common, no longer localized. This is a "cold" Blue Spruce environment whose major habitat type has been described as Blue Spruce/Forest Fleabane (DeVelice et al., 1986; Fitzhugh et al., 1987). Understories usually contain numerous shrub species as well as a luxuriance of herbs. Common plants include Rocky Mountain Maple, Mountain Lover, Creeping Mahonia, Twinflower (*Linnaea borealis*), Thimbleberry (*Rubus parviflora*), Wintergreens (*Ramischia secunda* and several species of *Pyrola*), Forest Ricegrass (*Oryzopsis asperifolia*), Fringed Brome, Foeny Sedge (*Carex foenea*), Pale Geranium (*Geranium richardsonii*), Forest Fleabane, and Meadowrue. This listing is deceptive, because the understory contains numerous species. Botanical diversity is very high, including some rare and sensitive plant species (New Mexico Native Plant Protection Advisory Committee, 1984).

Other Blue Spruce forests in New Mexico are found in special environments. Some stands occur on windy, exposed ridges or upper slopes (but still within a cold air inversion layer). These "dry" Blue Spruce forests have been classified and described as the Blue Spruce/Kinnikinnick habitat type found in northern mountains (DeVelice et al., 1976). An understory of common juniper (*Juniperus communis*) and Kinnikinnick (*Arctostaphylos uva-ursi*) and a rather sparse herbaceous cover help define this habitat type. The "wet" Blue Spruce forests occur on aquic and aquic intergrade soils along streamsides. Willows (especially *Salix bebbiana*), Thinleaf Alder (*Alnus tenuifolia*), occasional Narrowleaf Cottonwood (*Populus angustifolia*), and Red Ozier Dogwood (*Cornus stolonifera*) are diagnostic of streamside Blue Spruce forests. Such forests have high recreation, wildlife, and livestock values and are very heavily utilized. Flooding is also a frequent occurrence. Certain uses, especially sustained pressures from recreation or livestock grazing, can reduce the botanical diversity that is otherwise very high.

Recreation and livestock grazing are strong influences on streamside forests. More generally, however, wildfires, logging, and mistletoe parasitism have contributed to the widespread stand conditions observed today. Wildfires, generally prior to about 1900, resulted in an abundance of Aspen, Ponderosa Pine, and Douglas-fir. Recurrent fires in the "warm," grassy Blue Spruce forests created open, parklike stands with a high content of large yellow pines (Ponderosa Pine with yellowish bark which develops in trees older than 150 years) and Douglas-firs. The mechanism of fire succession and stand development is analogous to that of Ponderosa Pine/bunchgrass ecosystems (Moir and Dieterich, 1988), except that Douglas-fir becomes a second "fire climax" tree. Recurrent, low-intensity fires kept Aspen, Blue Spruce, and White Fir at low densities. Large stumps of pine and Douglas-fir and remnant, large-limbed, old yellow pines (whose form indicates growth in an open rather than crowded tree environment), plus old records all give evidence to the historical, parklike structure of these "warm" Blue Spruce forests.

Things changed after 1900, when both logging and fire suppression became widespread. Early logging was mostly high-grade: take the best, high-volume trees, and leave the rest. There was no silviculture. Fires were viewed as destructive; their influence on the ecology and development of stands was unknown. The results of high-grading and fire suppression (which continued to the 1980s) are the present stand conditions. These are typically densely stocked with fire-sensitive species (trees easily killed by low-intensity fires): Aspen and shade-tolerant Blue Spruce and firs. Stands are mostly sapling and smaller pole-sized trees (trees with stem diameters between 4–14 in. at breast height). Many trees were present as residual seedlings or small saplings after high-grade logging. No longer subject to shading by dominant Ponderosa Pine or Douglas-firs (which were harvested) nor to thinning by fires, thickets developed rapidly. Larger, cull or defect trees, rejects from high-grade logging, were often infected by mistletoe. The nearly thickets also became infected (susceptible trees include Ponderosa Pine, Douglas-fir, and Blue Spruce). As overstocked stands became weakened from both mistletoe parasitism and tree competition, they became vulnerable to insect predation, primarily by Western Spruce Budworm (whose feeding trees, in order of preference, are White Fir, Douglas-fir, and Blue Spruce), but also from other insects and diseases (USFS, 1988; Brown et al., 1986; Swetnam and Lynch, 1989).

The above account describes typical conditions within the warm Blue Spruce environment. The history and stand development in cold Blue Spruce forests is similar, except that Ponderosa Pine was absent or minor, and therefore high-grade logging concerned mostly Douglas-fir. Aspen is more important after fire or logging. Many Aspen stands originated from pre-1900 fires and from overstory harvests of conifers. At present, Aspens are being replaced by mostly Blue Spruce and Douglas-fir, trees characteristic of a mid-successional, mixed Aspen-conifer stage.

Western Spruce Budworm (*Choristoneura occidentalis*) is less significant as a tree defoliator in Blue Spruce forests than in the White Fir–Douglas-fir series (estimates of budworm susceptibility are given in USFS, 1986, 1987a). Blue Spruce itself seems to withstand epidemics with less damage, a fortunate circumstance for a tree of scenic corridors. Perhaps lowered night temperatures help reduce budworm metabolism rates and damaging effects. However, other host trees, especially in seedling and sapling sizes, can suffer high mortality during budworm episodes (Brookes et al., 1985). Ecologically, the budworm functions as a thinning agent, with healthy trees surviving and unhealthy trees (including those weakened by mistletoe) dying back or succumbing (Wulf and Cates, 1987).

White Fir–Douglas-fir Series

Major trees in forests of this series are White Fir, Douglas-fir, Southwestern White Pine, and Aspen. At high elevations or within locally cold environments in the mixed-conifer region, temperatures are too low for Ponderosa Pine, even in early stages of forest succession. Mean annual precipitation is about 28 in./yr. (71 cm/yr.), and mean annual air temperature is about 37°–40°F (3°–4°C). Soil temperatures are at the coldest limits of the frigid soil temperature regime (Carleton et al., 1974). Observations from spring visits indicate that winter snow is deeper and lasts longer than in the White Fir–Douglas-fir–Ponderosa Pine series (described later), but measurements of exact snow amounts are lacking.

Gambel Oak is minor or absent, but other shrubs or deciduous trees and herbs, such as Rocky Mountain Maple (*Acer glabrum*), Huckleberry (*Vaccinium myrtillus*), Forest Willow (*Salix scouleriana*), and Forest Fleabane (*Erigeron eximius*), are good indicators of the cold and wet limits at which the White Fir–Douglas-fir series is found in the mixed-conifer region. Here, too, Aspen becomes of major importance following stand replacing, holocaust fires. For example, the Aspen at Aspen Basin above Santa Fe originated from a fire around 1891. And many of the Aspen groves in the vicinity of Cloudcroft and from Cloudcroft to Sunspot, in the Sacramento Mountains, seem to have originated from fires sometime within the railroad logging era (Glover, 1984). Many of these Aspen sites in the Sacramento Mountains are in a White Fir/Rocky Mountain Maple habitat type (Alexander et al., 1984). Douglas-fir and White Fir are replacing the Aspen in most of the present stands, since Aspen is primarily an early and midseral tree (Aspen forests are described in a later section) (Pl. 11).

Mixed-conifer forests of these cold environments seem to be centers of Western Spruce Budworm epidemics (indices of budworm susceptibilities by habitat types are given in USFS, 1987a and b). Perhaps this is because it is an optimum environment for White Fir (a host tree), and overstocked stands of host trees lack adequate chemical defense mechanisms against budworm feeding (Fellin et al., 1990).

The White Fir/Rocky Mountain Maple and White Fir/Forest Fleabane habitat types are common and widespread examples of cold White Fir forests. The former is primarily shrubby in understory, and the latter is luxuriant with herbs; but the same plant species are found in both associations. The habitat types seem to be nearly identical environments, with perhaps only subtle features of soil texture, drainage, or depth causing shifts from shrub to herb dominance (DeVelice et al., 1987). Some common shrubs include Rocky Mountain Maple, Forest Willow, Oceanspray, Mountain Lover, Ninebark, Snowberry, and Waxflower. Among the wealth of herb species are Canada Violet, Ragweed Sage (*Artemisia franserioides*), Strawberries (*Fragaria americana, F. ovalis*), Forest Fleabane, Pale Geranium (*Geranium richarsonii*), Fringed Brome, Forest Fescue (*Festuca sororia*), and Foeny Sedge. A

few noteworthy herbs of limited locations include the Burnet Groundsel (*Senecio sanguisorboides*), Sacramento Mountain Groundsel (*Senecio sacramentanus,* mostly in the Sacramento Mountains), Cardamine Groundsel (*Senecio cardamine* in the Mogollon Mountains) and Elk Sedge (*Carex geyeri*) in the northern San Juan Mountains.

Some White Fir habitat types of special environments are the White Fir/Huckleberry (cold air basins), White Fir/Beardless Wild Rye (cobbly, alaskite parent materials in the Capitan Mountains), White Fir/Big-toothed Maple (streamside terraces and well-watered lower slopes in southern New Mexico), White Fir/Bedstraw (streamside terraces in northern New Mexico), and White Fir/Oceanspray (scree slopes throughout the state). These have all been described, with keys for identification, in the various habitat type publications (see references).

Douglas-fir–Southwestern White Pine Series

There is an environment in which Douglas-fir and Southwestern White Pine are found in forest stands, but without White Fir. This environment is often at high elevational limits of the mixed-conifer region (that is, 9,300–10,000 ft., somewhat lower on north-facing slopes). Perhaps, low temperature extremes exceed tolerances for White Fir but not for hardy ecotypes of Douglas-fir, and Pines. Mean annual precipitation is about 28 in./yr. (71 cm/yr.), and mean annual air temperature is around 37°F (3°C). Although summers are cool and winters cold, the soil temperature regime is within the definition of frigid, although very near the frigid-cryic boundary.

In this series, Ponderosa Pine may be associated with Douglas-fir as a seral tree, reproducing and surviving only in the early or middle stages of succession, although mature trees may persist into later stages. Ponderosa can sometimes establish itself in suitably large openings (for example, after a fire or clear cut), where environments are somewhat warmer than during the later stages of succession. Reasons for Ponderosa Pine occurrence without White Fir are conjectural. Nevertheless, the overall cold environment of this series is further suggested by Engelmann Spruce (*Picea engelmannii*) or Corkbark Fir (*Abies lasiocarpa* var. *arizonica*), sometimes found as accidentals. Southwestern White Pine is usually a common seral or late-seral tree, and Aspen (*Populus tremuloides*) is another seral tree that can be locally common.

These cold, high-elevation Douglas-fir forests may have an essentially shrubby understory, including such species as Rocky Mountain Maple (*Aber glabrum*), Ninebark, Oceanspray, Waxflower (*Jamesia americana*), and Snowberry. Or the forests may be luxuriant with herbaceous species. Common herbs include Fringed Brome, Mutton Grass, Mountain Trisetum (*Trisetum spicatum* ssp. *montanum*), Foeny Sedge (*Carex foenea*), Forest Fleabane (*Erigeron eximius*), Meadowrue (*Thalictrum fendleri*), Canada Violet (*Viola canadensis*), and Vetch (*Vicia americana*). These herbaceous forests are on deeper and perhaps more finely textured soils than the shrub-dominated stands. Sometimes there will be a very tall stratum of Rocky Mountain Maple as well as a luxuriant herb cover.

These high-elevation Douglas-fir forests occur from southwestern New Mexico (Mogollon and San Francisco Mountains) into central New Mexico (San Mateo, Magdalena, and Jemez Mountains), with very local occurrences in the northern part of the state. Habitat types have been described as Douglas-fir/Fringed Brome and Douglas-fir/Ninebark (Alexander et al., 1987; Fitzhugh et al., 1987; Moir and Ludwig, 1979; USFS, 1987a).

White Fir–Douglas-fir–Ponderosa Pine Series

Forests of this type are the most widespread mixed conifer forests in New Mexico. Both White Fir (*Abies concolor*) and Douglas-fir are coclimax trees (each species is able to reproduce and survive to maturity in advanced successional stages including old growth). Important associated trees are Southwestern White Pine, Ponderosa Pine, and Aspen (*Populus tremuloides*).

Mean annual precipitation is about 27 in./yr. (69 cm/yr.), but can be as low as 23 in./yr. (58 cm/yr.) (for example, Sunspot, New Mexico), and mean annual air temperature is 39°–42°F (4°–6°C).

The most widespread and common forest in this series is the White Fir/Gambel Oak habitat type. It occurs in most mountain ranges throughout the state where elevations are high enough to attain the White Fir climate. Gambel Oak is the major associated understory shrub or low tree, but other common shrubs can include New Mexico Locust, Snowberry, Creeping Mahonia (*Berberis repens*), Oceanspray, Waxflower, Mountain Lover (*Pachistima myrsinites*), and Wood Rose (*Rosa arizonica*). Total cover of Oak and other shrubs in mature stands can range from 5% of the ground area to over 70%. The understory also includes numerous herbaceous species whose combined cover is inversely related to the oak and coniferous overstory. Common associated herbs include Ross Sedge (*Carex rossii*), Vetch (*Vicia americana*), Peavines (*Lathyrus arizonica* and *L. leucanthus*), Yarrow (*Achillea millefolium* var. *lanulosa*), Meadowrue (*Thalictrum fendleri*), and herby Sages (*Artemisia ludoviciana* and *A. franserioides*). Grassy phases of the White Fir/Gambel Oak forest have good representation (over 5% cover) of such grasses as Arizona Fescue, Mountain or Screwleaf Muhlys, Kentucky Bluegrass (*Poa pratensis*), Muttongrass, and Fringed Brome (*Bromus ciliatus*).

Major disturbances that initiate succession include stand-replacing fires and logging operations. Fire succession has been studied by Hanks and Dick-Peddie (1974) in the White (Sacramento) Mountains. For the first year or two after fire, the cleared stands are dominated by herbaceous species, but Gambel Oak is already beginning to sprout from rootstock. For the next 40–100 years or so, Gambel Oak and associated shrubs exercise more or less complete dominance of the site. Within the oak thickets conifers have difficulty regenerating. Gradually, however, Ponderosa Pine, Southwestern White Pine, Douglas-fir, or White Fir seedlings do become established, although growth is slow under the canopy of oaks, New Mexico Locust, and other tall shrubs or deciduous trees. Scattered conifers emerge above the oaks about 50–100 years after the fire. This oak-conifer stage can persist for another 50 or more years, with gradual infilling of the site by more emergent conifers. A closed stand of replacement conifers may not fully develop until 100–200 years after the fire; the time depends upon numerous factors, such as shrub densities, conditions immediately after the fire, availability of tree-seed sources, soil conditions, livestock and wildlife use of the site after fire, climate, and site treatment by humans. Whatever the factors affecting the variability of tree restocking rates, the process of succession is nevertheless very slow once the oaks and associated shrubs gain dominance. Therefore, foresters often replant conifers within the first year in the hopes that coniferous growth will keep up with the oaks. In mid to late succession, characterized by conifer dominance in closed stands, the Gambel Oak, now shaded, dies back. Mature and old-growth stands contain this Oak as low shrubs sprouting from long-lived rootstock. Upon the next fire (or clearing created by logging), the Oaks and associated shrubs will again respond.

Because strand-replacing fires were once frequent in the White Fir/Gambel Oak environment, it is common today to find patchy mosaics of stands in various of the successional stages described above. Oak-dominated stages adjoin oak-conifer and closed-conifer stages. The pattern of these mosaics is determined partly by the erratic burning behavior of the fire, which might crown out here, burn at low intensity in the understory there, and miss entirely still other areas (Jones, 1974). The pattern is also affected by silvicultural prescriptions accompanying timber harvests. Heavy cuts favor oaks, and lighter cuts, especially those preceded by understory prescribed burns, favor conifers (USFS, 1986).

In addition to the White Fir/Gambel Oak habitat type, there are many other forest habitat types within the series. Forests with essentially shrubby understories are the White Fir/Kinnikinnick, White Fir/Snowberry, and White Fir/New Mexico Locust habitat types (DeVelice et al., 1986; USFS, 1987a). These are forests of specialized environments; such as ridgetops (White Fir/Kinnikinnick) or soils high in volcanic ash or cinders (White Fir/New Mexico Locust). Forests with grassy understories have been described as White Fir/Arizona Fescue and White Fir/Screwleaf Muhly habitat types (Fitzhugh et al., 1987; Alexander et al., 1987). Finally, there are forests whose mature and old-growth expressions have sparse understory. A weak shrub component and only scattered herbs (cover mostly less than 1%) is a diagnostic feature of the White Fir/Creeping Mahonia habitat type. This particular forest is usually hard to identify in earlier stages, because shrubs and herbs may become abundant if the tree overstory is logged or burned.

In all of these habitat types, both Ponderosa Pine and Aspen are present at sites where high intensity fires have occurred or where ground fires have slowed or prevented replacement by more shade-tolerant White Fir and Douglas-fir.

Periodic outbreaks of Western Spruce Budworm are another cause of retarded succession in these essentially warm, dry White Fir environments. Host trees for the budworm are White Fir and Douglas-fir, whereas trees of early and midsuccessional stages (Pines, Gambel Oak, and Aspen) are not defoliated by this particular insect. Many seedlings and saplings of host trees are killed in epidemic outbreaks, whereas healthy, mature trees typically survive. Repeated budworm defoliations over several outbreak episodes not only retard succession (by favoring seral trees), but can result as well in a single-storied structure of surviving mature or old-age Douglas-fir and White Fir, with gaps dominated by Aspen, Southwestern White Pine, or Ponderosa Pine. Forests most susceptible to Western Spruce Budworm outbreaks tend to be those already unhealthy from drought stress, mistletoe parasitism, overstocking, root diseases, or poor site conditions (Fellin et al., 1990; USFS, 1988). In 1987, about 266,000 acres of Western Spruce Budworm defoliation was detected, mostly in northern New Mexico (USFS, 1988). Mosaics of contrasting successional stages in the landscape of White Fir–Douglas-fir region are now considered by some forest ecologists to be results both of insects and past fires (Swetnam and Lynch, 1989; Swetnam, 1990). Budworm episodes since the 1940s are more severe, extensive, and synchronistic, a result of recent fire suppression and timber harvests (which removed the pines and left an often abundant regeneration of host trees). Extensive spraying with chemical insecticides has been shown to be futile (Fellin et al., 1990). Foresters today are attempting to reduce budworm susceptibilities by using timber harvests with improved silvicultural techniques (Brookes et al., 1985).

Douglas-fir–Limber Pine– Bristlecone Pine Series

This series features forests of ridgetops, upper south-facing slopes, and other sites where high insolation, wind exposure, or soil factors combine to produce large evaporative stress on tree canopies. Douglas-fir is usually found with Limber Pine (*Pinus flexilis*), Bristlecone Pine (*Pinus aristata*), and Ponderosa Pine. These forests occupy dry, topographic sites, where exact climatic data are unavailable.

The Douglas-fir and pines are interpreted as coclimax trees, because each exhibits reproductive success and survival to maturity in advanced stages of succession, including old growth. Typically, the stands are open, often with broad, large-limbed crowns that indicate tree growth in sunny environments. Doubtless fires were an important factor of the open, parklike stands, for many of the larger trees reveal fire scarred bases. However, fire frequencies have not been measured.

Grasses are conspicuous in the understory of several habitat types within this series. Principal species are Arizona Fescue, Fringed Brome, Mutton Grass, Slender Wheatgrass (*Agropyron trachycaulum*), Junegrass, and Mountain Muhly. Some common forbs among the grasses include Mountain Parsley (*Pseudocymopteris montanus*), Cinquefoil (*Potentilla hippiana*), Sticky Geranium (*Geranium caespitosum*), Meadow Fleabanes (*Erigeron subtrinervis* and *E. formosissimus*), and Harebell (*Campanula rotundifolia*). There may be scattered shrubs here and there, such as Snowberry, Currants (*Ribes* spp.), or Oceanspray, but the understory appears essentially grassy.

Forests in this series have been described as Limber and Bristlecone Pine phases of the Douglas-fir/Arizona Fescue and Douglas-fir/Mountain Muhly habitat types (USFS, 1987a). DeVelice et al. (1986) describe a Bristlecone Pine/Arizona Fescue habitat type in northern New Mexico primarily on steep southerly and westerly upper slopes at elevations from 8,600–10,000 ft. (2,620 to 3,100 m). The Terrestrial Ecosystems Survey for the Carson National Forest has a Limber Pine/Douglas-fir mapping unit (#300) on very steep slopes with cobbly soils (Typic Dystrochrepts) and rather shrubby understories (USFS, 1987c).

Many of today's stands may have developed from montane grasslands (see Chapter 7). Conditions for tree establishment in grassy meadows have been postulated by several researchers. Allen (1989) suggests that gradual entry of conifers or Aspen into montane grasslands may reflect a long fire-free period and a recent history of grazing by domestic livestock. Some ecologists describe climatic "windows" that result in abundant conifer seed crops and favorable conditions for their germination and survival. Trees get started on microsites in these dense, herb-dominated meadows, where disturbance by pocket gophers, hoof action by large grazing animals, or other accidents eliminate the grasses and provide seed-bed for conifers. The process of tree establishment into these meadows via disturbed microsites may take many years. But when a tree reaches sufficient size in a meadow, it provides opportunities for still more trees to become established (Franklin et al., 1970; Moir, 1967; Dye and Moir, 1977).

Douglas-fir–Gambel Oak Series

Mixed coniferous forests in cold-winter climates may not contain White Fir (except possibly as an accidental). Douglas-fir, Ponderosa Pine, and Southwestern White Pine (*Pinus strobiformis*) share dominance, and the pines can persist well into late succession and climax stages. This series is found in the driest cold-winter portion of the mixed-conifer region. Forests are usually localized in the landscape, such as on steep, south-facing slopes, exposed ridgetops, or bouldery escarpments.

The biological indicators of cold winter months are broadleafed, deciduous trees and shrubs. The principal species is Gambel Oak. The mean annual air temperature is about 41°F (5°C), and soil temperatures have been indicated as frigid (Carleton et al., 1974). Winter temperatures are too low for most evergreen broadleafed shrubs. Mean annual precipitation is about 25 in./yr. (64 cm/yr.).

Trees sometimes associated with Douglas-fir and Ponderosa Pine in these cold winter climates include Southwestern White or Limber Pines (depending upon geography), Pinyon Pine (*Pinus edulis*), One-seed, Rocky Mountain or Alligator Junipers (depending on geography), and usually Gambel Oak (which is more typically shrubby). A wide array of deciduous or semideciduous shrubs may be found, including Gambel and Wavyleafed Oaks, Oceanspray (*Holodiscus dumosus*), Skunkbush Sumac, Wax Currant (*Ribes cereum*), Mountain Mahogany, Ninebark (*Physocarpus monogynus*), and Snowberry. A few low, evergreen shrubs such as Creeping Mahonia or Mountain Lover may occur, but taller evergreen,

broadleafed shrubs are absent. Sometimes grasses are well represented, such as Arizon Fescue, Bromes (*Bromus anomalus* and *B. ciliatus*), Mountain and Screwleaf Muhlys, Mutton Grass, Junegrass (*Koeleria macrantha*), and Squirreltail (*Sitanion hystrix*).

Douglas-fir–Gambel Oak forests occur throughout New Mexico, as local topography and exposures dictate. Three grassy habitat types (Douglas-fir/Arizona Fescue, Douglas-fir/Screwleaf Muhly, and Douglas-fir/Mountain Muhly), and four shrubby habitat types (Douglas-fir/Wavyleaf Oak, Douglas-fir/Gambel Oak, Douglas-fir/Creeping Mahonia, and Douglas-fir/Oceanspray) have been described (DeVelice et al., 1986; Fitzhugh et al., 1987; Alexander et al., 1987b; USFS, 1986, 1987a).

Succession within the Douglas-fir–Gambel Oak series is commonly initiated by fire or timber harvest. At lower elevations, stand-replacing fires may bring about oak dominance or mixed shrubs in early and midsuccession. Pinyon and Junipers may be present in these stages. Some of the habitat types with grassy understories go through fire succession similar to that described by Moir and Dieterich (1988) in the Ponderosa Pine/bunchgrass habitat types. The main difference is that Douglas-fir would be included with Ponderosa Pine in middle and later stages.

Timber harvests also influence succession. Generally, heavier cuttings create openings that favor Oaks, Aspen, various other shrubs, or grasses. Lighter cuttings favor coniferous regeneration, especially Douglas-fir, as well as shade tolerant shrubs and herbs (such as sedges). Effects of timber harvest techniques upon tree regeneration and understory shrubs and herbs have been summarized in U.S. Forest Service guides (USFS, 1986, 1987a and b). But stages and rates of succession have not been studied in any detail.

Douglas-fir–Silverleaf Oak Series

Some Douglas-fir forests exist in mild winter climates. Mean annual precipitation is about 28–

29 in./yr. (71–74 cm/yr.) and mean annual air temperature is around 46°F (8°C). Mild winters probably result in a mesic soil-temperature regime, but this has not been fully documented. Instead, the evergreen nature of many associated plants is evidence of enough warmth that at least some photosynthesis takes place during cold months.

Common trees of this mild-winter climate are Douglas-fir, Ponderosa Pine, Arizona Pine (*Pinus arizonica*), Southwestern White Pine (*Pinus strobiformis*), and sometimes Pinyons (*Pinus edulis, P. discolor*) and Alligator Juniper. White Fir (*Abies concolor*) is absent or accidental. However, broadleafed, evergreen species characterize this series: principally Silverleaf and Netleaf Oaks (*Quercus hypoleucoides* and *Q. rugosa,* respectively). Associated evergreen species include Gray Oak, Wright's Silktassel, and Yuccas. Deciduous species are also present, including Skunkbush Sumac (*Rhus trilobata*), New Mexico Locust, and Gambel Oak. Shrubs such as Buckbrush (*Ceanothus fendleri*) and Mountain Mahogany (*Cercocarpus montanus*) are also found as faculative evergreens (retaining functional leaves during mild winters, losing them in cold winters). Herbs may or may not be well expressed, because the tree canopy density strongly influences ground cover. Some of the more common herbs are Longtongue Muhly, Mutton Grass, Sedges (*Carex geophila* and *C. rossii*), New Mexico Groundsel (*Senecio neomexicanus*), and Fleabane (*Erigeron delphinifolius*).

Mixed-conifer forests of mild winters are limited to southern New Mexico: the Animas Mountains, portions of the San Francisco and Mogollon mountains, southern parts of the Black Range, and the Sacramento Mountains. The two described habitat types are Douglas-fir/Silverleaf Oak and Douglas-fir/Wavyleaf Oak (USFS, 1986). In the Animas Mountains a crown fire within the former was mentioned by Wagner (1978), who observed conifer snags above a Silverleaf Oak woodland canopy. In 1989 a fire holocaust there eliminated most conifers on the summit; Douglas-fir and Arizona Pine survived mostly in wet, north-facing draws (of the Douglas-fir/Gambel Oak series).

The Douglas-fir/Wavyleaf Oak forest occurs on soils derived from limestone and less than 0.5 m depth to bedrock or course-textured soils low in water-holding capacity. Stands occur at low elevations (6,800–7,800 ft.; 2,060–2,390 m) in the Sacramento Mountains on the hottest, driest sites in the mixed conifer zone. This habitat type features Douglas-fir, Southwestern White Pine, and Ponderosa Pine all as coclimax trees, with Pinyon Pine and Alligator Juniper as seral trees. Wavyleaf Oak, Mountain Mahogany, and Skunkbush Sumac (*Rhus tribolata*) are major plants of the conspicuous shrub layer.

LOWER MONTANE CONIFEROUS FOREST

At elevations generally below about 8,500 ft. (2,600 m; see footnote 3, p. 58) F are relatively warm and dry forests dominated by the genus *Pinus* (Fig. 5.3C). The principal trees are Ponderosa Pine (*Pinus ponderosa*), Chihuahua Pine (*Pinus leiophylla*), Pinyons (*Pinus edulis, P. discolor*), Junipers (*Juniperus* spp.), and several Oaks (*Quercus* spp.). The climate is borderline for forests. Warm air and soil temperatures allow a potential growing season of around 180 days. However, available water in upper portions of the soil profile is deficient during some of the warm season (ustic soil moisture regime). The hot months, May and June, are also driest.

Ponderosa Pine–Gambel Oak Series

A widespread Ponderosa Pine forest is found within a cold winter climate. Mean precipitation averages 22–24 in./yr. (56–61 cm/yr.), and mean annual air temperature is about 41°F (5°C). The soil temperature regime is frigid (see footnote 1, page 48). Associated broadleafed shrubs and trees are deciduous in winter. These Ponderosa Pine forests often have grassy understories, composed of Arizona Fescue, Mountain Muhly, Screwleaf Muhly (*Muhlenbergia virescens*), Pine Dropseed, Mutton Grass, Squirreltail (*Sitanion hystrix*), Kentucky Bluegrass (*Poa pratensis*), and Sedges. Some sites, apparently with soils of higher gravel and/or cobble content, have Gambel Oak (*Quercus gambelii*), New Mexico Locust (*Robinia neomexicana*), Snowberry (*Symphoricarpos oreophila*), Wood Rose (*Rosa arizonica*), or other deciduous species as well as grasses and other herbs. At some ridgetop locations in central and northern New Mexico, Kinnikinnick (*Arctostaphylos uva-ursi*) and Common Juniper (*Juniperus communis*) display conspicuous matlike ground cover.

These Ponderosa Pine forests of cold winters are common in New Mexico, extending into Colorado and Arizona. Habitat types include the grassy pine forests: Ponderosa Pine/Arizona Fescue, Ponderosa Pine/Mountain Muhly, and Ponderosa Pine/Screwleaf Muhly. Habitat types with shrubby understories are Ponderosa Pine/Gambel Oak and Ponderosa Pine/Kinnikinnick.

Succession in the grassy pine forest habitat type has been thoroughly documented. Recurrent, low-intensity fires occurred at 2–10 year intervals prior to about 1900, when fire suppression actions were taken. These fires helped maintain open, parklike vistas, with here and there small patches of tree regeneration or small grassy openings (Cooper, 1960, 1961). Since about 1900, most fires have been put out. As a consequence, there is increased density of pine regeneration, often in dense thickets, abundance of dwarf mistletoe (a tree parasite), and suppressed herbaceous understory (for a review, see Moir and Dieterich, 1988). The year 1919 is well known in the Southwest for exceptional density of pine regeneration. Today many forest managers are again reverting to burning, using prescribed fires to help bring about pine forests with well spaced trees (rather than crowded thickets), invigorated

understories, and fewer stands infected with mistletoe.

Succession in Ponderosa Pine/Gambel Oak is characterized by a long period of oak dominance. New Mexico Locust and other shrubs are often associated with Gambel Oak. This period of oak dominance can result from stand-replacing fires, as well as from clear-cutting Ponderosa Pine. In addition, seed tree and light shelterwood cutting also favor oaks and related shrubs over Ponderosa Pine (USFS, 1986). As succession progresses into later stages, Ponderosa Pine will eventually overtop the oak and New Mexico Locust. Once shaded, these species become low understory shrubs. The oak can persist well into late succession and climax stages as mere sprouts from long-lived rootstock. When the next opening is created, sprouts from this rootstock will again develop into a dense shrub or tree canopy in a few years. Some proportion of successional oak in the landscape is known to have important wildlife and visual benefits. But slow rates of pine establishment and growth in oak thickets can result in too much oak brush in relation to coniferous stages in districts where heavy pine harvests have taken place too rapidly.

Ponderosa Pine–Silverleaf Oak Series

This series is found where winter temperatures are mild. There is about 24–35 in./yr. (61–64 cm/yr.) of mean annual precipitation, and annual air temperatures average around 49°F. The soil temperature regime is mesic (see footnote 1, p. 48). Ponderosa Pine is associated with evergreen oaks. Species such as Silverleaf, Netleaf, and Arizona White Oak can occur as trees in open sites or as shrubs beneath the shade of taller pines. Some stands may have appreciable cover or density of Alligator Juniper. Herbs are usually scarce or poorly represented, although grasses, such as Longtongue Muhly and Mutton Grass, are often found.

These pine-oak forests of mild winters are limited to southwestern New Mexico on the southern flanks of the Mogollon Mountains and southern Arizona. Habitat types have been classified as Ponderosa Pine/Silverleaf Oak and Ponderosa Pine/Netleaf Oak (Muldavin et al., 1990).

Little is known about these forests, which are very localized in the state and which intergrade quickly to Douglas-fir/Silverleaf Oak forests described in the upper montane conifer forest section above.

Ponderosa Pine–Pinyon Pine– Gambel Oak Series

The dry fringe of Ponderosa Pine forests is usually characterized by cold winters. Mean annual precipitation is about 20 in./yr. (51 cm/yr.), and mean annual air temperature is about 43°F (6°C). The soil temperature regime is frigid (Carleton and Brown, 1974). This climate is indicated by associated deciduous broadleafed shrubs or trees, including Gambel and Wavyleafed oaks (*Quercus gambelii, Q. undulata*), Snowberry (*Symphoricarpos oreophila*), and Skunkbush Sumac (*Rhus tribolata*). Common associated evergreen trees are Pinyon Pine, One-seed Juniper, Rocky Mountain Juniper, and Alligator Juniper. Some of the more common understory herbs include Grama grasses, Little Bluestem (*Schizachyrium scoparium*), Big Bluestem (*Andropogon gerardii*), Mutton Grass (*Poa fendleriana*), Mountain Muhly (*Muhlenbergia montana*), Pine Dropseed (*Blepharoneuron tricholepis*) and Deervetch (*Lotus wrightii*) (Pl. 12).

The geography of this fringe pine forest of cold winter climates extends from south-central and western New Mexico (portions of Catron County) northward into Colorado. Habitat types include Ponderosa Pine/Blue Grama, Ponderosa Pine/Gambel Oak (Pinyon Pine phase), Ponderosa Pine/Indian Ricegrass, Ponderosa Pine/Low Sagebrush, and Ponderosa Pine/Wavyleaf Oak (DeVelice et al., 1986; Fitzhugh et al., 1987; Alexander et al., 1987; Alexander et al., 1984).

Severe disturbances in forests of this series often lead to prolonged midsuccessional dominance by oaks and/or junipers. An example is the so-called spring burn of 1955, near Mayhill in the Sacramento Mountains, where today Wavyleaf Oak (with or without Alligator or One-seed Junipers) forms extensive shrubby thickets within the Ponderosa Pine/Wavyleaf Oak habitat type. Another example, in the Zuni Mountains, is the almost total removal of Ponderosa Pine during railroad logging days. At low elevations where Ponderosa Pine forest formerly occurred, there are today Pinyon-Juniper woodlands. These woodlands may represent a prolonged seral stage of the Ponderosa Pine/Blue Grama habitat type. One can suggest that under currently warming climates, this woodland stage represents the present or future vegetation potential. Details of successional stages and rates in cold winter, "fringe" Ponderosa Pine forests have not been worked out. However, the successional status of major trees and shrubs and their general responses to silvicultural treatments are described in U.S. Forest Service publications (USFS, 1986, 1987a and b).

Ponderosa Pine–Pinyon Pine– Gray Oak Series

This series occurs under a mild winter climate indicated by the presence of evergreen oaks and associated broadleafed evergreen shrubs and trees (Carleton and Brown, 1983). Mean annual precipitation averages about 22 in./yr. (56 cm/yr.), and mean annual air temperature is about 52°F (11°C). The soil-temperature regime is mesic (see footnote 1, p. 48). Common trees associated with Ponderosa Pine in these mild winter climates include Pinyon Pine (*Pinus edulis*), One-seed and Alligator Junipers (*Juniperus monosperma, J. deppeana*), and such evergreen oaks as Gray Oak, Emory Oak, and Arizona White Oak (*Quercus grisea, Q. emoryi, Q. arizonica*). Some common shrubs and herbs include Wright's Silktassel (*Garrya wrightii*), Yuccas (especially *Yucca baccata*), Beargrass (*Nolina microcarpa*), Longtongue Muhly (*Muhlenbergia longiligula*), Bluestems (*Schizachyrium cirratum* and *S. sco-*

parium), Pinyon Ricegrass (*Piptochaetium fimbriatum*), and Gramas (*Bouteloua* spp.).

The geography of this mild winter climate is that of southwestern New Mexico, extending northward on south-facing slopes to about the San Mateo Mountains, southwest of Socorro. Habitat types of this low-elevation, dry pine forest include Ponderosa Pine/Gray Oak and Ponderosa Pine/Blue Grama (Gray Oak phase) (Fitzhugh et al., 1987).

Fires were frequent in these dry, warm forests prior to about 1880. Although not studied in New Mexico, a comparable forest in Arizona (Ponderosa Pine/Arizona White Oak habitat type) revealed (from fire scars) that low intensity surface fires occurred somewhere within the 215-acre study site in sixty-seven of the years between 1770-1870 (Dieterich and Hibbert, 1990). Since about 1880, fires have been mostly suppressed. Timber and fuelwood cutting, grazing by domestic livestock, wildfire suppression, and burning by prescription are contemporary activities that have produced many changes in forest structure since 1880. Today these pine-evergreen oak habitats exist under a wide variety of early and middle successional stages, ranging from oak-juniper thickets to young blackjack pine stands (a blackjack is a Ponderosa Pine usually less than about 150 years old, as indicated by its blackish bark). Studies on the variety of successional stages do not yet exist.

Chihuahua Pine Series

Forests dominated by Chihuahua Pine (*Pinus leiophylla*) are also found within the mild winter climate. Annual precipitation averages about 22 in./yr. (56 cm/yr.), and mean annual air temperature is about 52°F (11.1°C). Mean annual soil temperature around 51°F (10.6°C) indicates a mesic soil temperature regime.

Mild winters are indicated by many associated broadleafed, evergreen trees and shrubs. These include Manzanita (*Arctostaphylos pungens*),

Toumey Oak (*Quercus toumeyi*), Wright's Silktassel, Emory and Arizona White Oaks (*Quercus emoryi, Q. arizonica*), and occasional Silverleaf Oak (*Quercus hypoleucoides*). Associated conifers are Border Pinyon (*Pinus discolor*) and Alligator Juniper. Common evergreen rosette shrubs are Yuccas (especially *Yucca schottii*), Agave (*Agave parryi*), and Beargrass (*Nolina microcarpa*). An herbaceous flora can be well represented. Common species are Bullgrass and Longtongue Muhlys (*Muhlenbergia emersleyi, M. longiligula*), Mutton Grass, Sedge (*Carex geophila*), Beggartick Grass (*Aristida orcuttiana*), Pinyon Ricegrass, Texas Bluestem (*Schizachyrium cirratum*), wild beans (*Phaseolus* spp.), herby Sages (*Artemisia ludoviciana, A. carruthii*), and Pennyroyal (*Hedeoma hyssopifolium*).

Chihuahua Pine forests in New Mexico are restricted to the Peloncillo (Willging, 1987) and Big Lue Mountains. The major habitat type is Chihuahua Pine/Arizona White Oak (Muldavin et al., 1990).

Principal influences in the Chihuahua Pine forest are fires, fuelwood cutting, and livestock grazing. Fire frequency has not been studied. However, stand-replacing fires can be expected to encourage oak and Alligator Juniper sprouting and seed germination of such shrubs as Manzanita and Deerbrush (*Ceanothus greggii, C. fendleri*). Chihuahua Pine, with its persistent serotinous and semi-serotinous cones, is a fire-adapted conifer and exhibits shade tolerance in an understory of Oaks and Junipers. Thus, midsuccession appears to be dominated by oaks and Junipers, and late succession by emergence of Chihuahua Pine above the Oak-Juniper canopy. Border Pinyon also exhibits shade tolerance and has been suggested as a late successional or climax pine (Muldavin et al., 1990; USFS, 1987b).

Fuelwood cutting stimulates oak and Alligator Juniper sprouting. Some maintenance of seed germinating shrubs may take place in hot, cleared microsites, but the stimulatory effect of high-temperature seed germination, induced by fire, is absent. Therefore, Manzanita and Deerbrush would gradually decline.

The present-day composition of these Chihuahua Pine forests represents a long history of domestic livestock grazing in most areas (Moir, 1979). Livestock grazing doubtless changed the vegetation according to livestock management practices (livestock densities and season of grazing). In the past, yearlong cattle grazing and simple deferred-rotation pasture systems tended to reduce the palatable herbaceous component and favor poisonous or armed understory species, such as herby Sages, Snakeweed and Catclaw Mimosas (*Mimosa biuncifera, M. dysocarpa*). In turn, these changes in understory composition probably influenced asexual reproduction of less palatable rhizomatous plants, such as the herby Sages, and mat forming, low grasses, such as Blue Grama (Neilson, 1986).

ASPEN DISTURBANCE FOREST

Forests of more or less pure Aspens (*Populus tremuloides*) are common in New Mexico (although their present extent is considerably less than it was in the early 1900s). These forests are very important for scenic beauty, wildlife, and other purposes. Most Aspen stands originate from a fire holocaust within a coniferous forest. After such events, Aspen is a prolific sprouter from rootstock that remained alive during the previous coniferous forest regime. The ecology of Aspen as a fire sprouter has been thoroughly summarized by DeByle and Winokur (1985).

Other Aspen groves originate from montane and subalpine meadows. Aspen becomes established in these grassy meadows rather quickly when fires are excluded or when the meadows are heavily grazed by livestock (Allen, 1989). One suggestive clue to a meadow origin of Aspen is the abundance of Thurber or Arizona Fescues under the Aspen canopy. These grasses require full meadow conditions for reproduction, but individual grass clumps can persist for years under the partial shade of an Aspen canopy. Another source of evidence for Aspens taking over high-elevation meadows are repeat air photographs (Allen, 1989).

Aspen can also originate on scree, debris slopes, or active talus. These cobbly, fragmental soils can support open and persistent stands, which may have encroached via rootstock from trees on adjacent finer-textured soils.

Generally, Aspen has been considered a seral tree in the Southwest (Moir and Ludwig, 1979). It is mostly a minor seral tree in the Douglas-fir and White Fir–Douglas-fir–Ponderosa Pine series. Here Aspen forms only localized clones, with little tendency for clones to spread and coalesce. Grasses, Oaks, or purely coniferous stands otherwise restrict the Aspen to established microsites. Aspen becomes more extensive and continuous in the remainder of the mixed conifer and in most forests of the lower and mid-elevation subalpine region. At upper elevations of the subalpine, Aspen sharply declines as a seral tree; clones above about 11,500 ft. (3,500 m) are rare.

Despite its importance as a seral tree, surprisingly little is known about Aspen succession (DeByle and Winokur, 1985). There may be certain stands in northern New Mexico where succession to conifers is slow enough (stands support less than 20–30 conifers per acre after 100 years of Aspen occupancy) that Aspen can be considered climax, for all practical purposes (Powell, 1988). On the other hand, Aspen is subject to numerous insect and disease attacks. Fungus damage in mature stands includes White Tree Rot, Sooty-bark Cankers, Flamulina Butt Rot, *Cyptosphaeria* cankers, and several root rots.

Damaging insects are Aspen Tortrix and Western Tent Caterpillar. Complexes of insects and disease include Ink Spot, Marsonnia Leaf Spot, and stem burls caused by Aspen Gall. Damage from big-game and domestic livestock browsing can weaken clones and ramets (individual suckers, or stems from a common rootstock), making them susceptible to fungal diseases. These and other causes of decline and mortality in Aspen groves are often quite evident in mature stands (Powell, 1988; USFS, 1988).

As a consequence of these built-in agents of Aspen decline, groves or ramets older than about 150 years are rare. More likely, stands simply fall apart within a short time after maturity. Numerous down Aspen logs are evidence of the rapid transition to coniferous dominance. In a matter of only one or two decades, there can be progression from an almost total Aspen overstory, to mixed Aspen-conifer overstory, and finally to a mostly coniferous overstory, where persisting Aspen ramets account for less than about 5% of canopy dominance. The exact timing of such succession depends on many factors, including the genetics of the particular Aspen clones, soils and other environmental conditions, stand conditions immediately following the originating fire, wildlife and livestock browsing, and the rapidity of coniferous establishment in the Aspen understory (DeByle and Winokur, 1985; Crouch, 1986).

Aspen has been observed to persist in coniferous stands of late succession or even near-climax stages. This persistence has two contrasting forms. Sprouts from rootstock emerge above the ground, grow for at most a few years, and then are browsed or die back. Such sprouting can go on for years, provided that the rootstock complex itself can maintain a reserve of carbohydrates. This reserve is provided by the second of the persisting Aspen forms in coniferous stands: the occasional dominant or upper canopy level ramet. These trees act as photosynthetic pumps, recharging the rootstock with carbohydrate reserves year after year. Should these large trees die, Aspen suckering would cease in a few years (the exact number of years is unknown).

It is not unusual to observe Aspen borders in open meadows or along roadsides, where behind the border is an essentially coniferous forest. In this case, the productive ramets are the border trees, and the dependent rootstock for suckering occur within the coniferous interior. Here forest succession is most likely taking place as a process of meadow replacement. Rates of meadow replacement by trees seem to be very rapid (Allen, 1989). Large areas of former Thurber Fescue grasslands in the Jemez Mountains, Mount Taylor area, and in the Pecos Wilderness are now closed Aspen stands or closed stands of mixed Aspen-conifer. A common community type is Aspen/Thurber Fescue (for a description, see Powell, 1988).

In the Upper Cruces Basin of the San Juan Mountains, a reverse process has taken place. Former Aspen groves have died out and montane and subalpine grasslands prevail. Remnant live trees and numerous downed Aspen logs are evidence of groves that have fallen apart and reverted to herbaceous communities. A long history of sheep and livestock grazing, big-game browsing, subsequent soil erosion, and possibly fire are probable causative factors, although the exact mechanisms remain obscure. A somewhat similar observation of Aspen senescence in groves adjoining grassy meadows of the Valle Vidal in the Cimarron Range was made by Williams and Moir (1984). Here soil erosion was not so evident, but browsing of stems and suckers by a large elk herd was obvious.

Forest managers are attempting to maintain Aspen using fire prescriptions and silvicultural techniques. Clear-cutting small patches (ca. 10–50 acres) may or may not be followed by a residue or slash burn. Results have been only partially successful, and in some cases there was little Aspen response. One known factor of failure to produce an Aspen grove by this technique is the attraction these small areas have for domestic livestock and large game browsers, such as deer or elk. If the wider landscape is largely devoid of palatable browse, these treatments serve as attractants when prolific suckering takes place. Animals quickly damage or even eliminate the suckers. Other failures may result as a consequence of an already high background of soil pathogens and attacking insects, or from Aspen rootstock low in carbohydrate reserves or mechanically damaged during treatment.

Large gaps created in mature or old-growth forests are probably a means of extending the longevity of Aspen into late stages of coniferous forests in the absence of fire. But little is known about how large a forest gap must be in order for an Aspen sucker to have the possibility of attaining canopy dominance and thus recharging or perpetuating the clonal rootstock. Forests that are in advanced stages of succession and have especially long intervals between fires, for example high-elevation forests in the corkbark fir–engelmann spruce fir series, may lose Aspen within a few years after the last tall ramets die. Forests within the Pecos Wilderness seem to illustrate this possibility. In some places, Aspen persists among midsuccessional stands of spruce and corkbark fir; elsewhere it is absent, even in young forests of spruce or corkbark fir which originated from fire. Another case of a high-elevation forest (above 10,000 ft., or 3,050 m) that seems to have lost Aspen because of infrequent fires or insufficient gap replacement is found in the vicinity of Sierra Blanca Peak (Dye and Moir, 1977). There are several large, downed Aspen logs in some stands, but no large living trees. In some old-growth stands there is little recent evidence of Aspen sprouting. It is doubtful whether clearcutting or fire replacement would result in an Aspen stage.

There is much yet to be learned about Aspen and its longevity and role in succession. It may well be that an alternation of forest generations (a coniferous generation, then an Aspen generation) is one of nature's mechanisms for maintaining forest health and productivity. Elsewhere, back-to-back generations of Aspen may be possible (Powell, 1988). Foresters in the Southwest have some excellent beginning studies (Johnston and Hendzel, 1985; Powell, 1988; DeByle and Winokur, 1985) and some background of practical experience. They have the scale and frequency of the natural disturbances that created Aspen stands to compare with the scale and frequency of silvicultural treatments intended to create and sustain Aspen in smaller patches. Eventually a solution will emerge between the size and frequency of forest disturbance and the size of the landscape required to maintain a proportion of the Aspen stage of succession in an otherwise coniferous landscape.

Table 5.1. Major Tundra and Forest Vegetation Types in New Mexico. (Associations adapted from Forest Service habitat types.) MS = mixed shrubs, MG-F = mixed grasses and forbs, S = sparse shrubs or herbs, * = riparian forest. A climatic gradient code ( ) has been used for series: C = cold winters, M = mild winters; 8 = tundra, 7 = subalpine forest, 6 = mixed conifer forest, 5 = pine forest; local climates given as: –1 = warm-dry, 0 = modal, +1 = cool-wet.

TUNDRA VEGETATION
  ALPINE TUNDRA
    Fellfield Series (C8, 0, –1)
      Potentilla sierrae-blancae

    Sedge Series (C8, 0, –1)
      Carex rupestris
      Kobresia myosuroides

    Alpine Avens Series (C8, 0, –1)
      Geum rossii–Polygonum bistortoides
      Geum rossii–Sibbaldia procumbens

    Willow Series (C8, snowmelt)
      Salix nivalis (Salix petrophila)/MF

FOREST VEGETATION
  SUBALPINE CONIFEROUS FOREST
    Bristlecone Pine Series (C7, +1)
      Pinus aristata/*Ribes montigenum*/S

    Engelmann Spruce–Bristlecone Pine Series (C7, +1)
      Picea engelmanni–Pinus aristata/S/
        Mertensia ciliata
      Picea engelmannii–Pinus aristata/*Vaccinium myrtillus*/*Polemonium pulcherrimum*

    Corkbark Fir–Engelmann Spruce Series (C7, 0)
      Abies lasiocarpa–Picea engelmannii/MS/
        Senecio sanguisorboides
      Abies lasiocarpa–Picea engelmannii/S/Moss
      Abies lasiocarpa–Picea engelmannii/
        Vaccinium myrtillus/MG-F
      Abies lasiocarpa–Picea engelmannii/S/
        Lathyrus arizonicus
      Abies lasiocarpa–Picea engelmannii/S/
        Mertensia ciliata

    Corkbark Fir–Engelmann Spruce–White Fir Series (C7, –1)
      Abies lasiocarpa–Picea engelmannii–Abies concolor/*Vaccinium myrtillus*/*Linnaea borealis*

Abies lasiocarpa–Picea engelmannii–Abis concolor/*Vaccinium myrtillus–Rubus parviflorus*/S
Abies lasiocarpa–Picea engelmannii–Abis concolor/MS/*Erigeron eximus*

    Engelmann Spruce–Douglas-fir Series (C7, –1)
      Picea engelmannii–Pseudotsuga menziesii/
        Vaccinium myrtillus/MG-F
      Picea engelmannii–Pseudotsuga menziesii/
        MS/*Lathyrus arizonicus*
      Picea engelmannii–Pseudotsuga menziesii/
        MS/Moss
      Picea engelmannii–Pseudotsuga menziesii/
        MS/*Elymus triticoides*
      Picea engelmannii–Pseudotsuga menziesii/
        MS/*Erigeron eximius*

    Engelmann Spruce–Limber Pine Series (C7, –1)
      Picea engelmannii–Pinus flexilis/
        Arctostaphylos uva-ursi/MG-F

  UPPER MONTANE CONIFEROUS (MIXED CONIFER) FOREST
    Blue Spruce Series (C6, cold air)
      **Picea pungens*/*Cornus stolonifera*/MG-F
      Picea pungens/*Arctostaphylos uva-ursi*/
        MG-F
      Picea pungens/S/*Festuca arizonica*
      Picea pungens/MS/*Carex foenea*
      Picea pungens/MS/*Erigeron eximius*
      **Picea pungens*/MS/*Poa pratensis*

    White Fir–Douglas-fir Series (C6, +1)
      Abies concolor–Pseudotsuga menziesii/*Acer glabrum*/MG-F
      Abies concolor–Pseudotsuga menzeisii/
        Vaccinium myrtillus/MG-F
      **Abies concolor–Pseudotsuga menzeisii*/*Acer grandidentatum*/MG-F
      Abies concolor–Pseudotsuga menzeisii/MS/
        Erigeron eximius

Table 5.1. (continued).

Abies concolor–Peudotsuga menzeisii/MS/
Elymus triticoides

Douglas-fir–Southwestern White Pine Series
(C6, +1)
Pseudotsuga menzeisii–Pinus strobiformis/
MS/*Bromus ciliatus*
Pseudotsuga menzeisii–Pinus strobiformis/
Physocarpus monogynus/MG-F

White Fir–Douglas-fir–Ponderosa Pine Series
(C6, 0)
Abies concolor–Pseudotsuga menzeisii–Pinus
ponderosa/Arctostaphylos uva-ursi/MG-F
Abies concolor–Pseudotsuga menzeisii–Pinus
ponderosa/Berberis repens/S
Abies concolor–Pseudotsuga menzeisii–Pinus
ponderosa/Robinia neomexicana/MG-F
Abies concolor–Pseudotsuga menzeisii–Pinus
ponderosa/Symphoricarpos oreophilus/
MG-F
Abies concolor–Pseudotsuga menzeisii–Pinus
ponderosa/Quercus gambelii/MG-F
Abies concolor–Pseudotsuga menzeisii–Pinus
ponderosa/Festuca arizonica
Abies concolor–Pseudotsuga menzeisii–Pinus
ponderosa/S/Muhlenbergia virescens

Douglas-fir–Limber Pine (Bristlecone Pine)
Series (C6, −1)
Pseudotsuga menzeisii–Pinus flexilis (Pinus
aristata)/S/Festuca thurberi
Pseudotsuga menzeisii–Pinus flexilis (Pinus
aristata)/MS/Muhlenbergia montana

Douglas-fir–Gambel Oak Series (C6, −1)
Pseudotsuga menzeisii–Quercus gambelii/
Physocarpus monogynus/MG-F
Pseudotsuga menzeisii–Quercus gambelii/
Quercus gambelii/MG-F
Pseudotsuga menzeisii–Quercus gambelii/
Quercus undulata/MG-F
Pseudotsuga menzeisii–Quercus gambelii/S/
Festuca arizonica
Pseudotsuga menzeisii–Quercus gambelii/
MS/*Muhlenbergia montana*

Douglas-fir–Silverleaf Oak Series (M6, −1)
Pseudotsuga menzeisii–Quercus
hypoleucoides/MG-F

Lower Montane Coniferous Forest
Ponderosa Pine–Gambel Oak Series
(C5, +1, 0)
Pinus ponderosa–Quercus gambelii/Quercus
gambelii/MG-F
Pinus ponderosa–Quercus gambelii/
Arctostaphylos uva-ursi/MG-F
Pinus ponderosa–Quercus gambelii/Quercus
undulata/MG-F
Pinus ponderosa–Quercus gambelii/MS/
Festuca arizonica
Pinus ponderosa–Quercus gambelii/MS/
Muhlenbergia montana

Ponderosa Pine–Silverleaf Oak Series
(M5, +1, 0)
Pinus ponderosa–Quercus hypoleucoides/
Quercus hypoleucoides/MG-F
Pinus ponderosa–Quercus hypoleucoides/
Quercus rugosa/S

Ponderosa Pine–Pinyon Pine–Gambel Oak
Series (C5, −1)
Pinus ponderosa–Pinus edulis–Quercus
gambelii/Quercus gambelii/MG-F
Pinus ponderosa–Pinus edulis–Quercus
gambelii/Quercus undulata/MG-F
Pinus ponderosa–Pinus edulis–Quercus
gambelii/Artemisia arbuscula/MG-F
Pinus ponderosa–Pinus edulis–Quercus
gambelii/MS/*Bouteloua gracilis*

Ponderosa Pine–Pinyon Pine–Gray Oak Series
(M5, −1)
Pinus ponderosa–Pinus edulis–Quercus
grisea/Quercus grisea/MG-F
Pinus ponderosa–Pinus edulis–Quercus
grisea/MS/*Bouteloua gracilis*

Chihuahua Pine Series (M5, −1)
Pinus leiophylla/Quercus hypoleucoides/
MG-F

Successional–Disturbance Forest
Aspen Successional Series (C6, C5, +1, 0)
Populus tremuloides/MS/MG-F

Table 5.2. Major Plants Comprising Tundra Vegetation in New Mexico.

Scientific Name	Common Name
CUSHION PLANTS	
Eritrichium nanum	Forget-Me-Not
Minuartia obtusiloba	Alpine Sandwort
Phlox caespitosa	Alpine Phlox
Potentilla sierrae-blancae	Sierra Blanca Cinquefoil
Silene acaulis	Alpine Campion
Tonestus pygmaeus	Cushion Yellow Aster
Trifolium dasyphyllum	Shaggyleaf Clover
Trifolium nanum	Dwarf Clover
PROSTRATE WOODY PLANTS	
Salix nivalis	Arctic Willow
Salix petrophila	Snow Willow
GRASSES AND GRASSLIKE PLANTS	
Agropyron trachycaulum	Slender Wheatgrass
Carex elynoides	sedge
Carex rupestris var. *drummondiana*	sedge
Deschampsia caespitosa	Tufted Hairgrass
Festuca brachyphylla	Alpine Fescue
Juncus drummondii	rush
Kobresia bellardii	Kobresia
Luzula spicata	Spike Woodrush
Poa arctica	Arctic Bluegrass
Trisetum spicatum spp. *spicatum*	Spike Trisetum
FORBS	
Artemisia scopulorum	Alpine Sage
Bessya alpina	Alpine Bessya
Caltha leptosepala	Marsh Marigold
Castilleja haydenii	Hayden Paintbrush
Castilleja occidentalis	Western Yellow Paintbrush
Clementsia rhodantha	Rose Crown
Erigeron coulteri	Coulter Fleabane
Erigeron melanocephalus	Blackheaded Daisy
Geum rossii	Alpine Avens
Hymenoxys acaulis	
Ligularia soldanella	Alpine Nodding Groundsel
Lloydia serotina	Alpine Lily
Polygonum bistortoides	Bistort
Potentilla diversifolia	Cinquefoil
Primula ellisiae	Ellis Primula
Rhodiola integrifolia	King's Crown
Rydbergia grandiflora	Old Man of the Mountain
Saxifraga chrysantha	Yellow Saxifrage
Saxifraga flagellaris	Spreading Saxifrage
Senecio atratus	Scree Groundsel
Sibbaldia procumbens	Sibbaldia
Trifolium brandegei	Spreading Clover
Veronica wormskjoldii	Speedwell

Table 5.3. Major Plants Comprising Forest Vegetation in New Mexico. * = common weedy plants of little indicator value except past disturbance.

Scientific Name	Common Name
TREES AND SHRUB-TREES	
Acer glabrum	Rocky Mountain Maple
Abies concolor	White Fir
Abies lasiocarpa	Subalpine Fir, Corkbark Fir (var. *arizonica*)
Juniperus deppeana	Alligator Juniper
Picea engelmannii	Engelmann Spruce
Picea pungens	Blue Spruce
Pinus aristata	Bristlecone Pine
Pinus edulis	Rocky Mountain Pinyon Pine
Pinus flexilis	Limber Pine
Pinus leiophylla	Chihuahua Pine
Pinus ponderosa	Ponderosa Pine
Pinus strobiformis	Southwestern White Pine
Populus tremuloides	Aspen
Pseudotsuga menziesii	Douglas-fir
Robinia neomexicana	New Mexico Locust
Quercus arizonica	Arizona White Oak
Quercus gambelii	Gambel Oak
Quercus grisea	Gray Oak
Quercus hypoleucoides	Silverleaf Oak
SHRUBS	
Agave parryi	Parry Agave
Arctostaphylos uva-ursi	Bearberry, Kinnikinnick
Artemisia arbuscula	Black Sage, Low Sage
Artemisia tridentata	Big Sagebrush
Berberis repens	Creeping Mahonia
Ceanothus fendleri	Fendler Buckbrush
Cercocarpus montanus	Mountain Mahogany
Garrya wrightii	Wright Silktassel
Holodiscus dumosus	Rock Spiraea
Hymenoxys richardsonii	Pingue
Jamesia americana	Cliffbush, Waxbush
Juniperus communis	Common Juniper
Lonicera arizonica	Arizona Honeysuckle
Lonicera utahensis	Utah Honeysuckle
Pachistima myrsinites	Mountain Lover
Physocarpa monogynus	Ninebark
Potentilla fruticosa	Shrubby Cinquefoil*
Quercus undulata	Wavyleaf Oak
Rhus glabra	Smooth Sumac
Rhus radicans	Poison Ivy
Rhus trilobata	Skunkbush

Table 5.3. (continued).

Scientific Name	Common Name
Ribes cereum	Wax Currant
Ribes inerme	
Ribes montigenum	Mountain Currant
Ribes pinetorum	Prickly Currant
Ribes wolfii	Mapleleaf Currant
Rosa woodsii	Woodrose
Rubus parviflorus	Thimbleberry
Rubus strigosa	Raspberry
Salix glauca var. *glabrescens*	Smoothleaf Willow
Salix scouleriana	Forest Willow
Sambucus caerulea	Blue Elderberry
Sambucus microbotrys	Rocky Mountain Red Elderberry
Sambucus neomexicana	New Mexican Elderberry
Shepherdia canadensis	Buffaloberry
Sorbus dumosa	Mountain Ash
Symphoricarpos oreophilus	Snowberry
Vaccinium myrtillus	Myrtleleaf Huckleberry
Vaccinium scoparium	Grouse Whortleberry

GRASSES AND GRASSLIKE PLANTS

Scientific Name	Common Name
Andropogon gerardii	Big Bluestem
Agropyron smithii	Western Wheatgrass
Agropyron trachycaulum	Slender Wheatgrass
Aristida orcuttiana	Beggartick Grass
Blepharoneuron tricholepis	Pine Dropseed
Bouteloua curtipendula	Sideoats Grama
Bouteloua gracilis	Blue Grama
Bromus anomalus	Nodding Brome
Bromus ciliatus	Fringed Brome
Carex bella	Pretty Sedge
Carex foenea	Foeny Sedge
Carex geophila	Earth Sedge
Carex rossii	Ross Sedge
Danthonia parryi	Parry Oatgrass
Elymus glaucus	Blue Ryegrass
Festuca arizonica	Arizona Fescue
Festuca sororia	Forest Fescue
Festuca ovina	Sheep Fescue
Festuca thurberi	Thurber Fescue
Koeleria pyramidata	Junegrass
Luzula parviflora	Millet Woodrush
Muhlenbergia longiligula	Longtongue Muhly
Muhlenbergia montana	Mountain Muhly
Muhlenbergia virescens	Screwleaf Muhly

Table 5.3. (continued).

Scientific Name	Common Name
Oryzopsis asperifolia	Roughleaf Ricegrass, Forest Ricegrass
Piptochaetium fimbriatum	Pinyon Ricegrass
Poa fendleriana	Muttongrass
Poa pratensis	Kentucky Bluegrass
Poa reflexa	
Schizachyrium scoparium	Little Bluestem
Sitanion hystrix	Bottlebrush Squirreltail
Trisetum spicatum spp. *montanum*	Spikeoats

FORBS
Achillea lanulosa	Western Yarrow*
Aconitum columbianum	Monkshood
Actea arguta	Baneberry
Agastache pallidiflora	Giant Hyssop
Allium cernuum	Nodding Onion
Allium geyeri	Geyer Onion
Antennaria parvifolia	Silver Pussytoes
Arenaria fendleri	Sandwort
Aquilegia spp.	Columbine
Artemisia ludoviciana	Louisiana Wormwood
Asclepias speciosa	Showy Milkweed*
Astragalus humistratus	Enemaweed
Bidens bipinnata	Spanish Needles
Bidens tenuisecta	Sticktight
Campanula rotundifolia	Bellflower
Castilleja linariaefolia	Wyoming Paintbrush
Cerastium brachypodium	Mouseear Chickweed
Cirsium undulatum	Wavyleaf Thistle
Commelina dianthifolia	Dayflower
Crepis acuminata	Tapertip Hawksbeard
Crepis intermedia	
Epilobium angustifolium	Fireweed*
Erigeron delphinifolius	New Mexico Fleabane
Erigeron eximius	Forest Fleabane
Erigeron formosissimus	Meadow Fleabane
Erigeron subtrinervis	Threenerve Fleabane
Eriogonum jamesii	Buckwheat
Eriogonum racemosum	Redroot Wildbuckwheat
Eupatorium herbaceum	Desert Thoroughwort
Eupatorium maculatum	Joe-Pye-Weed
Euphorbia fendleri	Spurge
Euphorbia spathulata	Spurge
Fragaria americana	Wild Strawberry
Fragaria ovalis	Wild Strawberry

Table 5.3. (continued).

Scientific Name	Common Name
Galium boreale	Northern Bedstraw
Galium fendleri	Fendler Bedstraw
Galium microphyllum	Littleleaf Bedstraw
Galium triflorum	Sweetscented Bedstraw
Gentiana heterosepala	
Gentiana strictiflora	
Geranium caespitosum	Purple Geranium, Sticky Geranium
Geranium richardsonii	Pale Geranium
Geum strictum	
Gnaphalium macounii	Winged Cudweed
Goodyera oblongifolia	Rattlesnake Plantain
Hackelia hirsuta	
Hackelia pinetorum	
Hedeoma hyssopifolium	False Pennyroyal
Hedeoma oblongifolium	
Helenium hoopesii	Orange Sneezeweed*
Helianthus nuttallii	Nuttall Sunflower
Hieracium fendleri	
Heterotheca villosa	Golden Aster
Ipomopsis aggregata	Skyrocket*, Scarlet Trumpet Flower
Lathyrus arizonica	Arizona Peavine
Lathyrus graminifolius	Grassleaf Peavine
Ligusticum porteri	Chuchupate, Osha
Linnaea borealis	Twinflower
Linum lewisii	Western Blue Flax
Lithospermum multiflorum	Puccoon
Lobelia cardinalis	Cardinal Flower
Lotus wrightii	Yelloweyed Pea, Deervetch
Lupinus argenteus	Silvery Lupine
Marrubium vulgare	Hoarhound*
Mentha arvensis	Mint
Mertensia franciscana	Franciscan Bluebells
Mirabilis decipiens	
Mirabilis linearis	
Oenothera hartweggii	
Oenothera hookeri	Late Primrose
Oreochrysum parryi	Parry Goldenweed
Osmorhiza depauperata	Sweet Cicely
Oxalis stricta	Sheepsour
Oxytropis lambertii	Lambert Locoweed
Oxytropis sericea	Silverleaf Locoweed
Pedicularis racemosa	White-flowered Lousewort
Pedicularis grayi	Gray Lousewort
Penstemon barbatus	Narrowleaf Penstemon

Table 5.3. (continued).

Scientific Name	Common Name
Penstemon virgatus	Wand Penstemon
Pericome caudata	
Phacelia neomexicana	New Mexican Scorpionweed
Polemonium foliosissimum	
Polemonium pulcherrimum	Jacobs Ladder
Polygonum sawatchense	Knotweed, Smartweed*
Potentilla hippiana	Horse Cinquefoil
Potentilla pulcherrima	Beauty Cinquefoil
Potentilla thurberi	Red Cinquefoil
Pseudocymopterus montanus	Mountain Parsley
Pteridium aquilinum	Bracken Fern*
Pyrola chlorantha	Wintergreen
Ramischia secunda	Sidebells Wintergreen
Ratibida columnaris	Prairie Coneflower
Scrophularia montana	Mountain Figwort
Senecio amplectens	Timberline Groundsel
Senecio cardamine	Cardamine Groundsel
Senecio neomexicanus	New Mexico Groundsel
Senecio sanguisorboides	Burnet Groundsel
Silene laciniata	Cutleaf Campion
Smilacina racemosa	Broadleaf False Solomonseal
Smilacina stellata	Stary Smilac
Solanum nigrum	Black Nightshade
Solidago missouriensis	Missouri Goldenrod
Solidago multiradiata	Alpine Goldenrod
Solidago sparsiflora	Fewflowered Goldenrod
Swertia radiata	Deersears
Taraxacum officinale	Common Dandelion*
Thalictrum fendleri	Meadowrue
Thermopsis pinetorum	Big Goldenpea, Golden Banner
Tradescantia pinetorum	Spiderwort
Tragopogon dubius	Salsify
Trifolium dasyphyllum	Alpine Clover
Urtica gracilis	Nettle
Valeriana edulis	Thickleaf Valerian
Verbascum thapsus	Mullein*
Verbena macdougalii	
Vicia americana	American Vetch
Viola canadensis	Canada Violet
Zygadenus elegans	Death Camas
Zygadenus paniculatus	Death Camas

CHAPTER 5 REFERENCES

ALPINE TUNDRA AND CONIFEROUS FORESTS

Alexander, B. G., F. Ronco, Jr., E. L. Fitzhugh, and J. A. Ludwig, 1984. A classification of forest habitat types of the Lincoln National Forest, New Mexico. USDA For. Serv. Gen. Tech. Report RM-104, 29 p.

Alexander, B. G., F. Ronco, Jr., A. S. White, and J. A. Ludwig, 1987. A classification of forest habitat types of the northern portion of the Cibola National Forest, New Mexico. USDA For. Serv. Gen. Tech. Report RM-143, 35 p.

Allen, C. D., 1989. Changes in the landscape of the Jemez Mountains, New Mexico. Ph.D. dissertation, Univ. California, Berkeley, CA.

Andrews, Tom, 1983. Subalpine meadow and alpine vegetation of the upper Pecos River. Final Report USDA For. Serv. Southwestern Region, Contr. RM-51-82, 54 p., Albuquerque, NM (Range Management Libr. #3873).

Baker, W. L., 1983. Alpine vegetation of Wheeler Peak, New Mexico, U.S.A.: gradient analysis, classification, and biogeography. Arctic & Alpine Res. 15: 223–240.

Berg, N. H., 1988. Krummholz snowdrifts: hydrologic implications at a Colorado treeline site. pp. 141–144. In: Management of subalpine forests: building on 50 years of research. USDA For. Serv. Gen. Tech. Report RM-149, vii, 253 p.

Brookes, M. H., J. J. Colbert, R. G. Mitchell, and R. W. Stark (Tech. Coors.), 1985. Managing trees and stands susceptible to western spruce budworm. USDA For. Serv. and Coop. State Research Serv. Tech. Bull. 1695, vi, 111 p.

Brown, D., S. M. Hitt, and W. H. Moir (eds.), 1986. The path from here: integrated forest protection for the future. Integrated Pest Management Working Group. 7 chapters + exhibits, U.S. Govt. Printing Off., Washington, DC, 303 p.

Carleton, O. and G. Brown, 1983. Primary climatic gradients in New Mexico and Arizona. pp. 85–90. In: Proceedings of the workshop on Southwestern habitat types, April 6–8, 1983, Albuquerque, New Mexico. USDA For. Serv. Southw. Region, Albuquerque, NM, 110 p.

Carleton, O., L. Young, and C. Taylor, 1974. Climosequence study of the mountainous soils adjacent to Santa Fe, New Mexico, USDA For. Serv. Southw. Region, Albuquerque, NM, 24 p.

Cooper, C. F., 1960. Changes in vegetation, structure, and growth of southwestern pine forests since white settlement. Ecol. Monogr. 30: 129–164.

Cooper, C. F., 1961. Pattern in ponderosa pine forests. Ecology 42: 493–499.

Crouch, G. L., 1986. Aspen regeneration in 6- to 10-year-old clearcuts in southwestern Colorado. USDA For. Serv. Res. Note RM-467, 4 p.

DeByle, N. V. and R. P. Winokur (eds.), 1985. Aspen: ecology and management in the western United States. USDA For. Serv. Gen. Tech. Report RM-119, 283 p.

DeVelice, R. L., J. A. Ludwig, W. H. Moir, and F. Ronco, Jr., 1986. A classification of forest habitat types of northern New Mexico and southern Colorado. USDA For. Serv. Gen. Tech. Report RM-131, 59 p.

Dieterich, J. H., 1983. Fire history of southwestern mixed conifer: a case history. Forest Ecol. & Management 6: 13–31.

Dieterich, J. H. and A. R. Hibbert, 1990. Fire history in a small ponderosa pine stand surrounded by chaparral. pp. 168–173. In: J. S. Krammes (Tech. Coordinator), Proceedings effects of fire management of Southwestern natural resources, 15–17 Nov. 1988, Tucson, Arizona. USDA For. Serv. Gen. Tech. Report RM-191, 293 p.

Dye, A. J. and W. H. Moir, 1977. Spruce-fir forest at its southern distribution in the Rocky Mountains, New Mexico. Amer. Midl. Nat. 97: 133–146.

Fellin, David, William H. Moir, James F. Linnane, and Ann M. Lynch, 1990. Western spruce budworm in Arizona and New Mexico: outbreak history and effects of direct suppression programs. Manuscript on file, USDA Forest Serv., Rocky Mountain Forest & Range Exp. Station, Ft. Collins, CO.

Ferguson, C. W., 1968. Bristlecone pine: science and esthetics. Science 159: 839–846.

Fitzhugh, E. L., W. H. Moir, J. A. Ludwig, and F. Ronco, Jr., 1987. Forest habitat types in the Apache, Gila, and part of the Cibola National Forests, Arizona and New Mexico. USDA For. Serv. Gen. Tech. Report RM-145, 116 p.

Franklin, J. F., W. H. Moir, G. W. Douglas, and C. Wiberg, 1970. Invasion of subalpine meadows by trees in the Cascade Range, Washington and Oregon. Arctic & Alpine Res. 3: 215–224.

Franklin, J. F., T. Spies, D. Perry, M. Harmon, and A. McKee, 1986. Modifying douglas-fir management regimes for nontimber objectives. pp. 373–379. In: C. D. Oliver, D. P. Hanley, J. A. Johnson (eds.), Douglas-fir: stand management for the future: proceedings of a symposium, June 18–20, 1985, Seattle, Washington. Contribution 55, College of Forest Resources, Univ. Washington, Seattle.

Glover, V. J., 1984. Logging railroads of the Lincoln National Forest, New Mexico. USDA For. Serv. Southw. Region, Cultural Resources Management Report 4, Albuquerque, NM, iv, 65 p.

Glover, V. J. and J. P. Hereford, Jr., 1986. Zuni Mountain railroads, Cibola National Forest, New Mexico. USDA For. Serv. Southw. Region, Cultural Resources Management Report 6, iv, 88 p.

Hanks, J. P. and W. A. Dick-Peddie, 1974. Vegetation patterns of the White Mountains, New Mexico. Southw. Nat. 18: 372–382.

Hansen-Bristow, K. J., 1986. Influence of increasing elevation on growth characteristics at timberline. Can. J. Bot. 64: 2517–2523.

Harris, L. D., 1983. The fragmented forest, island biogeography theory and the preservation of biotic diversity. Univ. Chicago Press, Chicago and London, xviii, 211 p.

Johnston, B. C. and L. Hendzel, 1985. Examples of aspen treatment, succession, and management in western Colorado. USDA For. Serv. Rocky Mountain Region, Lakewood, CO, 164 p.

Jones, J. R., 1974. Silviculture of southwestern mixed conifer and aspen: the status of our knowledge. USDA For. Serv. Res. Paper RM-122, 44 p.

Jones, R. A., 1981. Summary of snow survey measurements for Arizona and pertinent portions of New Mexico, 1938–1980. USDA Soil Conserv. Serv., Salt River Valley Water Users Assoc., and Ariz. Dept. Water Resources, Phoenix, 179 p.

LaMarche, V. C., Jr., D. A. Graybill, H. C. Fritts, and M. R. Rose, 1984. Increasing atmospheric carbon dioxide: tree ring evidence for growth enhancement in natural vegetation. Science 225: 1019–1021.

Leaf, C. F., 1975. Watershed management in the Rocky Mountain subalpine zone: the status of our knowledge. USDA For. Serv. Res. Paper RM-137, 31 p.

Marr, J. W., 1961. Ecosystems of the east slope of the Front Range in Colorado. Univ. Colo. Studies, Series in Biology No. 8, Univ. Colo. Press, Boulder, 134 p.

Marr, J. W., 1977. The development and movements of tree islands near the upper limit of tree growth in the southern Rocky Mountains. Ecology 58: 1159–1164.

Maser, C., J. M. Trappe, and R. A. Nussbaum, 1978. Fungal-small mammal inter-relationships with emphasis on Oregon coniferous forests. Ecology 59: 799–809.

Moir, W. H., 1967. The subalpine tall grass, *Festuca thurberi*, community of Sierra Blanca, New Mexico. Southw. Nat. 12: 321–328.

Moir, W. H., 1979. Soil-vegetation relationships in the central Peloncillo Mountains, New Mexico. Amer. Midl. Nat. 102: 317–331.

Moir, W. H. and J. H. Dieterich, 1988. Old-growth ponderosa pine from succession in pine-bunchgrass habitats in Arizona and New Mexico. Natural Areas J. 8: 17–24.

Moir, W. H. and J. A. Ludwig, 1979. A classification of spruce-fir and mixed conifer habitat types in Arizona and New Mexico. USDA For. Serv. Res. Paper RM-207, 47 p.

Moir, W. H. and J. A. Ludwig, 1983. Methods of forest habitat type classification. pp. 1–5. In: Proceedings of the workshop on Southwestern habitat types, April 6–8, 1983, Albuquerque, New Mexico. USDA For. Serv. Southw. Region, Albuquerque, NM, 110 p.

Moir, W. H. and H. Smith, 1970. Occurrence of the American salamander, *Aneides hardyii,* in tundra habitat. Arctic & Alpine Res. 2: 115–116.

Muldavin, E., F. Ronco, Jr., and E. F. Aldon, 1990. Consolidated stand tables and data base for southwestern forest habitat types. USDA For. Serv. Gen. Tech. Report RM-190, 51 p. + computer diskettes.

Neilson, R. A., 1986. High resolution climatic analysis and southwest biogeography. Science 232: 27–34.

New Mexico Native Plant Protection Advisory Committee, 1984. A handbook of rare and endangered plants of New Mexico. Univ. New Mexico Press, Albuquerque, NM, xviii, 291 p.

Powell, D. C., 1988. Aspen community types of the Pike and San Isabel National Forests in south-central Colorado. USDA For. Serv. Rocky Mt. Region, R2-ECOL-88-01, 254 p.

Ronco, F., Jr., 1970. Influence of high light intensity on survival of planted engelmann spruce. Forest Sci. 16: 331–339.

Soil Survey Staff, 1990. Keys to soil taxonomy, fourth ed., Soil Management Support Services tech. monograph no. 19, Blacksburg, Virginia, 422 numbered p.

Swetnam, T. W., 1990. Fire history and climate in the Southwestern United States. pp. 6–17. In: J. S. Krammes (Tech. Coordinator), Proceedings—effects of fire management of Southwestern natural resources, 15–17 Nov. 1988, Tucson, Arizona. USDA For. Serv. Gen. Tech. Report RM-191, 293 p.

Swetnam, T. W. and A. M. Lynch, 1989. A tree-ring reconstruction of western spruce budworm history in the southern Rocky Mountains. Forest Sci. 35: 962–986.

U.S. Forest Service (USFS), 1986. Forest and woodland habitat types (plant associations) of southern New Mexico and central Arizona (north of the Mogollon Rim), 2nd ed., USDA For. Serv. Southw. Region, Albuquerque, NM, 134 p.

U.S. Forest Service (USFS), 1987a. Forest and woodland habitat types (plant associations) of northern New Mexico and northern Arizona, 2nd ed., USDA For. Serv. Southw. Region, Albuquerque, NM, 160 p. + insert.

U.S. Forest Service (USFS), 1987b. Forest and woodland habitat types (plant associations) of Arizona south of the Mogollon Rim and southwestern New Mexico, 2nd ed., USDA For. Serv. Southw. Region, Albuquerque, NM, 155 p. + insert.

U.S. Forest Service (USFS), 1987c. Terrestrial ecosystems survey of the Carson National Forest. USDA For. Serv. Southw. Region, Albuquerque, NM, 552 p. + map.

U.S. Forest Service (USFS), 1988. Annual Southwestern Region pest conditions report—1987. Forest Pest Management Rep. R-3 88-2, USDA For. Serv. Southw. Region, Albuquerque, NM, 15 p.

U.S. Forest Service (USFS), 1989. General ecosystems survey. USDA For. Serv. Southw. Region, Watershed & Air Management, Albuquerque, NM, February 1989, 72 p. + appendices.

Wagner, W. L., 1978. Floristic affinities of Animas Mountain, southwestern New Mexico, M.S. thesis, Univ. New Mex., Albuquerque, xi, 180 p.

Wardle, P., 1968. Engelmann spruce (*Picea engelmannii* Engelm.) at its upper limits on the Front Range, Colorado. Ecology 49: 483–495.

Willging, R. C., 1987. Status, distribution, and habitat use of Gould's turkey in the Peloncillo Mountains, New Mexico. M.S. thesis, New Mex. State Univ., Las Cruces, xii, 95 p.

Williams, J. and W. H. Moir, 1984. Monitoring the effects of elk and cattle on high elevation meadows, Valle Vidal, Carson National Forest. Unpublished report to Carson National Forest, Taos, NM, 13 p. + data notebook + photos.

Wulf, N. W. and R. G. Cates, 1987. Site and stand characteristics. Chap. 7 (pp. 90–115). In: Martha H. Brookes et al. (Tech. Coors). Western spruce budworm. USDA For. Serv. and Coop State Res. Serv. Tech. Bull. 1694, vi, 174 p.

6

WOODLAND AND
SAVANNA VEGETATION

WOODLAND

Woodland vegetation differs from forest vegetation in two ways. First and most importantly, canopies of individual woodland trees rarely touch or overlap. Secondly, woodland tree species are generally of smaller stature than forest tree species. There is great variation in the degree of openness of woodlands.

Woodland vegetation on the North American continent varies from thorn woodland in western Mexico, to oak woodland on the western edge of the eastern deciduous forest in the United States and Canada, to pine-oak woodland in Texas, Arizona, California, and central Mexico, to pinyon-juniper woodland in the Rocky Mountains of Colorado and New Mexico and through the intermountain region of the Southwest to the eastern slopes of the Sierra Nevada Range, in southeastern California. Woodland vegetation covers approximately 23% of the land area of New Mexico, or roughly 4.2 million hectares (10.6 million acres; see Figure 6.1A and vegetation map insert). Over 90% of this is coniferous woodland and the rest is mixed woodland. Table 6.1 is a classification of woodland and savanna vegetation types in New Mexico.

CONIFEROUS WOODLAND

Coniferous Woodland is extensive in the southwestern United States and is considered to have evolved in the Great Basin from the Madro-Tertiary geoflora (Brown, 1982). Much of the area covered today by grassland and scrubland vegetation in the southwestern United States was covered by pinyon-juniper woodland 10,000 years ago. Woodlands of the Southwest were at their lowest elevations during the end of the Pleistocene (12,000–10,000 ybp). Baker (1983) found that, "Deserts of the Southwest were replaced during late-glacial time by woodlands of pinyon pines and juniper." According to Spaulding et al. (1983), "Throughout the Southwest, woodland and forest vegetation expanded at the expense of desert scrub and steppe." Again, referring to pinyon-juniper woodland, "On rocky slopes, it ranged from close to the basin floors upward into a transition with communities dominated by montane trees between 1500 and 2000 m elevation." Spaulding et al. state that evergreen oaks, including *Quercus turbinella,* were included in the woodlands. Following is a summary of the history of pinyon-juniper woodland vegetation in the Southwest, from an outline by Moir (1983):

1. Late Pleistocene (22,000–11,000 ybp): pinyon-juniper woodland at 550–1,525 m (1,800–5,000 ft.) elevation.

2. Early Holocene (11,000–8,000 ybp): junipers replace pinyons at lower elevations, creosotebush present in juniper woodlands.

3. Mid-Holocene (ca. 8,000–4,000 ybp): woodland species retreat north or upward in elevation, rocky slopes become drier, grassland belt develops at 1,200–2,000 m (4,000–6,500 ft.), especially in New Mexico and Texas.

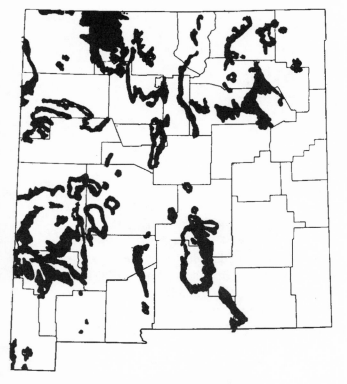

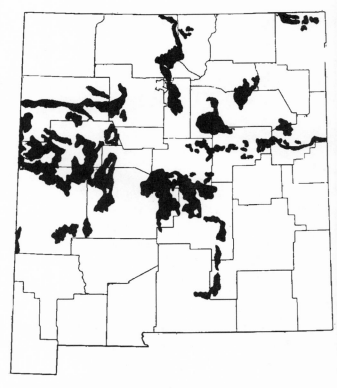

A. Distribution of Coniferous and Mixed Woodland.

B. Distribution of Juniper Savanna.

Figure 6.1 Distribution of Woodland and Savanna Vegetation in New Mexico. See New Mexico County Map on page xiv for reference, if needed, on county names.

4. Late Holocene (ca. 4,000–150 ybp): xeric margins expand; desert scrub communities develop at 1,200–2,000 m, especially in Arizona. Pinyon-juniper woodlands occupy approximately their present locations.

Except for management studies, relatively little research has been conducted on pinyon-juniper woodland vegetation, probably because this vegetation type has less value to humans than coniferous forest vegetation (valued for its tim-

ber) and grassland (for grazing by domestic livestock).

Coniferous woodland trees belong to two genera: *Pinus* (pine) and *Juniperus* (juniper). The pine species comprise a relatively short-needled, small-statured group, called Pinyon Pines. Pinyon and juniper tree species are generally the least shade tolerant and require the least amount of available moisture of all the conifers. The following tree species dominate Coniferous Woodland in the southwestern United States (* = not found in New Mexico):

Pines

Pinus aristata	Brislecone Pine
Pinus edulis	Colorado Pinyon
Pinus discolor	Border Pinyon
Pinus monophylla	One-leafed Pinyon
Pinus quadrifolia	Four-leafed Pinyon

Junipers

Juniperus ashei	Mexican Juniper
Juniperus californica	California Juniper
Juniperus deppeana	Alligator Juniper
Juniperus erythrocarpa	Redberry Juniper
Juniperus flacida	Drooping Juniper
Juniperus monosperma	One-seed Juniper
Juniperus osteosperma	Utah Juniper
Juniperus pinchotii	Redberry Juniper
Juniperus scopulorum	Rocky Mountain Juniper

Usually only one species of pinyon is found in a Coniferous Woodland stand, although occasionally two or even three species of juniper may be found in a stand.

CONIFEROUS AND MIXED WOODLAND

Except for Bristlecone Pine woodland, Coniferous Woodland in New Mexico is pinyon-juniper and mixed woodland (Table 6.1). There are small amounts of Bristlecone Pine woodland on high, dry ridges and talus slopes (DeVelice et al., 1986). These sites are similar to the rocky outcrop (escarpment) sites of lower elevations, where pinyon-juniper woodland can sometimes be found. In New Mexico, the only series in the Bristlecone Pine type is *Pinus aristata/Ribes montigenum*/S*. This is considered a habitat type by the Forest Service.

PINYON-JUNIPER WOODLAND

On the lower, drier boundary of Pinyon-Juniper Woodland vegetation, individual trees are often widely spaced. When their spacing is such that they are merely scattered in grass, the vegetation is referred to as savanna rather than woodland. Vegetation is usually called savanna if there are fewer than 320 individual trees per hectare (130/acre). Savanna is a transitional zone between woodland and grassland. The boundary between woodland and grassland may, however, be abrupt and without a savanna. On the upper, more moist boundary of Pinyon-Juniper Woodland vegetation, woodland trees may be closely spaced, resulting in almost a closed canopy. Kennedy (1983) found an average tree density of 420/h (170/a) in the *Pinus edulis–Juniperus monosperma*/S/*Stipa columbiana* communities of the Jicarilla and Sacramento mountains in south-central New Mexico. The Forest Service in New Mexico indicates that in closed *Pinus edulis–Juniperus/Artemisia tridentata*/MG-F communities of north-central and northwestern New Mexico, tree densities may be 690 plus or minus 120 individuals per hectare (Moir and Carleton, 1987).

It is not uncommon to find seral Pinyon-Juniper Woodland vegetation as a result of past disturbance of coniferous forest. In New Mexico, the disturbed forest has usually been ponderosa pine forest. The presence of young ponderosa pines in pinyon-juniper woodland could signify the successional nature of the stand. In the mountains around the Plains of San Agustin, Potter (1957) found pinyon-juniper woodland dominated by Colorado Pinyon (*P. edulis*) and Alligator Juniper (*J. deppeana*). These stands were located at elevations of from 7,000 to 9,000 ft. (2,100–2,700 m). The small numbers of young Ponderosa Pine that Potter found in these stands could indicate ecotonal vegetation between coniferous woodland and montane coniferous forest or, more likely, imply a woodland sere on disturbed ponderosa pine sites (Pl. 13).

Dittmer (1951) summarized the major features of the coniferous woodland in New Mexico as follows:

Few mountain areas of the southwest are so dry and so hot that they will not have a pinon-juniper association. The one-seeded junipers occupy the lower areas of this belt while the pinon pines grow better in the upper margins. Above this belt and extending down into it slightly in the northern areas is *Juniperus scopulorum,* while in the southern area the alligator-bark juniper is more prominent. This association ordinarily would make good grazing land, because the blue grama grass occupies most of the soil between trees. However, as is so common throughout the southwest, overgrazing has hit this belt as hard as the grasslands themselves.

Moir and Carleton (1987) propose the following three elevational subzones for the woodland life zone of Region 3 (Arizona and New Mexico, the lowest of these is our savanna zone).

1. The aridic (warm, dry) juniper savannas
   Tree cover: 5–30%
   Height of tallest trees: <5 m

2. Typical or model open woodlands
   Tree cover: 30–50%
   Height of tallest trees: 4–8 m

3. Mesic (cool, wet) closed woodlands
   Tree cover: 50–80%
   Height of tallest trees: 7–13 m

In portions of its range, the vegetation below Pinyon-Juniper Woodland may be Montane Scrub or Desert Grassland, instead of Plains-Mesa Grassland vegetation. In south-central and southwestern Colorado and north-central and northwestern New Mexico, Montane Scrub vegetation, dominated by Big Sagebrush (*Artemisia tridentata*), has often replaced the grasslands that once bordered the woodlands on their lower side. On steep slopes, Pinyon-Juniper Woodland may comprise a narrow band between the forest above and the grassland or scrubland below.

Pinyon-Juniper Woodland is not restricted to mountain slopes. It was and still can be found on mesas, if they are located at an elevation where moisture availability is sufficient. There are thousands of hectares of Pinyon-Juniper vegetation on mesas in the central and south-central part of the state (see vegetation map insert).

Pinyon-Juniper Woodland may be found on rocky outcrops surrounding an elevated mesa where the precipitation is only enough to support grassland. This juxtaposition of woodland and grassland vegetation results because such escarpments provide relatively high moisture conditions in cracks and fissures where run-off accumulates. Training material for Region 3 of the Forest Service states that, "A scarp is any slope between adjacent landforms of sharply contrasting elevations. It is formed by geologic erosion and slope retreat via the processes of hillslope and channel erosion." And, "Here we limit scarp woodlands to slopes steeper than 40% and with cobbly-bouldery soils having much discontinuity because of rock outcrop or bare rock exposure" (USDA Forest Service, 1986). Scarp Pinyon-Juniper Woodland vegetation is common along breaks associated with the rivers of northeastern and east-central New Mexico. Scarp woodland stands are often too small in area to be included on our maps.

Colorado Pinyon (*Pinus edulis*) is by far the most common pinyon of the Pinyon-Juniper Woodland vegetation of New Mexico. It is replaced by Border Pinyon (*P. discolor*) in the southwestern corner of the state. A one-needle variety, *P. edulis* var. *fallax* (Little, 1968) is found on the southeastern portion of the Colorado Plateau in west-central New Mexico. This variety is not to be confused with One-Leafed Pinyon (*P. monophylla*) of the eastern California and Great Basin Pinyon-Juniper Woodland vegetation. Woodin and Lindsey (1954) note that One-seed Juniper (*J. monosperma*) is not important in Colorado or Mexico. Rocky Mountain Juniper (*J. scopulorum*) replaces One-seed Juniper in Colorado, while Alligator Juniper (*J. deppeana*) replaces it in Mexico. Woodin and Lindsey also state that pinyon densities are higher than juniper densities in the upper portion of the pinyon-juniper woodland zone. In the central portion of their range, the pinyons and junipers tend to exhibit equal densities. These observations are similar to those of Dittmer (1951) and the Forest Service (1986, 1987a and b). Generally, Pinyon Pines have higher available-moisture requirements than junipers. Consequently, Colorado Pinyon has higher densities than juniper in the higher, more moist portions of this vegetation zone and is found in minor amounts in the lower, drier portions. For example, Donart, Sylvester, and Hickey (1978b) have a pinyon-juniper series (CW1) and a juniper-pinyon series (CW2) that reflect a reversal in relative pinyon and juniper densities.

One-seed Juniper (*Juniperus monosperma*) is the most widespread juniper in New Mexico. It may share dominance with Rocky Mountain Juniper in the northern third of the state; with Utah Juniper (*J. osteosperma*) and Rocky Mountain Juniper in the northwestern corner; with Alligator Juniper (*J. deppeana*) in south-central and southwestern areas; and with Redberry Juniper (*J. pinchotii*) or Mexican Juniper (*J. ashei*) in the southeast. Redberry Juniper tends to be a shrub and in New Mexico is more commonly a member of montane scrub than of pinyon-juniper woodland vegetation. Junipers are often erroneously called cedars; so-called cedar posts are all from junipers.

There are differences in available-moisture requirements among juniper species, reflected in their different relative positions in the vegetation, documented by Kennedy (1983) in the Guadalupe, Sacramento, and White mountains; Naylor (1964) in the Sandia Mountains; Robinson (1969) in the Guadalupe Mountains; and Fletcher (1978) in the Datil Mountains. Alligator Juniper (*J. deppeana*) requires more moisture and appears to be more shade tolerant than other junipers. Alligator Juniper may also be a member of Ponderosa Pine communities in New Mexico. Large living Alligator Juniper relicts can be found today on logged sites of past pine forest in the Sacramento and

White mountains of south-central New Mexico. Kennedy (1983) described a *Pinus ponderosa–Juniperus deppeana*/S/*Muhlenbergia dubia* habitat type from the Guadalupe and Sacramento mountains. This type is included in the forest vegetation classification in Table 5.1. Kennedy found that the average elevation of thirty-one plots dominated by Pinyon and One-seed Juniper in the Lincoln National Forest was 1,980 m (6,540 ft.), while the average elevation of thirteen plots dominated by Pinyon and Alligator Juniper was 2,100 m (6,950 ft.).

The most common pinyon-juniper type (Table 6.1) in south-central and possibly all of New Mexico is *Pinus edulis–Juniperus monosperma*/S/*Bouteloua gracilis*. Kennedy (1983) identified this type from the Guadalupe, Jicarilla, Capitan, and Sacramento mountains. Barnes (1986) identified this type in the mountains near Los Alamos. Barnes also found a type near Los Alamos with Mountain Muhly (*Muhlenbergia montanus*) as the dominant grass instead of Blue Grama. Another common and widespread type found in south-central New Mexico on lower drier sites is the *P. edulis–J. monosperma*/*Rhus trilobata*/*Muhlenbergia pauciflora* type. In these communities, juniper has a much higher density than pinyon. At higher elevations, the *P. edulis–J. deppeana*/S/S type is common from the Sacramento Mountains to the Datil Mountains of west-central New Mexico. In the Peloncillo Mountains, Alligator Juniper and One-seed Juniper share dominance with Border Pinyon (*P. discolor*), rather than with Colorado Pinyon (*P. edulis*). In these types, Mexican Pinyon is usually many times denser than junipers. The shrub layer in this type is dominated by Toumey Oak (*Q. toumeyi*) and ground cover is sparse (S). The pinyon-juniper associations having Rocky Mountain Juniper as dominant or codominant with One-seed Juniper and Colorado Pinyon are found in the north-central part of the state, while communities with Utah Juniper (*J. osteosperma*) as the dominant juniper occur in north-central and northwestern New

Mexico. These northern New Mexico pinyon-juniper communities usually have Big Sagebrush (*A. tridentata*) as the dominant shrub. In the south-central mountains there is a pinyon-juniper type with Mountain Mahogany (*Cercocarpus montanus*) dominating the shrub layer and Big Bluestem (*Andropogon gerardii*) the dominant grass (Kennedy, 1983).

Some pinyon-juniper community types have an understory shrub layer dominated by one shrub species. Others have a number of shrub species dominating the understory. About 40% of the pinyon-juniper communities in the classification for New Mexico have virtually no shrub layer as indicated by "S" for sparse. The shrub layers of some types may be dominated by Wavyleafed Oak (*Quecus undulata*), Gambel Oak (*Q. gambelii* [shrubby form]), or Toumey Oak (*Q. toumeyi*). Other shrubs that may be important in pinyon-juniper understory are Big Sagebrush, Mountain Mahogany, and Skunkbush (*Rhus trilobata*). Table 6.2 lists major plants comprising woodland and savanna vegetation in New Mexico. Many shrubs, forbs, and grasses typical of the vegetation types bordering pinyon-juniper woodland vegetation are also found in woodlands in New Mexico; the species listed in Table 6.2 are only those most commonly found in this type.

MIXED WOODLAND

In New Mexico there are areas where the dominant trees in woodland vegetation are not restricted to conifers but may include broadleaf tree species as codominants. In west and north-central New Mexico, Gambel Oak (*Q. gambelii*) may be a codominant in woodland (see Table 6.1). Donart, Sylvester, and Hickey (1978b) found this type of woodland in the San Francisco Mountains, southwest of Reserve; in the Burro Mountains, northwest of Lordsburg; in the Tularosa Mountains, west of the Plains of San Agustin; in the Nacimiento and Jemez mountains, west, south, and east of Jemez Springs; and in the mountains north, south, and east of Zuni, in northwestern New Mexico. Gambel Oak, the dominant oak of the Mixed Woodlands mentioned, is deciduous.

There are woodlands in New Mexico where the dominant nonconiferous trees are evergreen. The term "encinal" has been used for these woodland types (Brown, 1982), and we have used it in the classification of mixed woodland vegetation. One of these encinal mixed woodland types is found in the Guadalupe Mountains, in south-central New Mexico. The broadleaf, evergreen codominant tree is Texas Madrone (*Arbutus xalapensis*). The rest of the encinal communities in the state have oaks as the broadleaf, evergreen tree component. Evergreen oaks that may dominate these types are Arizona White Oak (*Q. arizonica*), Gray Oak (*Q. grisea*), and Emory Oak (*Q. emoryi*). Arizona White Oak–Gray Oak hybrids are common in southwestern New Mexico. *Q. grisea* in parentheses in the classification (Table 6.1) indicates that Gray Oak may be the dominant oak. Pointleaf Manzanita (*Arctostaphylos pungens*) and Wright Silktassel (*Garrya wrightii*) is frequently the dominant shrub in encinal woodland. The coniferous trees Border Pinyon and Alligator Juniper have relatively low densities in these communities. In New Mexico, these oak-dominated encinal mixed woodlands are restricted to the southwestern corner. Perez-Garcia (1978) reports large areas covered by oak woodland just south of New Mexico, in the Sierra Madre Occidental of northern Chihuahua, Mexico. These woodlands have the same dominant tree species as those of southwestern New Mexico.

SAVANNA

JUNIPER SAVANNA

Savanna is not normally considered a major vegetation type in the United States. However, vegetation of widely scattered low trees in a grass matrix is widespread and covers many hectares in New Mexico. Woodlands in New Mexico are usually dominated by two or more species of conifers or oaks. The transition from woodland to grassland involves a marked decrease in the density of trees, accompanied by reduction to a single tree species, usually a juniper. This ecotone between woodland and grassland can be broad and extensive in New Mexico; it has been expanding during this century. The omission or partial recognition of this large ecotone results in vegetation maps that may be confusing and appear to be incompatible. On the vegetation map "Biotic Communities of the Southwest" (Brown and Lowe, 1980), the woodland boundaries were drawn to encompass all areas containing junipers, even areas where the junipers were scattered. The map "Potential Vegetation of New Mexico" (Donart, Sylvester, and Hickey, 1978b) added cross-hatching to grassland and woodland, if juniper densities were low enough to be savanna-like. If all of the cross-hatched areas of this map are added to the woodland areas, the two maps appear to reflect similar views of the vegetation. Because of the confusion caused by partially or totally ignoring this extensive savanna ecotone, we have chosen to classify its vegetation (Table 6.1), map it (Figure 6.1B and vegetation map insert), and discuss it as a major vegetation type (Pl. 14).

Most of the savanna in New Mexico is juniper savanna. One-seed Juniper (*J. monosperma*) is usually the juniper species of juniper savanna vegetation in New Mexico. In southern New Mexico, Redberry Juniper (*J. erythrocarpa*) is also found as a savanna juniper. Cully and Knight (1987) report some juniper savanna from San

Juan and Rio Arriba counties, in northwestern New Mexico, where Utah Juniper (*J. osteosperma*) is the savanna juniper.

There are a few savanna areas in New Mexico where oak, instead of juniper, is the dominant. Wagner (1977) describes some stands of oak savanna in the Animas Mountains, in southwestern New Mexico, as lower encinal vegetation. The tree was Arizona White Oak (*Quercus arizonica*), and there were some Pointleaf Manzanita (*Arctostaphylos pungens*) shrubs scattered in a mixture of grass species. Oak savanna with Gray Oak (*Q. grisea*) can be found in southern mountains, such as the Organ Mountains.

In the northwestern quarter of the state, where Big Sagebrush (*Artemisia tridentata*) is commonly a dominant shrub, there may also be Juniper Savanna communities where Bigelow Sagebrush (*Artemisia bigelovii*) or Shadscale (*Atriplex confertifolia*) are dominant shrubs in savanna communities. Campbell (1981–1985) found Bigelow Sagebrush to be the dominant shrub of many juniper savanna communities in San Juan County. Cully and Knight (1987) and Donart, Sylvester, and Hickey (1978b) found shadscale to be a dominant shrub of juniper savanna communities in McKinley and Sandoval counties.

There are a few areas in central New Mexico where the shrub Wavyleaf Oak (*Q. undulata*) is scattered with junipers; most of the savanna vegetation in the state, however, has little or no shrub component. Under present conditions in New Mexico, juniper savanna vegetation is likely to continue to increase in area. The "natural," or climax, nature of juniper savanna vegetation in New Mexico is debatable and is discussed below.

DISTURBANCE AND SUCCESSION

Human-induced disturbance is varied and widespread in woodland vegetation in the state.

Typical disturbances are changes in fire frequencies, grazing by domestic livestock, clearing for livestock grazing, and harvesting of trees, mostly for fuelwood. Often woodland sites have been subjected to a combination of these disturbances. When these disturbances are not chronic, they usually initiate succession. Types and rates of succession are dictated in part by climate, soils, initial conditions before disturbance, severity of disturbance, and genetic opportunities (plant ecotypes). Some generalized stages of succession in woodland vegetation have been suggested: herb to herb-shrub to seedling juniper to juniper-shrub or juniper-herb to juniper-pinyon woodland (open or closed) to mature pinyon-juniper woodland (Erdman, 1970). Shott (1984) looked at succession in pinyon-juniper woodland in south-central New Mexico. He found that the eventual dominant woody species become established after disturbance as follows: first the understory shrub, Wavyleafed Oak (*Q. undulata*), followed by Oneseed Juniper (*J. monosperma*), and lastly, Colorado Pinyon (*P. edulis*). Stands of mature pinyon-juniper are difficult to find in New Mexico; when they are found, it is often questionable to assume that the species aggregations in the various layers are typical (natural). Naylor (1964) noticed, in the Sandia Mountains, that the understory shrub layer of pinyon-juniper stands contained species that are members of montane scrub vegetation. These were mixed with young pinyons and junipers. He also found pinyon and juniper seedlings in the ground cover. It is likely that if they were left undisturbed, there would be a reduction or even a disappearance of some of the understory shrub species in these communities. When there is a conspicuous shrub layer, there is a tendency to suppose that the stand is in a middle to late stage of succession. Consequently, some of the community types in the classification (Table 6.1) may not be true climax types. On the other hand, pinyon-juniper types were only named if stands are commonly found containing young or seedlings of the dominant species. This condition suggests stability, because replacement of dominant-species individuals should maintain the species' relative positions in the communities.

There has been considerable interest in the expansion of woodland onto grassland sites, undoubtedly due to the importance of ranching in the state. After an analysis of vegetation patterns from repeat photography, Sallach (1986) asserted that much of the recent increase of pinyon-juniper woodland on grassland in the mountains of New Mexico is actually a return of woodland to sites that had previously been woodland; the trees had been removed to improve the sites for livestock. If this assertion is correct, it would appear that most of the "invasion" of grassland by conifers is limited to junipers and is a transition to savanna rather than to woodland. The increase in juniper was already apparent in the early part of this century. Watson (1912) stated for the "Cedar Formation" (our Savanna):

Next comes the formation of which *Juniperus monosperma* is the dominant plant. East of Albuquerque it is confined strictly to the mountains, but where the mesa rises higher (6500 ft. or over) it stretches out over the plains. In the Estancia Valley it seems to be spreading at the expense of the prairie, as considerable areas are dotted over with young trees where there are no signs of old ones.

In 1932, Emerson maintained that hardy grama grass roots out-compete pinyon and juniper roots in the upper portions of the soil profiles in northeastern New Mexico; these are the roots of tree seedlings and the horizontal roots of mature trees. Cottam and Stewart (1940) stated that juniper moves both lower and higher when grass competition is reduced.

It is not uncommon to see early succession toward juniper savanna, where junipers are advancing into grassland areas via erosion gullies. Savanna vegetation has a higher moisture requirement than grassland vegetation and grassland vegetation has a higher requirement than scrubland vegetation. Under heavy grazing, scrubland may advance from the lower, dry side into grassland, and savanna may advance from the higher (mesic) side. As pointed out in Chapter 3 and shown in Figure 3.6, erosion due to over-grazing results in a decrease in infiltration (moisture entering the soil where it falls). Prior to erosion, with a high degree of infiltration under grass cover, the soil moisture is suitable for grass but not for junipers. With run-off instead of in-filtration, erosion creates miniature drainways, leaving much of the soil surface drier. The newly created catchments of the miniature drainways hold sufficient water for junipers. As erosion continues, more and more of the previously grass-covered landscape becomes dissected with gullies and juniper density increases. The expansion of juniper savanna onto previous grassland is so common that it is debatable whether there is or ever has been any natural, or climax, juniper savanna vegetation in New Mexico.

Table 6.1. Woodland and Savanna Vegetation Types in New Mexico. * = adapted from habitat types of USDA, Forest Service (1986, 1987a and b) and Barnes (1986).

WOODLAND AND SAVANNA VEGETATION
CONIFEROUS WOODLAND
  Bristlecone Pine Series
    *Pinus aristata/Ribes montigenum/S**
    *Pinus aristata/Festuca arizonica**
    *Pinus aristata/Festuca thurberi**

  Colorado Pinyon–One-seed Juniper Series
    Pinus edulis–Juniperus monosperma/
      Artemisia tridentata/MG-F
    Pinus edulis–Juniperus monosperma/
      Artemisia tridentata/Bouteloua gracilis
    Pinus edulis–Juniperus monosperma/
      Quercus undulata (Q. turbinella)/
      Bouteloua gracilis
    Pinus edulis–Juniperus monosperma/Rhus
      *trilobata/Muhlenbergia pauciflora**
    Pinus edulis–Juniperus monosperma/
      Cercocarpus montanus/Andropogon
      *gerardii**
    Pinus edulis–Juniperus monosperma/MS/
      *Muhlenbergia montanus**
    Pinus edulis–Juniperus monosperma/S/
      Oryzopsis hymenoides
    Pinus edulis–Juniperus monosperma/S/
      *Bouteloua gracilis**
    Pinus edulis–Juniperus monosperma/S/
      Bouteloua curtipendula
    Pinus edulis–Juniperus monosperma/S/Stipa
      *columbiana**

Colorado Pinyon–Alligator Juniper Series
    Pinus edulis–Juniperus deppeana/MS/MG-F
    Pinus edulis–Juniperus deppeana/S/
      *Bouteloua gracilis**
    Pinus edulis–Juniperus deppeana/S/
      *Muhlenbergia dubia**
    *Pinus edulis–Juniperus deppeana/S/S**

Colorado Pinyon–Utah Juniper Series
    Pinus edulis–Juniperus osteosperma/MS/
      MG-F

Colorado Pinyon–Rocky Mountain Juniper Series
    Pinus edulis–Juniperus scopulorum/
      Artemisia tridentata/S

Colorado Pinyon–Mixed Juniper Series
    Pinus edulis–Juniperus monosperma–
      Juniperus scopulorum/Quercus gambelii/
      MG-F
    Pinus edulis–Juniperus deppeana–Juniperus
      scopulorum/S/MG-F
    Pinus edulis–Juniperus monosperma–
      Juniperus osteosperma/MS/MG-F

Border Pinyon–Mixed Juniper Series
    Pinus discolor–Juniperus deppeana–
      Juniperus monosperma/Quercus toumeyi/S

Table 6.1. (continued).

MIXED WOODLAND

   Colorado Pinyon–Oak–Juniper Series

     Pinus edulis–Quercus gambelii–Juniperus monosperma/MS/MG-F

     Pinus edulis–Quercus gambelii–Juniperus monosperma–Juniperus scopulorum/MG/MG-F

   Encinal Series

     Pinus edulis–Juniperus monosperma–Arbutus xalapensis/MS/MG-F

     Quercus arizonica (Q. grisea)–Quercus hypoleucoides–Pinus discolor–Juniperus deppeana/S/MG-F

     Quercus arizonica (Q. grisea)–Quercus emoryi–Pinus discolor–Juniperus deppeana/Nolina microcarpa/Bouteloua gracilis

SAVANNA

   Arizona White Oak Series

     Quercus arizonica/Arctostaphylos pungens/MG

One-seed Juniper Series

     Juniperus monosperma/Artemisia tridentata/S

     Juniperus monosperma/Artemisia bigelovii/Bouteloua gracilis–Hilaria jamesii

     Juniperus monosperma/Quercus undulata (Q. turbinella)/MG

     Juniperus monosperma/Atriplex confertifolia/Oryzopsis hymenoides

     Juniperus monosperma/S/*Oryzopsis hymenoides–Hilaria jamesii*

     Juniperus monosperma/S/*Hilaria belangeri–Bouteloua* spp.

     Juniperus monosperma/S/*Bouteloua* spp.

     Juniperus monosperma/S/*Andropogon gerardii–Bouteloua curtipendula*

One-seed Juniper–Rocky Mountain Juniper Series

     Juniperus monosperma–Juniperus scopulorum/Artemisia tridentata/MG-F

Utah Juniper Series

     Juniperus osteosperma/Artemisia tridentata/MG

Table 6.2. Major Plants Comprising Woodland and Savanna Vegetation in New Mexico.

Scientific Name	Common Name	Region Found
TREES		
Arbutus xalapensis	Texas Madrone	southeastern
Juniperus ashei	Mexican Juniper	southeastern
Juniperus deppeana	Alligator Juniper	widespread
Juniperus erythrocarpa	Redberry Juniper	southern
Juniperus monosperma	One-seed Juniper	widespread
Juniperus osteospema	Utah Juniper	northwestern
Juniperus scopulorum	Rocky Mountain Juniper	northern
Pinus aristata	Bristlecone Pine	northern (high elevations)
Pinus discolor	Border Pinyon Pine	southwestern
Pinus edulis	Colorado Pinyon Pine	northern, central, south-central
Quercus arizonica	Arizona White Oak	southwestern
Quercus emoryi	Emory Oak	southwestern
Quercus gambelii	Gambel Oak	widespread
Quercus grisea	Gray Oak	central, southern

Table 6.2 (continued).

Scientific Name	Common Name	Region Found
SHRUBS		
Amelanchier oreophila	Mountain Serviceberry	northern
Arctostaphylos pungens	Pointleaf Manzanita	southwestern
Artemisia bigelovii	Bigelow Sagebrush	northwestern
Artemisia tridentata	Big Sagebrush	northern
Atriplex canescens	Fourwing Saltbush	widespread
Atriplex confertifolia	Shadscale	northwestern
Berberis fremontii	Fremont Barberry	northern
Berberis haematocarpa	Algerita	central, southern
Cercocarpus montanus	Mountain Mahogany	widespread
Certoides lanata	Winterfat	widespread
Chrysothamnus nauseosus	Rubber Rabbitbrush	widespread
Fendlera rupicola	Cliff Fendlerbush	north-central, central, southern
Garrya wrightii	Wright Silktassel	southern
Gutierrezia sarothrae	Broom Snakeweed	widespread
Juniperus communis	Dwarf Juniper	widespread (high elevations)
Juniperus pinchotii	Pinchot Juniper	southeastern
Lycium pallidum	Pale Wolfberry	central, southern
Nolina microcarpa	Sacahuista	southern
Opuntia imbricata	Tree Cholla	central, southern
Quercus toumeyi	Toumey Oak	southwestern
Quercus turbinella	Scrub Liveoak	southern
Quercus undulata	Wavyleafed Oak	widespread
Philadelphus microphyllus	Mockorange	widespread
Rhus trilobata	Skunkbush	widespread
Ribes montigenum	Gooseberry Currant	widespread (high elevations)
Ribes spp.	currant	widespread
Robinia neomexicana	New Mexican Locust	widespread
Symphoricarpos oreophilus	Snowberry	widespread
Yucca baccata	Banana Yucca	central, southern
GRASSES		
Agropyron smithii	Western Wheatgrass	northern, central
Andropogon gerardii	Big Bluestem	eastern, southeastern
Aristida adscensionis	Sixweeks Threeawn	widespread
Aristida divaricata	Poverty Threeawn	central, southern
Aristida orcuttiana	Beggartick Grass	southern
Aristida purpurea	Purple Threeawn	widespread
Bouteloua curtipendula	Sideoats Grama	widespread
Bouteloua gracilis	Blue Grama	widespread
Bouteloua hirsuta	Hairy Grama	widespread
Eragrostis erosa	Chihuahua Lovegrass	southwestern
Eragrostis intermedia	Plains Lovegrass	widespread
Hilaria jamesii	Galleta	widespread

Table 6.2 (continued).

Scientific Name	Common Name	Region Found
Koeleria cristata	Junegrass	widespread
Lycurus phleoides	Wolftail	widespread
Muhlenbergia arenacea	Ear Muhly	widespread
Muhlenbergia emersleyi	Bullgrass	southern
Muhlenbergia montanus	Mountain Muhly	widespread
Muhlenbergia pauciflora	New Mexican Muhly	widespread
Muhlenbergia repens	Creeping Muhly	southern
Muhlenbergia porteri	Bush Muhly	widespread
Munroa squarrosa	False Buffalograss	widespread
Oryzopsis hymenoides	Indian Ricegrass	central, northern
Oryzopsis micrantha	Littleseed Ricegrass	northern
Piptochaetium fimbriatum	Pinyon Ricegrass	widespread
Poa fendleriana	Muttongrass	widespread
Schizachyrium scoparium	Little Bluestem	widespread
Sitanion hystrix	Bottlebrush Squirreltail	widespread
Sporobolus cryptandrus	Sand Dropseed	widespread
Stipa columbiana	Columbia Needlegrass	central, northern
Stipa neomexicana	New Mexican Porcupinegrass	widespread

FORBS

Scientific Name	Common Name	Region Found
Artemisia ludoviciana	Louisiana Wormwood	widespread
Aster praealtus	Tall Aster	southern
Berlandiera lyrata	Lyreleaf Greeneye	widespread
Calliandra humilis	Dwarf Calliandra	central, southern
Castilleja integra	Wholeleaf Paintbrush	widespread
Chaenactus douglasii	Douglas Falseyarrow	northern
Chamaesaracha coronopus		widespread
Chenopodium album	Lambsquarters	widespread
Chenopodium incanum		widespread
Descurainia obtusa		central, western
Draba aurea	Golden Draba	widespread
Erigeron spp.	fleabane	widespread
Galium microphyllum	Littleleaf Bedstraw	southern
Geranium caespitosum	Purple Geranium	widespread
Goodyera oblongifolia	Rattlesnake Plantain	northern
Hedeoma spp.	False Pennyroyal	central, southern
Ipomopsis aggregata	Skyrocket	widespread
Leucelena aricoides	White Aster	widespread
Machaeranthera spp.		widespread
Mirabilis multiflora	Four O Clock	widespread
Penstemon whippleanus	Whipple Penstemon	widespread (high elevations)
Polygonum sawatchense	Sawatch Knotweed	northern
Salsola kali	Russian Thistle	widespread

Table 6.2 (continued).

Scientific Name	Common Name	Region Found
Salvia subincisa		widespread
Thlaspi alpestre	Wild Candytuft	widespread
Sphaeralcea coccinea	Red Globemallow	widespread
Zinnia grandiflora	Rocky Mountain Zinnia	widespread

CHAPTER 6 REFERENCES

WOODLAND AND SAVANNA VEGETATION

Aldon, E. F., 1964. Ground-cover changes in relation to runoff and erosion in west-central New Mexico. U.S. Department of Agriculture. Forest Service Research Note RM-34. Albuquerque, New Mexico.

Armentrout, S. M., 1986. Vegetational gradients surrounding pinyon pine and one-seed juniper in south-central New Mexico. Thesis, New Mexico State University, Las Cruces.

Armentrout, S. M. and R. D. Pieper, 1988. Plant distribution surrounding Rocky Mountain pinyon pine and oneseed juniper in southcentral New Mexico. Journal of Range Management, 41: 139–143.

Axelrod, D. F., 1940. Historical development of the woodland climax in western North America. American Journal of Botany, 27: 21 (Abstract).

Baker, R. G., 1983. Holocene vegetational history of the western United States. pp. 109–127. In: H. E. Wright (ed.), Late Quaternary Environments of the United States: Vol. 2, S. C. Porter (ed.), The Holocene. University of Minnesota Press, Minneapolis.

Barnes, F. J., 1986. Carbon gain and water relations in pinyon-juniper habitat types. Dissertation, New Mexico State University, Las Cruces.

Brown, D. E., 1982. Biotic communities of the American Southwest—United States and Mexico. Desert Plants, 4: 332 pp. Boyce Thompson Southwestern Arboretum, Superior, Arizona.

Brown, D. E. and C. H. Lowe, 1980. Biotic communities of the southwest, (map). Rocky Mountain Forest and Range Experiment Station, U.S. Department of Agriculture, Forest Service. Tempe, Arizona.

Campbell, L., 1981–1985. Coal mine permit applications; Chap. VII, La Ventana; Chap. V, Black Diamond; Chap. V, Mentmore; Chap. VI, Black Lake; Chap. V, Crown; Chap. VI, South Hospah; and Chap. VI, Star Lake. Metric Corporation, Albuquerque, New Mexico.

Choate, G. A., 1966. New Mexico's forest resources. U.S. Department of Agriculture, Forest Service.

Christensen, E. M., 1958. A comparative study of the climates of mountain brush, pinyon-juniper, and sagebrush communities in Utah. Proceedings of the Utah Academy of Science, Arts, and Letters, 36: 174–175.

Cottam, W. P. and G. Stewart, 1940. Plant succession as a result of grazing and of meadow desiccation by erosion since settlement in 1862. Journal of Forestry, 38: 613–626.

Crawford, R. C., 1974. An ecological analysis of an oak-juniper community in the Franklin Mountains, El Paso County, Texas. Thesis, University of Texas at El Paso, El Paso.

Cully, A. and P. Knight, 1987. A handbook of vegetation maps of New Mexico counties. 135

pp. New Mexico Natural Resources Department, Santa Fe.

Daubenmire, R. F., 1942. Soil temperature versus drought as a factor determining the lower altitudinal limits of trees in the Rocky Mountains. Botanical Gazette, 105: 1–13.

DeVelice, R. L., J. A. Ludwig, W. H. Moir, and F. Ronco, Jr., 1986. A classification of forest habitat types of northern New Mexico and southern Colorado. USDA Forest Service General Technical Report RM-131.

Dittmer, H. J., 1951. Vegetation of the Southwest—past and present. The Texas Journal of Science, 3: 350–355.

Donart, G. B., D. Sylvester, and W. Hickey, 1978a. A vegetation classification system for New Mexico, U.S.A. pp. 488–490. In: Proceedings of the First International Rangeland Congress. Denver, Colorado.

———, 1978b. Potential natural vegetation—New Mexico. New Mexico Interagency Range Committee Report II, M7-P0-23846. USDA Soil Conservation Service. (map, scale 1:1,000,000).

Emerson, F. W., 1932. The tension zone between the grama grass and pinon-juniper associations in northern New Mexico. Ecology, 13: 347–358.

Erdman, J. A., 1970. Pinyon-juniper succession after natural fires on residual soils of Mesa Verde, Colorado. BYU Science Bulletin, Biological Series 11. Provo, Utah.

Fletcher, R. A., 1978. A floristic assessment of the Datil Mountains. Thesis, University of New Mexico, Albuquerque.

Groce, V. L., 1966. Vegetational characteristics of range sites on pinyon-juniper-grassland range in southcentral New Mexico. Thesis, New Mexico State University, Las Cruces.

Howell, J., Jr., 1941. Pinyon and juniper woodlands of the Southwest. Journal of Forestry, 39: 542–545.

Johnson, B., 1984. Plant associations (habitat types) of Region 2, Ed. 3.5, USDA Forest Service, Rocky Mountain Region, Lakewood, Colorado.

Kennedy, K. L., 1983. A habitat type classification of the pinyon-juniper woodlands of the Lincoln National Forest, New Mexico. pp. 54–61. In: W. H. Moir and L. Hendzel (technical coordinators), Proceedings of the Workshop on Southwestern Habitat Types, U.S. Department of Agriculture, Forest Service, Southwestern Region, Albuquerque, New Mexico.

Kohler, D. A., 1974. The ecological impact of feral burros on Bandelier National Monument. Thesis, University of New Mexico, Albuquerque.

Layser, E. F. and G. H. Schubert, 1979. Preliminary classification for the coniferous forest and woodland series of Arizona and New Mexico. USDA Forest Service Research Paper, RM-208. Albuquerque.

Little, E. L., Jr., 1977. Research in the pinyon-juniper woodland. U.S. Department of Agriculture. Forest Service, General Technical Report, RM-39: 8–19.

Ludwig, J. A. and W. G. Whitford, 1977. An evaluation of transmission line construction on pinyon-juniper woodland and grassland communities in New Mexico. Journal of Environmental Management, 5: 127–137.

Lymberly, G. A., 1979. Ecology of pinyon-juniper vegetation in south-central New Mexico. Thesis, New Mexico State University, Las Cruces.

Moir, W. H., 1979. Soil-vegetation patterns in the central Peloncillo Mountains, New Mexico. American Midland Naturalist, 102: 317–331.

———, 1982. A fire history of the high Chisos, Big Bend National Park, Texas. The Southwestern Naturalist, 27: 87–98.

———, 1983. A series vegetation classification for Region 3. pp. 91–95. In: Proceedings of the Workshop on Southwestern Habitat Types, USDA Forest Service, Southwestern Region, Albuquerque, New Mexico. 110 pp.

Moir, W. H. and J. O. Carleton, 1987. Classification of pinyon-juniper (P-J) sites on National Forests in the Southwest. pp. 216–226. In: Proceedings = Pinyon-Juniper Conference. USDA Forest Service General Technical Report INT-215, 581 pp.

Naylor, J. N., 1964. Plant distributions of the

Sandia Mountains area, New Mexico. Thesis, University of New Mexico, Albuquerque.

Parker, R. H., 1949. A biological study of the pinyon pine-juniper association at Juan Tabo in the Sandia Mountains of New Mexico. Thesis, University of New Mexico, Albuquerque.

Perez-Garcia, A., 1978. Ecology of the oak communities on the eastern foothills of the Sierra Madre Occidental in Chihuahua. Dissertation, New Mexico State University, Las Cruces.

Pieper, R. D., 1968. Vegetation on grazed and ungrazed pinyon-juniper grasslands in New Mexico. Journal of Range Management, 21: 51–52.

————, 1977. The southwestern USA pinyon-juniper ecosystem. U.S. Department of Agriculture. Forest Service, General Technical Report, RM-39: 1–6.

Potter, L. D., 1957. Phytosociological study of San Augustin Plains, New Mexico. Ecological Monographs, 27: 113–136.

Robinson, J. L., 1969. Forest survey of the Guadalupe Mountains, Texas. Thesis, University of New Mexico, Albuquerque.

Sallach, B. K., 1986. Vegetation changes in New Mexico documented by repeat photography. Thesis, New Mexico State University, Las Cruces.

Schott, M. R., 1984. Pinyon-juniper ecology with emphasis on secondary succession. Dissertation, New Mexico State University, Las Cruces.

Schott, M. R. and R. D. Pieper, 1986. Succession in pinyon-juniper vegetation in New Mexico. Rangelands, 8: 126–128.

Schott, M. R. and R. D. Pieper, 1987. Succession of pinyon-juniper communities after mechanical disturbance in southcentral New Mexico. Journal of Range Management, 40: 88–94.

Spaulding, W. G., E. B. Leopold, and T. R. Van Devender, 1983. Late Wisconsin Paleoecology of the American Southwest. pp. 259–293. In:

H. E. Wright (ed.), Late Quaternary Environments of the United States: Vol. 1, S. C. Porter (ed.), The Late Pleistocene. University of Minnesota Press, Minneapolis.

Tatschl, A. K., 1966. A floristic study of the San Pedro Parks Wild Area, Rio Arriba County. Thesis, University of New Mexico, Albuquerque.

USDA Forest Service, 1986. Forest and woodland habitat types (plant associations) of southern New Mexico and central Arizona (north of Mogollon Rim), 2nd ed., USDA Forest Service Southern Region, Albuquerque, New Mexico.

————, 1987a. Forest and woodland habitat types (plant associations) of northern New Mexico and northern Arizona, 2nd ed., ibid.

————, 1987b. Forest and woodland habitat types (plant associations) of southern Arizona (south of Mogollon Rim) and southwestern New Mexico, 2nd ed., ibid.

Wagner, W. L., 1977. Floristic affinities of the Animas Mountains, southwestern New Mexico. Thesis, University of New Mexico, Albuquerque.

Watson, J. R., 1912. Plant geography of northcentral New Mexico. Botanical Gazette, 54: 194–217.

West, N. E., K. H. Rea, and R. J. Tausch, 1975. Basic synecological relationships in pinyon-juniper woodlands. In: Gifford, G. E. and F. E. Burby (eds.). The pinyon-juniper ecosystem: A symposium. Utah State University, Logan.

West, N. E., R. J. Tausch, K. H. Rhea, and P. T. Tueller, 1978. Taxonomic determination, distribution, and ecological indicator values of sagebrush within the pinyon-juniper woodlands of the Great Basin. Journal of Range Management, 31: 87–92.

Woodin, H. E. and A. A. Lindsey, 1954. Juniper-pinyon east of the continental divide, as analyzed by the line-strip method. Ecology, 35: 473–489.

7

GRASSLAND VEGETATION

North American grasslands originated as a result of increasing aridity on the Great Plains during the late Oligocene and Miocene, approximately 20,000,000 years ago. This aridity caused the Arcto-Tertiary forest vegetation occupying the Great Basin to retreat north and Neotropical-Tertiary forests to move south and east, leaving only drought-tolerant members, primarily grasses and other herbaceous plants (Dix, 1964). Evidently, the major development of grassland took place in the late Miocene or even Pliocene. The dominant grasses had Arcto, Neotropical-, and Madro-Tertiary origins; the genera *Agropyron* (wheatgrass), *Koeleria* (junegrass), and *Poa* (bluegrass) from the north and *Bouteloua* (grama grass), *Sporobolus* (dropseed), and *Stipa* (needlegrass) from the south. During and after the Ice Age (Pleistocene), vegetation in the Great Basin moved and changed, resulting in grasslands that were virtually the same as those found by western Europeans in the sixteenth and seventeenth centuries. Post-Pleistocene drying caused the deciduous forests to move farther east, but many grasses and forbs adapted and stayed behind. Examples are *Andropogon* (bluestem), *Panicum* (Switchgrass), *Sorghastrum* (indian grass), and gayfeather *(Liatrus)*, one of the forbs.

The grasslands of New Mexico can be divided into subalpine-montane grassland, plains-mesa grassland, and the transitional, or ecotonal, type, called desert grassland. These units have been combined differently and given different names by other authors. Brown (1982) uses alpine and subalpine grassland and montane meadow for our subalpine-montane grassland; plains and Great Basin grassland for our plains-mesa grassland; and semidesert grassland for our desert grassland. Brown's classification had to accommodate the entire southwestern United States and northwestern Mexico, while ours has been tailored to be most useful for understanding the vegetation of New Mexico. Little of what Brown called alpine grassland occurs in New Mexico; that which does would best be described as riparian vegetation (see Chapter 9). Figure 7.1A–C and the vegetation map insert show where grassland occurs in New Mexico.

SUBALPINE-MONTANE GRASSLAND

These grasslands are found above or within subalpine and montane coniferous forest vegetation in the state. An excellent study of montane grasslands of the Jemez Mountains examines topographic features, areal extent, soil characteristics, and ecotones along with maintenance and floristic composition (Allen, 1984). Allen found rapid tree invasion into Thurber Fescue meadows, while Moir (1967) found that comparable meadows were generally not being claimed by trees at higher elevations in southern New Mexico. Many of Allen's generalizations concerning Jemez Mountains vegetation are applicable to this type of vegetation in the state. Subalpine-montane grassland usually occurs from above 2,700 m (8,900 ft.) up to 3,500 m (11,500 ft.), with an elevational spread of about 304 m (1,000 ft.) on any given mountain. The grasslands are "best developed on relatively smooth terrain" (Allen, 1984). Most meadows are found on southwestern

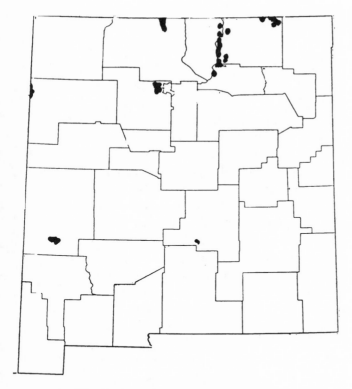

A. Distribution of Montane Grassland.

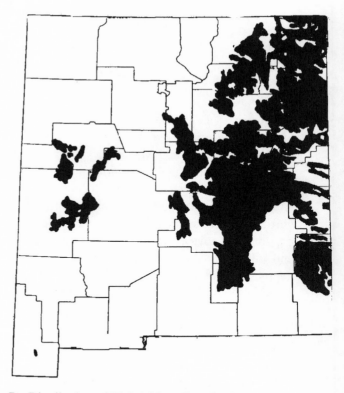

B. Distribution of Plains-Mesa Grassland.

Figure 7.1 Distribution of Grassland Vegetation in New Mexico. See New Mexico County Map on page xiv for reference, if needed, on county names.

exposures, but occasionally meadows can be found on exposures ranging from 140 to 300 degrees. The grasslands are most often on slopes of from 20% to 50%, but can also be found on level ridges or in valley bottoms. The size of these meadows is a function of the size of forest openings, smoothness of terrain, and degree of recent tree invasion. Subalpine-montane grasslands are generally less than 100 hectares (247 acres) in extent. The distribution of subalpine-montane grasslands large enough to be mapped is shown in Figure 7.1A (Pl. 15 and 16).

The meadows of higher slopes often have abrupt boundaries with subalpine coniferous forest; this is less true at lower altitudes, with montane coniferous forest. Allen and Moir both suggest that in areas of abrupt boundaries, the grassland may be climax. In the Sierra Blanca region, Moir (1967) found no conifer seedlings in any of 280 quadrants in mature Thurber Fescue meadows. It is unlikely that such meadows constitute a seral stage on the way to coniferous forest vegetation. At lower elevations, where there are broad transition ecotones between forest and grassland, the mead-

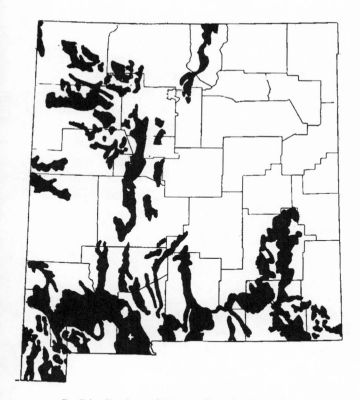

C. Distribution of Desert Grassland.

ows may be dependent upon fire. It is generally agreed by foresters and ecologists that under heavy grazing, trees may become established in montane grassland (Allen, 1984).

Forbs share dominance with grass in subalpine-montane grassland. They may contribute up to 35% of the live vegetation cover in these communities (Allen, 1984). For a classification of subalpine-montane grassland in New Mexico, see Table 7.1; for a list of major plants comprising this type of vegetation, see Table 7.2. Forb contribution to these communities is from many dif-

ferent species, rather than high densities of one or two species. The dominant grasses of subalpine-montane grassland are tall (up to 1 m) bunch grasses. Important genera are *Festuca* (fescue), *Danthonia* (oatgrass), *Deschampsia* (tufted hairgrass), *Koeleria* (junegrass), and some species of *Poa* (bluegrass) and *Muhlenbergia* (muhly). Where this grassland is found on small valley bottoms, the meadows or wet meadows, as they are sometimes called, are usually dominated by mixed sedge (*Carex*) species and one of the grasses Tufted Hairgrass (*D. caespitosa*), Junegrass (*K. cristata*), bluegrasses (*P. interior, fendleriana,* and *longiligula*), or species of danthonia (Table 7.1). Rushes (*Juncus*) and flat sedges (*Cyperus*) are also common in these wet meadows. Common forbs are Mountain Dandelion (*Agoseris aurantiaca*), fleabanes (*Erigeron* spp.), cinquefoils (*Potentilla* spp.), Blueflag (*Iris missouriensis*), and Bracken (*Pteridium aquilinum*). Figure 7.2B is a picture of a typical montane meadow.

A number of changes have been taking place in the state's montane grasslands as a result of heavy grazing. Thurber and Arizona Fescue, Danthonias, and Junegrass tend to be replaced by Kentucky Bluegrass (*Poa pratensis*). Fescue meadows rested from livestock in the Pecos Wilderness revert to Thurber Fescue and other native grasses, with *Poa pratensis* and rhizomatous sedges between clumps (Moir, personal communication, 1988). Reversion occurs fairly fast (2–4 years) at ca. 3,000–3,300 m (10,000–11,000 ft.). In many areas, the Kentucky Bluegrass meadows seem relatively stable, possibly due to soil change; it is unlikely that these meadows will completely return to their prelivestock composition. Therefore, Kentucky Bluegrass associations are classified with other subalpine-montane grassland associations, rather than in a succession-disturbance category (Table 7.1).

Moir (1967) found a general elevational gradient for dominance of the various fescues of the subalpine meadows in the Sierra Blanca region. Sheep Fescue (*F. ovina*) occurred at highest elevations, Thurber Fescue (*F. thurberi*) next, and Arizona Fescue (*F. arizonica*) was the lowest. Of course, there is considerable overlap, and one of the most common Subalpine-Montane Grassland associations in the state is actually the *Festuca thurberi–Festuca arizonica* association. The *Festuca arizonica–Muhlenbergia montana* association (Table 7.1) attests to the importance of Arizona Fescue at slightly lower elevations. Arizona Fescue–Mountain Muhly and Pine Dropseed (*Blepharoneuron tricholepis*)–Mountain Muhly communities are associated with montane coniferous forest rather than with the higher sub-alpine coniferous forest. Kentucky Bluegrass associations may be found with both forest types.

As in wet meadows, dry meadows contain many perennial and annual forb species. These slope and ridge meadows are a riot of color during late summer and early fall. Often the reds, blues, and yellows are all in bloom at the same time: Skyrocket (*Ipomopsis aggregata*) and paint brushes (*Castilleja* spp.); penstemon (*Penstemon* spp.), lupines (*Lupinus* spp.), and bluebell (*Campanula rotundifolia*); and cinquefoils (*Potentilla* spp.), Mountain Parsley (*Pseudocymopteris montanus*), and goldenrods (*Solidago* spp.).

PLAINS-MESA GRASSLAND

Plains-Mesa Grassland is the most extensive grassland in the state. It covers the eastern plains west to bajadas (erosion slopes of mountain ranges) of the Sangre de Cristo Mountains in the north and Sacramento Mountains farther south. This grassland covers suitable high mesas between mountain ranges across the state to the west, including the San Agustin Plains of the Colorado Plateau, in west-central New Mexico. There are small patches in northwestern New Mexico, but most areas occupied by this grassland in the early 1800s are now covered by desert Grassland (Figure 7.1B).

Plains-Mesa Grassland merges with savanna or woodland at its high (mesic) boundaries and with desert grassland or desert scrubland at its low (xeric) boundaries. In eastern New Mexico, dryland and irrigated farming have greatly reduced the amount of this vegetation (vegetation map insert). The reduction of this vegetation type is continuing under urbanization and grazing by domestic livestock; under grazing, succession at the high boundaries is in the direction of juniper savanna and on the low sides toward desert grassland or desert.

Plains-Mesa Grassland in climax condition is composed almost entirely of grasses. The few shrubs and forbs constitute less than 10% (Heerwagen, 1956). Major plants of the plains-mesa grassland in New Mexico are listed in Table 7.3. The area of transition from grassland to grassland-scrubland ecotone (desert grassland) vegetation is often subtle and extensive. The judgment as to when shrub densities are sufficient to consider a site as belonging to desert grassland rather than plains-mesa grassland can be a difficult one. Consequently, the number of shrubs listed in Table 7.3 is larger than might be expected for a grassland.

Blue Grama (*Bouteloua gracilis*) is a common denominator of the Plains-Mesa Grassland in New Mexico (Table 7.1). It codominates with Buffalograss (*Buchloe dachtyloides*) in the northeast and east-central plains (including a small outlier in the Animas Valley; see Figure 7.1B and vegetation map insert), with Western Wheatgrass (*Agropyron smithii*) on northern and northwestern mesas, or with Galleta (*Hilaria jamesii*) on mesas in the northern half of the state. On fine-textured soils in the north and northwest, Indian Ricegrass (*Oryzopsis hymenoides*) may share dominance with Blue Grama, or communities can be found with New Mexico Feathergrass (*Stipa neomexicana*) and Needle-and-Thread (*Stipa* comata), sharing dominance with Blue Grama. Areas can also be found dominated by Three-awns (*Aristida fendleriana* and *A. longiseta*), Sideoats Grama (*B. curtipendula*), and Blue Grama. Plate 17 shows a typical Plains-Mesa Grassland community.

There is a grass vegetation type common in southern New Mexico that is dominated by Tobosa Grass (*Hilaria mutica*). The areas where this type is found are often lower than the surrounding mesa and are miniature catchments, referred to as tobosa swales. Swales are small depressions that allow the water to spread; it either slowly exits to an arroyo or enters the soil before standing very long. Swales are common on the landscapes supporting plains-mesa and desert grassland vegetation. The following, from Shreve (1942b) refers to swales in northern Chihuahua, Mexico. "Other bolsons in the center of the plateau receive limited drainage from their surrounding hills and bajadas and are without a central playa, or lake bed. In these the floor of the bolson is nearly level and the soil is deep and fine in texture. In such situations are found the llanos, or nearly pure *Hilaria* grassland areas, which collectively cover large expanses in northern Chihuahua." Swale vegetation could be classed as a type of riparian vegetation, but it has usually been described as a distinct type of grassland. The distinctive feature of swale riparian vegetation is the consistent dominance by Tobosa Grass (*Hilaria mutica*). Saltgrass (*Distichlis stricta*) and/or the blue-green algae *Nostoc* spp. may be found between the dominant grass clumps. These tobosa swales often provide important winter pastures for livestock. Most of them have withstood livestock grazing, and the composition and structure of their vegetation today are virtually as they were in presettlement times. Occasionally swales will be codominated by Tobosa and Alkali Sacaton (*Sporobolus airoides*). Swale boundaries tend to be composed of deep sand that has accumulated as water flow slows and spreads as it enters the swale. These sandy borders support stands of

Soaptree Yucca (*Yucca elata*). As the grasses on the mesas surrounding swales diminished from grazing, the mesa sites opened up and erosion increased; the sandy boundaries of the swales then began to broaden as more sand dropped out of the erosion flow. This increase in area of sandy sites has resulted in an increase in Soaptree Yucca during the last 120 years (York and Dick-Peddie, 1969). Some swales have large amounts of Sacaton (*Sporobolus wrightii*) (Pl. 18).

Neuenschwander, Sharrow, and Wright (1975) investigated succession to a tobosa swale. They found that pioneers on a swale site were the lichen, *Lecidea crendas dealbata*, and the Blue-Green Alga, *Nostoc commune flagilliforme*, followed by Hog Potato (*Hoffmannseggia jamesii*), White Ragweed (*Hymenopappus robustus*), Mexican Drymary (*Drymaria holosteoides*), Alkali Sacaton (*Sporobolus airoides*), Burrograss (*Scleropogon brevifolius*), and Pigweed (*Amaranthus* spp.). Finally, the Dropseed, *Sporobolus auriculatus*, dominates before tobosa takes over.

There are some communities where Blue Grama is not dominant but is usually present. On sandy sites in the east, there are prairie pockets dominated by Little Bluestem (*Schizachyrium scoparium*) and Sideoats Grama. Indian Ricegrass–Mixed Dropseed (*Sporobolus* spp.) communities can be found on some sandy northern and central mesas. Mixed dropseed species may codominate with Black Grama (*B. eriopoda*) on dry slopes in the southern third of the state. Sideoats Grama–Black Grama–New Mexico Feathergrass communities may also be found.

Even though forbs do not constitute a major feature of plains-mesa grassland vegetation, the species that do occur have indicator value. For example, Red Globemallow (*Sphaeralcea coccinea*), Curly Cup Gumweed (*Grindelia squarrosa*), Spiny Goldenweed (*Haplopappus spinulosus*), coneflowers (*Ratibida* spp.), and Rocky Mountain Zinnia (*Zinnia grandiflora*) are all members of the plains-mesa grassland (Heerwagen, 1956). When they are found as part of desert grassland vegetation, these forbs may be considered relics of a prior Plains-Mesa Grassland. A few low shrubs may be scattered through plains-mesa grassland communities, including small Soapweed (*Yucca glauca*) in northeastern and east-central New Mexico. Estafiata or Fringed Sage (*Artemisia frigida*),is another low shrub sometimes scattered in this type of vegetation. To the west, Winterfat (*Ceratoides lanata*) and Bigelow Sagebrush (*Artemisia bigelovii*) may be scattered in plains-mesa grassland communities. Erosion gullies commonly have rabbitbrush (*Chrysothamnus* spp.) species along their edges or in their beds. The ubiquitous disturbance shrub Broom Snakeweed (*Gutierrezia sarothrae*) increases under heavy grazing in this grassland. Clumps of low-growing Honey Mesquite (*Prosopis glandulosa*) are occasionally found in grazed grasslands of the eastern plains.

Due to rain-shadow effect and only a minor amount of frontal activity, the Plains-Mesa Grassland vegetation of New Mexico represents the southwestern boundary of the continental grasslands. It is not surprising that when stocking rates similar to those in Texas and Oklahoma were used

on the relatively xeric and open grama grass (Black Grama), found on slopes with shallow soils, there was an extensive and rapid succession of these grasslands toward Desert Grassland or desert scrublands. Improved range-management practices resulting from continued rangeland research may not only retard succession to Desert Grassland from Grassland and from Desert Grassland to Desert Scrubland, but may in some areas even reverse the successional trends. The following is from Donart (1984):

Implementation of knowledge has resulted in Bureau of Land Management lands changing from 1.5 percent in good to excellent, 14.3 percent fair, and 84.2 percent in poor or bad condition in 1936 to 17 percent in good to excellent, 50 percent fair and 33 percent in poor or bad condition by 1975 (Box, 1978). Private rangelands for the United States have gone from 5 percent in excellent, 15 percent in good, 40 percent in fair and 40 percent in poor condition in 1963 to 12 percent in excellent, 28 percent in good, 42 percent in fair and 18 percent in poor condition in 1977 (Pendleton, 1979).

DESERT GRASSLAND

Desert Grassland vegetation has also been referred to as semidesert grassland (Brown, 1982); desert-grassland transition (Shreve, 1942a); desert savanna (Shantz and Zon, 1924); desert plains grassland (Weaver and Clements, 1938 and LeSuer, 1924); desert shrub grassland (Darrow, 1944); and grassland transition (Muller, 1947). Desert Grassland has been the most frequently used term (Shantz and Zon, 1924; Nichol, 1937; Benson and Darrow, 1944; Buechner, 1950; Castetter, 1956; Humphrey, 1958; and Lowe, 1964). The transitional nature of this vegetation is apparent from these names. Brown (1982) makes

the point that even though it is an ecotone based upon life-form, this vegetation type is a "distinctive separate biome." Brown links the desert grassland to habitats bordering or within the Chihuahuan Desert. However, it is difficult to determine recent successional-disturbance Desert Grassland from original Desert Grassland. Consequently, all of the areas of desert shrub-grassland vegetation are considered to be desert grassland and are mapped as such (Figure 7.1C and vegetation map insert). For example, areas in the northwest, codominated by sagebrush or saltbush shrubs and grasses, are considered Desert Grassland. West (1983) refers to this vegetation as "shrub-steppe." In the Peloncillo Mountains, Moir (1979) has called this vegetation "shrub steppe" and combined it with "grama steppe" to form a piedmont desert grassland. The grama steppe described by Moir in the Peloncillos was virtually without shrubs and would be in our Grama Grass Series of the Plains-Mesa Grassland category (Pl. 19).

Much of the Desert Grassland and desert scrubland in New Mexico occupies sites that were previously grassland. Whitfield and Anderson (1938) found that under intensive grazing, forbs and shrubs replace palatable grasses. The following comes from Campbell (1929): ". . . Jardine and Forsling ('22) indicate the ultimate result of injudicious grazing may be transformation of Grama Grass (*Bouteloua eriopoda*) range to the *Prosopis* dune type." In the absence of grazing by domestic livestock, Campbell and Bomberger (1934) observed that Black Grama suffers no more from drought than Snakeweed. They note that when insect damage is added, Snakeweed suffers more than Black Grama during a dry cycle. Canfield (1948) writes that under protection from livestock grazing, Bush Muhly (*Muhlenbergia porteri*) grows out in the open rather than in or under shrubs. In the absence of man-initiated disturbance, desert shrubs will not increase in density in Plains-Mesa or Desert Grassland. York and Dick-Peddie (1969) document unchanged boundaries between grassland and scrubland over the past 100 years on ungrazed sites in New Mexico. Stewart (1982) found stable desert grassland communities with dense Black Grama and scattered desert shrubs on ungrazed mesa sites in southern New Mexico. As indicated earlier (Chapter 3), fire appears to have been of little importance in Desert Grassland vegetation before settlement (Cable, 1967 and 1972; Buffington and Herbel, 1965; and Cornelius, 1988). The patchy nature of the vegetation and its disjunct occurrence on the basin and range landscape make carrying an extensive fire unlikely. Also, many of the desert shrubs, such as Sacahuista, Sotol, and *Mimosa biuncifera,* crown sprout after fire (Kittams, 1972).

Desert grassland vegetation may border Chihuahuan Desert scrub or Great Basin Desert scrub at its lower (drier) boundaries and plains-mesa grassland or montane scrub at its higher, more mesic boundaries. The dominant grass of desert grassland vegetation is Black Grama (*Bouteloua eriopoda*). Shrub and forb components are made up of many different species, as expected from the ecotonal position of this vegetation type. Most major shrub species of this type are also major species of the montane, Great Basin Desert, and Chihuahuan Desert scrub (see Tables 8.1, 3, and 4 in Chapter 8).

Plants considered possibly to be diagnostic (climax) members of Desert Grassland are identified in Table 7.4; plants considered to be diagnostic members of the Chihuahuan Desert scrub are identified in Table 8.6, in Chapter 8. Identifying species as diagnostic of these two vegetation types is a judgment made with the help of the literature, in particular Muller (1940) and Johnston (1977). The difficulty in recognizing climax stands of these vegetation types makes such designations arbitrary, and the reader may feel that there are omissions and/or erroneous inclusions. Plants that tend to be common to many vegetation types have little indicator value and are not considered diagnostic members of either type; examples are Spectaclepod (*Dithyrea wislizenii*), Russian Thistle (*Salsola kali*), White Horsenettle (*Solanum elaeagnifolium*), Buffalo Gourd (*Cucurbita foetidissima*), Goldenweed (*Haplopappus gracilis*), and Scorpionweed (*Phacelia corrugata*). The blurred distinction between desert grassland and Chihuahuan Desert scrub over a rather extensive zone precludes any absolute separation today.

Because much of the vegetation included as desert grassland has recently changed or may be changing, species considered to be diagnostic of this type can be helpful in understanding a community's past. If desert grassland forb species are common in a Chihuahuan Desert scrub community, the desert community is probably a recent one. If dominant shrubs of a desert grassland community are species that also dominate desert scrub communities, the vegetation on the site may be seral and on the move toward desert scrub. On the other hand, if shrub dominants of a desert grassland community are also major species of montane scrub vegetation, the community may constitute an ecotone beween montane scrub and plains-mesa grassland. Examples might be communities where Century Plant (*Agave parryii*), Turpentinebush (*Ericameria laricifolia*), Algerita (*Berberis haematocarpa*), or Evergreen Sumac (*Rhus choriophylla*) are the dominant shrubs.

The list of major plants comprising Desert Grassland vegetation in New Mexico (Table 7.4) is long, due to the transitional nature of this type and the inherently large number of forb species as major members. In the classification of desert grassland vegetation (Table 7.1), series and/or associations dominated by species other than those identified as being diagnostic of desert grassland are likely to be recent succession-disturbance types occupying prior plains-mesa grassland sites. It is debatable whether these associations comprise stable communities or whether succession will continue in the direction of desert scrubland. Also, if protected from grazing, could the vegetation of these desert grassland sites possibly return to plains-mesa grassland? In an analysis of grass in New Mexico's national forests after twenty-five years of protection, Potter and Krenetsky (1967) determined that forbs decrease under protection and increase under grazing. They also found that grass cover makes over threefold increases in montane grassland under protection, but there is little or no difference in grass cover between protected and grazed plots in desert grassland. Atwood (1987) found that Black Grama grass in southern New Mexico exhibits little difference in cover on plots protected from grazing for approximately fifty years from that on continuously

grazed plots. Desert grassland communities where Blue Grama dominates are examples of communities that have recently become members of desert grassland vegetation; in the not too distant past they were probably plains-mesa grassland communities. Gardiner (1951), in discussing desert grassland in south-central New Mexico, writes that, "shrubs are now [1949] in possession of sites upon which Black Grama grass hay was cut and baled within the memory of living residents."

The composition of Desert Grassland communities in New Mexico is highly variable (Table 7.1). The following are some general observations about desert grassland on the Jornada Experimental Range, in southern New Mexico (Little and Campbell, 1943). Shrubs are of greater importance than grass. Forbs are relatively important (+ 10%). Sandy soils are dominated by the shrubs Honey Mesquite (*Prosopis glandulosa*), Broom Snakeweed (*Gutierrezia sarothrae*),or Soaptree Yucca (*Yucca elata*); adobe soils by Tarbush (*Flourensia cernua*) or Allthorn (*Koeberlinia spinosa*); and gravelly slopes by Creosotebush (*Larrea tridentata*). No single forb species stands out in this vegetation type. Little and Campbell found that out of twenty-seven forb species, each with a contribution of 0.1% or more, only six had a contribution over 0.5%. Stickleaf (*Mentzelia albicaulis*) was highest, with a contribution of 2.5%. Stein (1977) characterizes the typical species composition of desert grassland vegetation on bajada plots in southern New Mexico. The species pattern by life-form is expressed as the percentage of plots in which a species was found, as follows:

shrub—Creosotebush, 70%; Broom Snakeweed, 41%; Long-Leaf Joint-Fir (*Ephedra trifurca*), 36%; Tarbush, 21%; and *Opuntia violacea*, 36%.

grasses—Fluff Grass (*Erioneuron pulchellum*), 97%; Black Grama, 39%; Six-Weeks Three-Awn (*Aristida adscensionis*), 38%; Purple Three-Awn (*A. purpurea*), 36%; and Sideoats Grama (*B. curtipendula*), 30%.

forbs—Desert Marigold (*Baileya multiradiata*), 92%; *Eriogonum rotundifolia*, 50: *Iva ambrosiaefolia*, 49%; *Eriogonum tricopes*, 44%; *Bahia absynthafolia*, 42%; *Zinnia acerosa*, 27%; and Desert Holly (*Perezia nana*), 24%.

Notice that only three species, a shrub (Creosotebush), a grass (Fluff Grass), and a forb (Desert Marigold) were found in most of the plots. This hints at the considerable variation found in communities of desert grassland vegetation.

Few major desert grassland forbs are found in the northwestern quarter of the state, where sheep grazing has occurred for many human generations. This might suggest that climax desert grassland is associated primarily with the Chihuahuan Desert. However, Black Grama, the major desert grassland grass, is common and occasionally is a grass dominant in northwestern desert grassland communities (Table 7.1). Generally, it can be assumed that desert grassland communities with shrub dominants of Big Sagebrush (*Artemisia tridentata*), Four-Wing Saltbush (*Atriplex canescens*), or Shadscale (*Atriplex confertifolia*), and cool-season grasses as grass dominants, are probably new desert grassland communities that had been plains-mesa grassland types prior to this century. The series containing associations with these dominants could have been placed in the Succession-Disturbance category, but with insufficient knowledge as to stability of the communities making up these associations, we have included them with the regular series and associations (Table 7.1).

A considerable area in west-central New Mexico is covered by sacahuista–galleta–blue grama vegetation. This is classified as desert grassland by virtue of the high shrub and forb densities, coupled with low grass density; it is obviously a new succession-disturbance desert grassland type. The *Juniperus monosperma–Bouteloua* spp.–*Hilaria belangeri*–MF association could be considered savanna vegetation, depending upon the density of the juniper.

Associations encompassing communities having black grama and sideoats grama as grass dominants, with Chihuahuan Desert or montane scrub shrub species as shrub dominants, might be examples of climax rather than disturbance desert grassland associations. Fluff Grass and Bush Muhly are also common grass dominants. The forb species in these communities will be primarily those designated as diagnostic in Table 7.4. Most of the communities in these associations are found on dry, gravelly piedmont slopes or bajadas of the basin and range mountains of southern New Mexico. These communities are common on the slopes of the mountain ranges on either side of the Rio Grande, as far north as the entrance of the Rio Puerco, north of Socorro. Figures 7.2F and G are pictures of typical desert grassland communities.

Shrubs that may be locally dominant in desert grassland communities are Mariola (*Parthenium incanum*), Spicebush (*Aloysia wrightii*), Thick-Leaved Yuccas (*Y. baccata, torreyi*, or *schottii*), agaves (*A. lechuguilla, palmeri*, or *parryi*), Sotol, and Ocotillo. Low shrubs that may be scattered through these communities include Feather Peabush (*Dalea formosa*), False Mesquite (*Calliandra eriophylla*), various prickly pears (*Opuntia* spp.), and *Barrel Cactus (Ferocactus wislizenii)*. Yerba de Pasmo (*Baccharis pterinoides*), Gatuño (*Mimosa dysocarpa*), and Range Ratany (*Krameria glandulosa*) may be locally abundant in these desert grassland communities. Communities of the *Prosopis glandulosa–Microrhamnus ericoides–Gutierrezia sarothrae/Muhlenbergia porteri–Euphorbia fendleri* Association (Table 7.1) are found in southeastern New Mexico (Secor et al., 1983). These Mesquite-Javalinabush-Snakeweed communities are found on shallow, sandy soils, in a mosaic with communities in the shinnery series of plains-mesa sand scrub (Table 8.3, Chapter 8).

Striking aspect dominance is common in desert grassland communities. Creosotebush–Black Grama–Fluff Grass communities often appear to be carpeted in the spring by the bright yellow Bladderpod (*Lesquerella gordonii*) and carpeted again in the fall by the orange-yellow Lemonweed (*Pectis papposa*). Interesting aspect dominants are the algae, *Nostoc* spp., found on exposed soil surfaces in some desert grassland vegetation.

When active, they form a dark green sheet over bare areas between grass clumps and play a nutrient-cycling role in mature communities (Neuenschwander, Sharrow, and Wright, 1975, and Stewart, 1982).

Table 7.1. Grassland Vegetation Types in New Mexico.

GRASSLAND VEGETATION
Sᴜʙᴀʟᴘɪɴᴇ-Mᴏɴᴛᴀɴᴇ Gʀᴀssʟᴀɴᴅ
  Mixed Sedge Series
    Carex spp.–*Deschampsia caespitosa*–MF
    Carex spp.–*Koeleria cristata*–MF
    Carex spp.–*Poa fendleriana–P. longiligula–*
      P. interior–MF

  Fescue Series
    Festuca thurberi–MF
    Festuca thurberi–Festuca arizonica–MF
    Festuca arizonica–Muhlenbergia montana–
      MF

  Pine Dropseed–Mountain Muhly Series
    Blepharoneuron tricholepis–Muhlenbergia
      montanus–MF

  Kentucky Bluegrass Series
    Poa pratensis–MG-F

Pʟᴀɪɴs-Mᴇsᴀ Gʀᴀssʟᴀɴᴅ
  Bluestem-Sideoats Series
    Schizachyrium scoparium–Bouteloua
      curtipendula–MF

  Grama-Buffalograss Series
    Bouteloua gracilis–Buchloe dactyloides–MF

  Grama-Feathergrass Series
    Bouteloua curtipendula–B. eriopoda–Stipa
      neomexicana–MF
    Bouteloua gracilis–Stipa neomexicana–S.
      comata–MF

  Grama-Galleta Series
    Bouteloua gracilis–Hilaria jamesii–MF
    Bouteloua eriopoda–Hilaria jamesii–MF
    Bouteloua belangeri–Hilaria jamesii–MF

  Grama-Threeawn Series
    Bouteloua gracilis–Aristida longiseta–MF
    Bouteloua curtipendula–B. gracilis–B.
      hirsuta–Aristida fendleriana–A. longiseta–
      MF

  Grama–Western Wheatgrass Series
    Bouteloua gracilis–Agropyron smithii–MF
    Bouteloua curtipendula–B. gracilis–B.
      hirsuta–Agropyron smithii–MF

  Grama–Indian Ricegrass Series
    Bouteloua gracilis–Oryzopsis hymenoides–
      MF

  Galleta–Indian Ricegrass–Needlegrass Series
    Hilaria jamesii–Oryzopsis hymenoides–Stipa
      comata–MF

  Indian Ricegrass–Dropseed Series
    Oryzopsis hymenoides–Sporobolus airoides–
      S. contractus–S. cryptandrus–MF

  Grama Grass Series
    Bouteloua gracilis–B. eriopoda–MF
    Bouteloua gracilis–B. eriopoda–B.
      curtipendula–MF (piedmont)

  Grama-Dropseed Series
    Bouteloua eriopoda–Sporobolus contractus–
      S. flexuosus–S. giganteus–MF

  Tobosa Series (swales)
    Hilaria mutica
    Hilaria mutica–Sporobolus airoides
    Hilaria mutica–Bouteloua gracilis

  Sacaton Series (swales)
    Sporobolus airoides–Nostoc commune

Table 7.1. (continued).

Sporobolus airoides–Distichlis stricta
Sporobolus giganteus–S. wrightii–
  Schizachyrium scoparium

DESERT GRASSLAND (ECOTONE)
Shrub–Alkali Sacaton Series
  Atriplex canescens/Sporobolus airoides–SMF
  Atriplex confertifolia/Sporobolus airoides–
    SMF
  Atriplex canescens–A. confertifolia/
    Sporobolus airoides–SMF

Shrub–Western Wheatgrass Series
  Artemisia tridentata/Agropyron smithii–MF

Shrub–Indian Ricegrass Series
  Artemisia tridentata/Oryzopsis hymenoides–
    MF
  Atriplex canescens–A. confertifolia/
    Oryzopsis hymenoides–SMF
  Gutierrezia sarothrae/Oryzopsis
    hymenoides–MF

Shrub-Galleta Series
  Artemisia tridentata/Hilaria jamesii–MF
  Artemisia bigelovii/Hilaria jamesii–MF
  Atriplex canescens/Hilaria jamesii–MF

Shrub–Blue Grama Series
  Artemisia frigida/Bouteloua gracilis–MF

Shrub–Black Grama Series
  Artemisia tridentata/Bouteloua eriopoda–MF
  Gutierrezia sarothrae/Bouteloua eriopoda–
    MF
  Chrysothamnus greenei/Bouteloua eriopoda–
    MF
  Ephedra trifurca–Gutierrezia sarothrae/
    Bouteloua eriopoda–MF

Shrub–Mixed Grass Series
  Artemisia tridentata/Hilaria jamesii–
    Bouteloua gracilis–Sporobolus airoides–S.
    cryptandra–MF
  Artemisia tridentata–Atriplex obovata/
    Hilaria jamesii–Bouteloua gracilis–SMF
  Artemisia tridentata–Gutierrezia sarothrae–
    Ephedra torreyana/Hilaria jamesii–
    Oryzopsis hymenoides–MF
  Atriplex obovata/Hilaria jamesii–Sporobolus
    airoides–Atriplex powellii–SMF
  Chrysothamnus greenei–Gutierrezia
    sarothrae/Hilaria jamesii–Bouteloua
    gracilis–MF
  Yucca elata/Bouteloua gracilis–Muhlenbergia
    porteri–MF
  Yucca elata/Hilaria mutica (swales)
  Prosopis glandulosa–Microrhamnus
    ericoides–Gutierrezia sarothrae/
    Muhlenbergia porteri–Euphorbia fendleri
  Larrea tridentata/Bouteloua eriopoda–
    Erioneuron pulchellum–Muhlenbergia
    porteri–MF
  Acacia neovernicosa/Bouteloua eriopoda–
    Panicum obtusum–MF
  Nolina microcarpa/Hilaria jamesii–
    Bouteloua gracilis–MF
  Dasylirion wheeleri–Ericameria laricifolia/
    Bouteloua curtipendula–B. hirsuta–MF
  Yucca baccata–Parthenium incanum/
    Bouteloua curtipendula–B. eriopoda–MF
  Yucca torreyi–Fouquieria splendens/
    Bouteloua curtipendula–B. eriopoda–MF
  Agave lechuguilla–Dasylirion leiophyllum–
    Yucca baccata/Bouteloua spp.–MF
  Juniperus monosperma/Bouteloua spp.–
    Hilaria belangeri–MF

Table 7.2. Major Plants Comprising Subalpine-Montane Grassland in New Mexico.

Scientific Name	Common Name
GRASSES AND GRASSLIKE PLANTS	
Agropyron subsecundum	Bearded Wheatgrass
Blepharoneuron tricholepis	Pine Dropseed
Bromus ciliatus	Hairy Brome
Carex bella	Beautiful Sedge
Carex festivella	Meadow Sedge
Carex foenea	Nebraska Sedge
Cyperus spp.	flat sedge
Danthonia california	California Oatgrass
Danthonia intermedia	Timber Danthonia
Danthonia spicata	Poverty Danthonia
Deschampsia caespitosa	Tufted Hairgrass
Eleocharis acicularis	Needle Spikerush
Elymus virginicus	Virginia Wildrye
Equisetum spp.	horsetail
Festuca arizonica	Arizona Fescue
Festuca ovina	Sheep Fescue
Festuca thurberi	Thurber Fescue
Juncus spp.	rush
Koeleria cristata	Junegrass
Muhlenbergia montana	Mountain Muhly
Poa fendleriana	Fendler Muttongrass
Poa interior	Inland Bluegrass
Poa longiligula	Longtongue Muttongrass
Poa pratensis	Kentucky Bluegrass
FORBS	
Achillea lanulosa	Western Yarrow
Agastache pallidiflora	Giant Hyssop
Agoseris aurantiaca	Orange Mountain Dandelion
Allium cernuum	Rocky Mountain Nodding Onion
Antennaria parvifolia	Pussytoes
Campanula rotundifolia	Bellflower
Castilleja spp.	indian paintbrush
Cerastium spp.	mouse-ear chickweek
Delphinium spp.	larkspur
Erigeron canus	Hoary Fleabane
Erigeron confusa	
Erigeron subtrinervis	Threenerve Fleabane
Erigeron superbus	
Geranium caespitosum	Purple Geranium
Helenium hoopesii	Orange Sneezeweed
Ipomopsis aggregata	Skyrocket

Iris missouriensis	Western Blueflag
Lathyrus arizonica	Arizona Peavine
Linum lewisii	Blue Flax
Lupinus spp.	lupine
Oxytropus lamberti	Crazyweed
Penstemon spp.	penstemon
Phacelia heterophylla	Scorpionweed
Potentilla concinna	Elegant Cinquefoil
Potentilla hippiana	Horse Cinquefoil
Potentilla pulcherrima	Beauty Cinquefoil
Potentilla sierrae-blancae	
Pseudocymopteris montanus	Mountain Parsley
Pteridium aquilinum	Bracken Fern
Rudbeckia hirta	Blackeyed Susan
Senecio neomexicanus	New Mexican Groundsel
Silene scouleri	Sours Catchfly
Sisyrinchium demissum	Blue-eyed Grass
Solidago decumbens	Dwarf Alpine Goldenrod
Stellaria longipes	Longstalk Starwort
Swertia radiata	Deer's-ears
Taraxacum spp.	dandelion
Thermopsis montana	Dwarf Alpine Goldenpea
Thlaspi fendleri	Wildcandytuft
Vicia americana	American Vetch

Table 7.3. Major Plants Comprising Plains-Mesa Grassland in New Mexico.

Scientific Name	Common Name
SHRUBS	
Artemisia frigida	Estafiata
Artemisia bigelovii	Bigelow Sagebrush
Artriplex canescens	Fourwing Saltbush
Certoides lanata	Winterfat
Chrysothamnus greenei	
Chrysothamnus nauseosus var. *bigelovii*	Bigelow Rabbitbrush
Corypantha vivipara	Pincushion Cactus
Echinocereus engelmannii	Engelmann Strawberry Cactus
Echinocereus fendleri	Fendler Strawberry Cactus
Ephedra torreyana	Mormon Tea
Eriogonum effusum	
Gutierrezia microcephala	Hairworm Snakeweed
Gutierrezia sarothrae	Broom Snakeweed
Lycium pallidum	Pale Wolfberry
Mammillaria spp.	mammillaria

Table 7.3. (continued).

Scientific Name	Common Name
Menodora scabra	Rough Menodora
Opuntia imbricata	Tree Cholla
Opuntia polyacantha	Prairie Pricklypear
Prosopis glandulosa	Honey Mesquite
Yucca elata	Soaptree Yucca
Yucca glauca	Small Soapweed

GRASSES

Agropyron smithii	Western Wheatgrass
Aristida barbata	Havard Threeawn
Aristida fendleriana	Fendler Threeawn
Aristida longiseta	Red Threeawn
Aristida purpurea	Purple Threeawn
Bouteloua curtipendula	Sideoats Grama
Bouteloua eriopoda	Black Grama
Bouteloua gracilis	Blue Grama
Bouteloua hirsuta	Hairy Grama
Bromus tectorum	Cheatgrass
Buchloe dactyloides	Buffalograss
Distichlis stricta	Saltgrass
Eragrostis oxylepis	
Hilaria jamesii	Galleta
Hilaria mutica	Tobosa
Lycurus phleoides	Wolftail
Muhlenbergia arenacea	Ear Muhly
Muhlenbergia cuspidata	Stonyhills Muhly
Muhlenbergia torreyi	Ring Muhly
Munroa squarrosa	False Buffalograss
Oryzopsis hymenoides	Indian Ricegrass
Panicum obtusum	Vinemesquite
Schedonnardus paniculatus	Tumblegrass
Schizachyrium scoparium	Little Bluestem
Setaria macrostachya	Plains Bristlegrass
Sitanion hystrix	Bottlebrush Squirreltail
Sorghastrum nutans	Yellow Indiangrass
Stipa comata	Needle and Thread
Stipa neomexicana	New Mexico Feathergrass
Sporoblus airoides	Alkali Sacaton
Sorobolus contractus	Spike Dropseed
Sporobolus cryptandrus	Sand Dropseed
Sporobolus flexuosus	Mesa Dropseed
Sporobolus giganteus	Giant Dropseed
Sporobolus nealleyi	Gypgrass
Trichachne californica	Arizona Cottontop

Table 7.3. (continued).

Scientific Name	Common Name
FORBS	
Astragalus bisulcatus	
Astragalus drummondii	Drummond Milkvetch
Astragulus lentiginosus	Blue Locoweed
Atriplex argentea	Silverscale
Atriplex powellii	
Bahia dissecta	
Calochortus ambiguus	Mariposalily
Eriogonum wrightii	
Gaura coccinea	Scarlet Gaura
Grindelia squarrosa	Curlycup Gumweed
Haplopappus gracilus	Spiny Goldenweed
Hoffmanseggia glauca	
Hymenoxys scaposa	Plains Rubberweed
Ipmoea leptophylla	
Ipomopsis longiflora	Pale Trumpets
Lappula redowskii	Stickseed
Lesquerella fendleri	Fendler Bladderpod
Leucelene ericoides	White Aster
Petalostemum candidum	White Prairieclover
Petalostemum purpureum	Purple Prairieclover
Polanisia trachysperma	Clammyweed
Psoralea lanceolata	Lemon Scufpea
Psoralea tenuiflora	Slender Scufpea
Ratibida columnifera	Prairie Coneflower
Ratibida tagetes	
Senecio spp.	groundsel
Sphaeralcea coccinea	Red Globemallow
Sphaeralcea fendleri	Fendler Globemallow
Thelesperma spp.	greenthread
Zinnia grandiflora	Rocky Mountain Zinnia

Table 7.4. Major Plants Comprising Desert Grassland in New Mexico. *=diagnostic (climax) member of Desert Grassland.

Scientific Name	Common Name
SHRUBS	
Acacia constrica	White Thorn
*Acacia neovernicosa	Viscid Acacia
*Agave lechuguilla	Lechuguilla
*Agave palmeri	Palmer Agave
Agave parryi	Parry Agave
*Aloysia wrightii	Spicebush
Artemisia bigelovii	Bigelow Sagebrush
Artemisia frigida	Estafiata
Artemisia tridentata	Big Sagebrush
Atriplex canescens	Fourwing Saltbush
Atriplex confertifolia	Shadscale
Atriplex obovata	
*Baccharis pterinoides	Yerba de Pasmo
Berberis haematocarpa	Algerita
*Calliandra eriophylla	False Mesquite
Ceratoides lanata	Winterfat
Chrysothamnus greenei	
*Dalea formosa	Feather Peabush
Dasylirion leiophyllum	Desert Spoon
*Dasylirion wheeleri	Sotol
Echinocereus spp.	hedgehog cactus
Ephedra torreyana	Mormon Tea
*Ephedra trifurca	Longleaf Jointfir
Ericameria laricifolia	Turpentinebush
*Eupatorium wrightii	Spreading Thoroughwort
*Ferocactus wislizenii	Barrel Cactus
Flourensia cernua	Tarbush
Fouquieria splendens	Ocotillo
*Gutierrezia microcephalum	
*Gutierrezia sarothrae	Broom Snakeweed
Gymnosperma glutinosum	Tatalencho
Haplopappus tenuisecta	Burroweed
*Isocoma wrightii	Jimmyweed
Koeberlinia spinosa	Allthorn
Krameria glutinosa	Range Ratany
Larrea tridentata	Creosotebush
Lycium pallidum	Pale Wolfberry
*Menodora scabra	Rough Menodora
*Microrhamnus ericoides	Javalinabush
Mimosa biuncifera	Catclaw Mimosa
*Minosa dysocarpa	Gatuno
Nolina microcarpa	Sacahuista

Table 7.4. (continued).

Scientific Name	Common Name
Opuntia imbricata	Tree Cholla
Opuntia leptocaulis	Pencil Cholla
Opuntia phaeacantha	Engelmann Pricklypear
Opuntia polyacantha	Plains Pricklypear
Opuntia spinosior	
Parthenium icanum	Mariola
Prosopis glandulosa	Honey Mesquite
Rhus choriophylla	Evergreen Sumac
Rhus microphylla	Littleleaf Sumac
Senecio longiloba	Wooly Groundsel
Yucca baccata	Banana Yucca
Yucca elata	Soaptree Yucca
Yucca glauca	Small Soapweed
Yucca schottii	Schott Yucca
Yucca torreyi	Torrey Yucca
Zizyphus obtusifolia	Graythorn

GRASSES
Aristida adscensionis	Sixweeks Threeawn
Aristida longiseta	Red Threeawn
Agropyron smithii	Western Wheatgrass
Bothriochloa barbinodis	Cane Beardgrass
Bouteloua breviseta	Chino Grama
Bouteloua curtipendula	Sideoats Grama
Bouteloua eriopoda	Black Grama
Bouteloua gracilis	Blue Grama
Bouteloua hirsuta	Hairy Grama
Elyonurus barbiculmis	Balsamscale
Eragrostis diffusa	Spreading Lovegrass
Eragrostis intermedia	Plains Lovegrass
Eragrostis neomexicana	New Mexico Lovegrass
Erioneuron pulchellum	Fluffgrass
Heteropogon contortus	Tanglehead
Hilaria belangeri	Curly Mesquite
Hilaria jamesii	Galleta
Hilaria mutica	Tobosa
Lycurus phleoides	Wolftail
Muhlenbergia porteri	Bushmuhly
Oryzopsis hymenoides	Indian Ricegrass
Panicum obtusum	Vine Mesquite
Scleropogon brevifolius	Burrograss
Schizachyrium scoparium	Little Bluestem
Setaria macrostachya	Plains Bristlegrass
Trichachne californica	Arizona Cottontop

Table 7.4. (continued).

Scientific Name	Common Name
Tridens muticus	Slim Tridens
Sporobolus airoides	Alkali Sacaton
Sporobolus contractus	Spike Dropseed
Sporobolus cryptandrus	Sand Dropseed
Sporobolus flexuosus	Mesa Dropseed
Sporobolus nealleyi	Gypgrass

FORBS
*Allionia incarnata	Trailing Four O Clock
*Amaranthus palmeri	Carelessweed
*Ambrosia acanthicarpa	Burweed
Artemisia dracunculoides	False Tarragon
Artemisia ludoviciana	Louisiana Wormwood
*Astragalus wootonii	Wooton Loco
Atriplex powelli	
*Bahia absynthafolia	Field Bahia
*Baileya multiradiata	Desert Marigold
Boerhaavia spp.	spiderling
Caesalpinia densiflora	Hogpotato
Caesalpinia depanocarpa	
*Cassia bauhunoides	Twinleaf
*Chamaesaracha spp.	
Chenopodium incanum	
Cordylanthus wrightii	Clubflower
*Cryptantha crassisepala	Plains Hiddenflower
*Cryptantha micrantha	Desertnut
*Cucurbita digitata	Cleftleaf Gourd
Cucurbita foetidissima	Buffalo Gourd
*Desmanthes cooleyi	Bundleflower
Dithyrea wislizenii	Spectaclepod
Dyssodia acerosa	Dogweed
Erigeron spp.	fleabane
*Eriogonum abertianum	
Eriogonum jamesii	Antelope Sage
*Eriogonum rotundifolia	
*Eriogonum tricopes	
Erodium cicutarium	Filaree
Euphorbia albomarginata	Rattlesnakeweed
Euphorbia fendleri	
*Evolvulus nuttallianus	
*Froelichia gracilis	Cottonweed
*Gaillardia pulchella	Blanket Flower
*Gnaphalium chilense	Cottony Everlasting
*Gomphrena caespitosa	Ballclover

Table 7.4. (continued).

Scientific Name	Common Name
Haplopappus gracilis	Goldenweed
Hedeoma spp.	false pennyroyal
Heterotheca latifolia	Telegraph Plant
Ipomopsis longiflora	Pale Trumpets
Ipomopsis pumila	
Iva ambrosiaefolia	Sumpweed
Lepidium lasiocarpum	Peppergrass
Lesquerella fendleri	Fendler Bladderpod
Lesquerella gordonii	Gordon Bladderpod
Leucelene ericoides	White Aster
Linum aristatum	
Lotus greenei	Deervetch
Lotus neomexicana	
Machaerantha tanacetifolia	Tahoka Daisy
Malacothrix fendleri	Desert Dandelion
Melampodium leucanthum	Rock Daisy
Mentzelia albicaulis	Whitestem Stickleaf
Nama hispidum	Purple Curlleaf
Nerisyrenia camporum	
Notholaena sinuata	Wavy Cloakfern
Oenothera caespitosa	Tufted Evening Primrose
Oenothera primiveris	Large Yellow Desert Evening Primrose
Pectis angustifolia	Limoncillo
Pectis papposa	Lemonweed
Penstemon fendleri	Fendler Beardtongue
Perezia nana	Desert Holly
Phacelia corrugata	Scorpionweed
Plantago purshii	Wooly Indianwheat
Portulaca oleracea	Common Purslane
Psilostrophe tagetina	Paperflower
Rumex hymenosepalus	Canaigre
Salsola kali	Russian Thistle
Schrankia occidentalis	Sensitive Brier
Selenocarpus diffusus	Spreading Moonpod
Sida leprosa	Scurfy Sida
Solanum elaeagnifolium	White Horsenettle
Sphaeralcea coccinia	Red Globemallow
Sphaeralcea incana	
Sphaeralcea subhastata	
Talinum aurantiacum	Flameflower
Tetraclea coulteri	
Thelesperma megapotamicum	Greenweed
Tidestromia lanuginosa	Mouse Ear

Table 7.4. (continued).

Scientific Name	Common Name
*Verbena bipinnatifida	Dakota Verbena
*Verbesina enceliodes	Crownbeard
Zephyranthes longiflora	Zephyr Lily
*Zinnia acerosa	
Zinnia grandiflora	Rocky Mountain Zinnia

CHAPTER 7 REFERENCES

GRASSLAND VEGETATION

Allen, C. D., 1984. Montane grassland in the landscape of the Jemez Mountains, New Mexico. Thesis, University of Wisconsin, Madison.

Atwood, T. L., 1987. Influence of livestock grazing and protection from livestock grazing on vegetational characteristics of *Bouteloua eriopoda* rangelands. Thesis, New Mexico State University, Las Cruces.

Banks, R. L., 1970. An ecological study of scaled quail in southeastern New Mexico. Thesis, New Mexico State University, Las Cruces.

Beavis, W. D., J. C. Owens, J. A. Ludwig, and E. W. Huddleston, 1982. Grassland communities of eastcentral New Mexico and density of the range caterpillar, *Hemileuca oliviae* (Lepidoptera: Saturniidae). The Southwestern Naturalist, 27: 335–343.

Benson, L. and R. A. Darrow, 1944. A manual of southwestern trees and shrubs. University of Arizona Biological Sciences Bulletin 6. 411 pp. Tucson.

Branscomb, B. L., 1956. Shrub invasion of a southern New Mexico desert grassland range. Journal of Range Management, 11: 129–132.

Brown, D. E., 1982. Biotic communities of the American Southwest–United States and Mexico.

Desert Plants, 4. 332 pp. Boyce Thompson Southwestern Arboretum, Superior, Arizona.

Buechner, H. K., 1950. Life history, ecology, and range use of the pronghorn antelope in Trans-Pecos Texas. American Midland Naturalist, 43: 257–354.

Buffington, L. C. and C. H. Herbel, 1965. Vegetation changes on a semi-desert grassland range from 1858–1963. Ecological Monographs, 35: 139–164.

Cable, D. R., 1967. Fire effects on semidesert grasses and shrubs. Journal of Range Management, 20: 170–176.

————, 1972. Fire effects in southwestern semi-desert grass shrub communities. Tall Timbers Fire Ecological Conference No. 12: 109–127.

Campbell, R. S., 1931. Plant succession and grazing capacity on clay soils in southern New Mexico. Journal of Agricultural Research, 43: 1027–1051.

Campbell, R. S. and E. H. Bomberger, 1934. The occurrence of *Gutierrezia sarothrae* on *Bouteloua eriopoda* ranges. Ecology, 15: 49–61.

Canfield, R. H., 1948. Perennial grass composition as an indicator of condition of southwestern mixed grass ranges. Ecology, 29: 190–204.

Castetter, E. F., 1956. The vegetation of New Mexico. New Mexico Quarterly, 26: 257–288.

Cornelius, J. M., 1988. Fire effects on vegetation of a northern Chihuahuan desert grassland. Dissertation, New Mexico State University, Las Cruces

Cottam, W. P. and G. Stewart, 1940. Plant succession as a result of grazing and of meadow desiccation by erosion since settlement in 1862. Journal of Forestry, 38: 613–626.

Cully, A. and P. J. Knight, 1987. A handbook of vegetation maps of New Mexico counties. 135 pp., New Mexico Natural Resources Department, Santa Fe.

Darrow, R. A., 1944. Arizona range resources and their utilization—1. Cochise County. University of Arizona Agricultural Experiment Station Bulletin, 103: 311–366. Tucson.

DeOliviera, R. R., 1961. Survey of the vegetation of the eastern base of San Augustin Peak and vicinity, San Augustin Mountains, Dona Ana County, New Mexico. Thesis, New Mexico State University, Las Cruces.

Dittberner, P. L., 1971. A demographic study of some semidesert grassland plants. Thesis, New Mexico State University, Las Cruces.

Dix, R. L., 1964. A history of biotic and climatic changes within the North American grassland. pp. 71–89. In: Grazing in terrestrial and marine environments. A symposium of the British Ecological Society, Bangor, 1962. Blackwell Scientific Publications, Oxford.

Donart, G. B., 1984. The history and evolution of western rangelands in relation to woody plant communities. pp. 1235–1258. In: National Research Council/National Academy of Sciences: Developing strategies for rangeland management. Westview Press, Boulder, Colorado. 2022 pp.

Donart, G. B., D. Sylvester, and W. Hickey, 1978. A vegetation classification system for New Mexico, USA. pp. 488–490. In: D. N. Hyder (ed.), Proceedings of 1st International Rangeland Congress, Denver, Colorado.

Dyksterhuis, E. J., 1951. Use of ecology on rangeland. Journal of Range Management, 4: 319–322.

Francis, R. E., 1987. Phytoedaphic communities of the Rio Puerco Watershed, New Mexico.

Gardiner, J. L., 1951. Vegetation of the creosotebush area of the Rio Grande valley in New Mexico. Ecological Monographs, 21: 397–403.

Heerwagen, A., 1956. Mixed prairie in New Mexico. pp. 284–300. In: J. E. Weaver and F. W. Albertson. Grasslands of the Great Plains. Johnson Publishing Company, Lincoln, Nebraska.

Hennessy, J. T., R. P. Gibbons, J. M. Tromble, and M. Cardenas, 1983. Vegetation changes from 1935 to 1980 in mesquite dunelands and former grasslands of southern New Mexico. Journal of Range Management, 36: 370–374.

Humphrey, R. R., 1958. The desert grassland—a history of vegetational change and an analysis of cause. Botanical Review, 24: 193–252.

Jameson, D. A., 1962. Effects of burning on galleta–black grama range invaded by juniper. Ecology, 43: 760–763.

Johnston, M. C., 1977. Brief resume of botanical, including vegetational features, of the Chihuahuan desert region with special emphasis on their uniqueness. pp. 335–362. In: R. H. Wauer and D. H. Riskind (eds.). Transactions of the symposium of the biological resources of the Chihuahuan Desert region, United States and Mexico. U.S. Department of the Interior, National Park Service Transactions and Proceedings Series No. 3. U.S. Government Printing Office, Washington, D.C.

Kittams, W. R., 1972. Effects of fire on the vegetation of the Chihuahuan desert region. Tall Timbers Fire Ecological Conference, No. 12: 427–444.

LeSuer, H., 1945. The ecology of the vegetation of Chihuahua, Mexico. North of parallel 28. 92 pp. University of Texas Publication 452, Austin.

Little, E. L., Jr. and R. S. Campbell, 1943. Flora of Jornada Experimental Range, New Mexico. American Midland Naturalist, 30: 626–670.

Lowe, C. H. (ed.), 1964. The vertebrates of Arizona. 270 pp. University of Arizona Press, Tucson.

Martin, S. C., 1972. Semidesert ecosystems. Journal of Range Management, 25: 317–319.

Meents, J. K., 1979. Avian community structure in Chihuahuan desert grasslands. Dissertation, New Mexico State University, Las Cruces.

Moir, W. H., 1967. The subalpine tall grass, *Festuca thurberi,* community of Sierra Blanca, New Mexico. The Southwestern Naturalist, 12: 321–328.

———, 1979. Soil vegetation patterns in the Central Peloncillo Mountains, New Mexico, USA. American Midland Naturalist, 102: 217–331.

Muller, C. H., 1940. Plant succession in the *Larrea-Flourensia* climax. Ecology, 21: 206–212.

———, 1947. Vegetation and climate of Coahuila, Mexico. Mandrono, 9: 33–57.

Neuenschwander, L. F., S. H. Sharrow, and H. A. Wright, 1975. Review of tobosa grass, *Hilaria mutica.* The Southwestern Naturalist, 20: 255–264.

Nichol, A. A., 1937. The natural vegetation of Arizona. University of Arizona Agricultural Experiment Station. Technical Bulletin 68: 181–222.

Potter, L. D. and J. C. Krenetsky, 1967. Plant succession with released grazing on New Mexico range lands. Journal of Range Management, 20: 145–151.

Secor, J. B., S. Shamas, D. Smeal, and A. L. Gennaro, 1983. Soil characteristics of two desert plant community types that occur in the Los Medaños area of southeastern New Mexico. Soil Science, 136: 133–144.

Shantz, H. L. and R. Zon, 1924. Natural vegetation. Atlas of American Agriculture. Part I, Section E (map). U.S. Department of Agriculture. Washington, D.C.

Shreve, F., 1942a. The desert vegetation of North America. Botanical Review, 8: 195–246.

———, 1942b. Grassland and related vegetation in Northern Mexico. Madrono, 6: 190–198.

Smith, S. D., 1975. The growth patterns, productivity, and phytosociology of soaptree yucca (*Yucca elata* Engelm.). Thesis, New Mexico State University, Las Cruces.

Stein, R. A., 1977. Vegetation and soil patterns in a Chihuahuan desert bajada. Thesis, New Mexico State University, Las Cruces.

Stewart, L. H., 1982. Desert grassland communities on Otero Mesa, Otero County, New Mexico. Thesis, New Mexico State University, Las Cruces.

Weaver, J. E. and F. W. Albertson, 1956. Grassland of the Great Plains. Johnson Publishing Company. Lincoln, Nebraska.

Weaver, J. E. and F. E. Clements, 1938. Plant Ecology. 601 pp. McGraw-Hill Book Company, New York.

West, N. E., 1983. Comparisons and contrasts between continents. pp. 461–472. In: N. E. West (ed.). Ecosystems of the World, 5: Temperate Deserts and Semi-Deserts. Elsevier Scientific Publishing Company, New York.

Wester, D. B. and H. A. Wright, 1987. Ordination of vegetation change in Guadalupe Mountains, New Mexico, USA. Vegetatio, 72: 27–33.

Whitfield, C. J. and H. L. Anderson, 1938. Secondary succession of the desert plains grasslands. Ecology, 19: 171–180.

Whitfield, C. J. and E. L. Beutner, 1938. Natural vegetation in the desert plains grassland. Ecology, 19: 26–37.

Whitson, P. D., 1974. Factors contributing to production and distribution of Chihuahuan Desert annuals. US/IBP Desert Biome Research Memorandum 75-11. Utah State University, Logan.

Wondzell, S. M., 1984. Recovery of desert grasslands in Big Bend National Park following 36 years of protection from grazing by domestic livestock. Thesis, New Mexico State University, Las Cruces.

York, J. C. and W. A. Dick-Peddie, 1969. Vegetation changes in southern New Mexico during the past hundred years. pp. 157–199. In: W. G. McGinnies and B. J. Goldman (eds.). Arid lands in perspective. University of Arizona Press, Tucson.

Zimmerman, U. D., 1967. The response of grassland to disturbance in northeastern New Mexico. Thesis, New Mexico State University, Las Cruces.

8

SCRUBLAND VEGETATION

Scrubland vegetation is dominated by shrub species. Included as shrubs are woody-based perennials, such as cacti and the group sometimes referred to as "rosette" shrubs, which includes century plant (*Agave* spp.), some of the yuccas (*Yucca* spp.), sacahuista (*Nolina microcarpa*), and sotol (*Dasylirion* spp.). Many of these woody-based perennials are also members of the scrubland-grassland ecotone vegetation called Desert Grassland.

Much scrubland vegetation consists of shrubs having low available-moisture requirements. This is the case with shrubs of the Great Basin Desert Scrub, Chihuahuan Desert Scrub, and most of the shrubs of the Montane Scrub. Shrub adaptations include water-retention (reduction of water loss) features. Consequently, shrubs of the Montane Scrub occupy the drier, more xeric mountain habitats, but would not survive in grasslands or deserts. Plants dominating plains-mesa sand scrub are able to obtain water and survive in deep sand. Conditions of low available moisture that give rise to this type of vegetation may result from one or a combination of factors. These include low annual precipitation, short growing season, hot dry atmosphere, desiccating winds, steep slopes, south to west facing exposures, shallow rocky soils, and sandy, gravelly, and saline soils.

In New Mexico, scrublands can be subdivided into Montane Scrub. Closed Basin Scrub, Plains-Mesa Sand Scrub, Great Basin Desert Scrub, and Chihuahuan Desert Scrub. Figures 8.1A–D are maps depicting major areas of scrubland vegetation in New Mexico; Table 8.1 is a classification of scrubland vegetation. Closed Basin Scrub is the only riparian vegetation in New Mexico having sufficient horizontal dimensions to be delineated at the scale used for our maps. Although mapped (Fig. 9.3 and vegetation map insert) as a scrubland, Closed Basin Scrub has been classified and described as riparian vegetation (see Chapter 9). In southern New Mexico, the basin-and-range nature of the landscape, coupled with post-Pleistocene erosion, accelerated by disturbance, results in complex vegetation mosaics (disjunctions). Consequently, it can be difficult to differentiate between Chihuahuan Desert scrub and Desert Grassland.

Due to the scattered and patchy nature of scrubland vegetation in the state, *Biotic Communities of the American Southwest–United States and Mexico* (Brown, 1982), *A Handbook of Vegetation of New Mexico Counties* (Cully and Knight, 1987), and *Potential Vegetation of New Mexico* (Donart, Sylvester, and Hickey, 1978) have been useful in determining occurrences and composition of this vegetation.

MONTANE SCRUB

This scrubland is found in situations where the available moisture is less than might be expected considering the altitude, latitude, and/or surrounding vegetation. Montane Scrub often constitutes a patch or strip within other, more extensive types of vegetation. These patches and strips of climax montane scrub reflect conditions such as a high, rocky, windswept knoll; a southwesterly

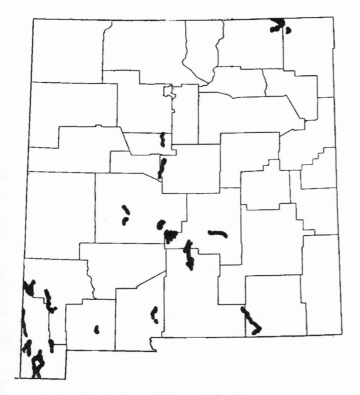 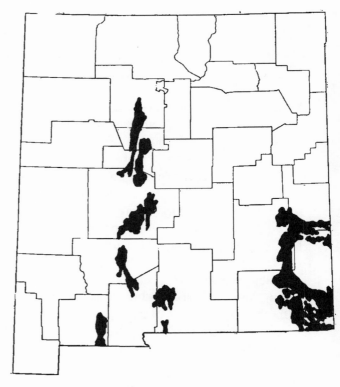

A. Distribution of Montane Scrub.   B. Distribution of Plains-Mesa Sand Scrub.

Figure 8.1 Distribution of Scrubland Vegetation in New Mexico. See New Mexico County Map on page xiv for reference, if needed, on county names.

facing slope; a rocky slope; or an escarpment of an exposed rock stratum. Some of the montane scrub found in New Mexico is extensive enough to be mapped at the scale we have used (Fig. 8.1A and vegetation map insert).

Brown (1982) separated what we are calling Montane Scrub into interior chaparral and Great Basin montane scrubland. We feel this is not warranted in New Mexico because a number of the major shrubs are common to all of the mountain scrubland areas in the state. Figure 8.2 dis-

plays regions of the state in which the dominant plants of the montane scrub occur. Note the relatively large number (eleven) of shrubs common to all Montane Scrub regions. In addition, areas where shrub species with restricted ranges codominate with widespread species, no unique combinations occur to justify two separate groups.

It is often difficult to detect whether a scrubland is climax or successional. Aspen (*Populus tremuloides*) is the primary large plant that follows disturbance in the higher forests (subalpine

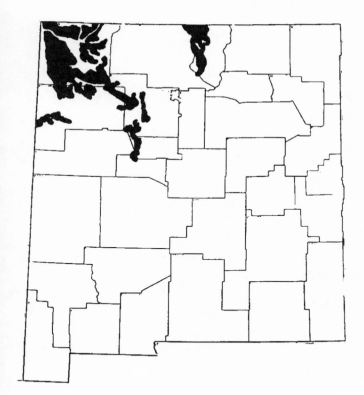

C. Distribution of Great Basin Desert Scrub.

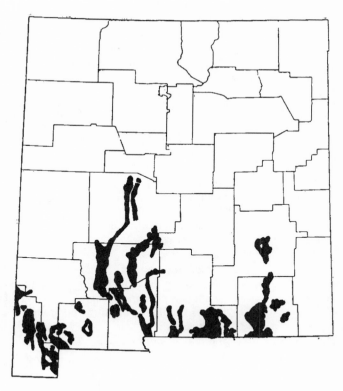

D. Distribution of Chihuahuan Desert Scrub.

coniferous forest and upper montane coniferous forest). Scrubland vegetation follows disturbance in lower elevation forests (lower montane coniferous forest and upper woodland), as after disturbance in ponderosa pine forest. These shrubs may persist even after relatively mature stands of pine have returned. In New Mexico, most of this montane scrubland successional stage is dominated by a shrub form of Gambel Oak (*Quercus gambelii*) and New Mexican Locust (*Robinia neomexicana*). Both of these plants are

members of other vegetation types, and in such types they take on a tree life-form. When these two species dominate mountain scrubland vegetation, it can be assumed that the vegetation is successional and does not represent true montane scrub. For this reason, Gambel Oak and New Mexican Locust are not included in any series of the montane scrub, but they are included in series of the successional-disturbance scrub.

Major plants of the Montane Scrub in New Mexico are listed in Table 8.2. Many of the dom-

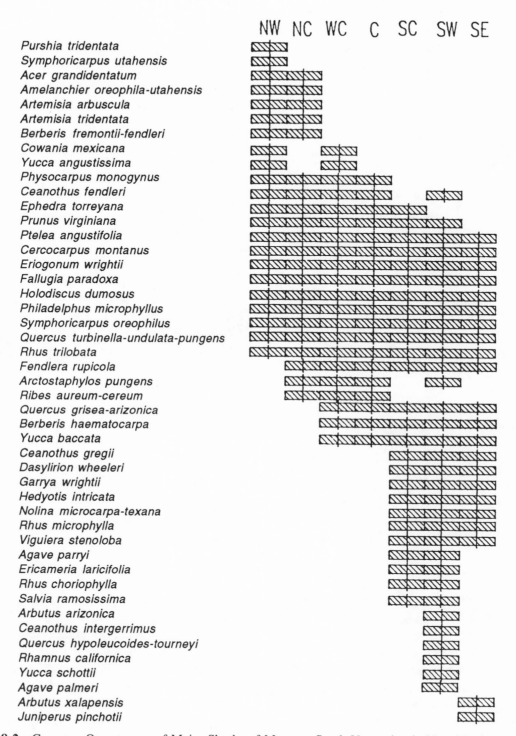

Figure 8.2 Common Occurrences of Major Shrubs of Montane Scrub Vegetation in New Mexico. northwest = NW, north-central = NC, west-central = WC, central = C, south-central = SC, southwest = SW, southeast = SE

inant shrubs of the higher montane scrub vegetation are also found as members of the shrub layer in montane coniferous forest and woodland vegetation. Some of these plants are Bigtooth Maple (*Acer grandidentatum*), Silverleaf Oak (*Quercus hypoleucoides*), Manzanita (*Arctostaphylos pungens*), Buckbrush (*Ceanothus fendleri*), Snowberry (*Symphoricarpos oreophilus*), and currants (*Ribes* spp.). Many dominant plants of the lower montane scrub are also members of desert grassland vegetation, particularly in the Rosette Scrub–Grama Grass Series. Examples are Sotol (*Dasylirion wheeleri*), Century Plant (*Agave parryi*), Banana Yucca (*Yucca baccata*), and Sacahuista (*Nolina microcarpa*).

Several semiriparian species are important in montane scrub; this is largely the reason for using a semiriparian category. Species considered to be "obligate riparian" are those virtually restricted to drainage systems. Species that can dominate in both riparian habitats and pseudoriparian habitats are considered here to be semiriparian species. Upon close inspection of the microhabitats where these semiriparian species grow in montane scrub, it can usually be found that the microhabitats contain more available moisture than the surrounding area. This additional moisture may be due to seeps or to small catchments of run-off from surrounding bare rocks. Such microhabitats are referred to as pseudoriparian, because they are not part of a true drainage system (for details see Chapter 9). Apache Plume (*Fallugia paradoxa*), Chokecherry (*Prunus virginiana*), Bigtooth Maple (*Acer grandidentatum*), and Hoptree (*Ptelea angustifolia*) are examples of semiriparian species (see Table 9.2). Watson (1912) noted that in mountains near Albuquerque, Apache Plume may form thickets up as high as 2,700 m (9,000 ft.).

Some of the information about montane scrub on various New Mexico mountains was obtained from the following sources: Animas (Wagner, 1977); Brazos Canyon (Standley, 1915); Cebolleta (Osborn, 1966); Guadalupe (Gehlbach, 1967;

Robinson, 1969; Wester, 1984; and Wester and Wright, 1987); Hachita (Moir, 1963); Jemez (Osborn, 1966; and Robertson, 1968); Manzano (Bedker, 1966); Organ (Dick-Peddie and Moir, 1970); Peloncillo (Moir, 1979); San Andres (Little and Campbell, 1943; and Von Loh, 1977); Sandia (Naylor, 1964); San Pedro Parks (Tatschl, 1966); and Zuni (Riffle, 1973). Major tree-shrubs and shrubs dominating the upper, middle, and lower montane scrub elevational (available moisture) zones in the state are shown in Table 8.3. In this table, many of the plants in the upper zone are also understory dominants of Woodland and Montane Coniferous Forest vegetation. Plants dominating the lower Montane Scrub zone may also be members of Desert Grassland or even Desert Scrub vegetation.

Mountain Mahogany (*Cercocarpus montanus*) is a major dominant of montane scrub and may share dominance with serviceberry (*Amelanchier* spp.) and/or Cliffrose (*Cowania mexicana*) in the northwestern mountains of the state; with Skunkbush and/or Scrub Oak (*Quercus turbinella* and *Q. undulata*) in north-central portions; or with Redberry Juniper (*Juniperus pinchotii*) and Desert Ceanothus (*Ceanothus greggii*) in the southeast. Texas Madrone (*Arbutus xalapensis*) is occasionally a member of this mountain scrub vegetation.

Montane scrub in the San Andres Mountains, in south-central New Mexico, contains examples of middle and lower elevational elements. In addition to the ubiquitous Mountain Mahogany (*Cercocarpus montanus*), the montane scrub of the San Andres Mountains contains, at its upper elevations, Buckbrush (*Ceanothus fendleri*), Skunkbush (*Rhus trilobata*), Gambel Oak (*Quercus gambelii*), Desert Rose (*Rosa stellata*), and Bluets (*Hedyotis intricata*). Von Loh (1977) identified a Mountain Mahogany–Bluets association in these mountains. Von Loh found the lower montane scrub dominated by Resinbush (*Viguiera stenoloba*), Algerita (*Berberis haematocarpa*), Shrub Liveoak (*Quercus turbi-*

nella), Sacahuista (*Nolina macrocarpa*), Desert Ceanothus (*Ceanothus greggii*), and Banana Yucca (*Yucca bacata*) (Pl. 20).

In southwestern New Mexico, vegetation referred to as upper encinal by Wagner (1977) can be considered montane scrub. Widespread dominants such as Mockorange (*Philadelphus micro-phyllus*), Buckbrush, and Mountain Mahogany are joined by species peculiar to this corner of the state, such as Arizona Madrone (*Arbutus arizonica*), Hoary Yucca (*Yucca schottii*), Century Plant (*Agave palmeri*), Shrubby Senna (*Cassia wislezeni*), Border Pinyon (*Pinus discolor*), and Toumey Oak (*Quercus toumeyi*).

PLAINS-MESA SAND SCRUB

Most of the areas supporting Plains-Mesa Sand Scrub are found in Woodland, Savanna, and Grassland vegetation (see vegetation map insert). Even relatively new dune areas (mesquite coppice dunes) are found on past mesa grassland sites. The major deep-sand areas are on the shoulders of old (post-Pleistocene) floodplains of the Pecos, Rio Grande, and San Juan rivers or their tributaries. One area is located on the west shoulder of the Rio Grande floodplain, from south of Albuquerque to the entrance of the Rio Salado, where it broadens out into a rather large dune field. There are a number of areas on the southern shoulder of the Rio Chama floodplain not far to the west of its confluence with the Rio Grande and a few on the shoulders of the drainages to the San Juan, in northwestern New Mexico. There is an extensive deep-sand area west of the Llano Estacado caprock and east of the Pecos River. White Sands National Monument occupies a small area in the southeast corner of a vast gypsum sand area in the Tularosa Basin (Fig. 8.1B).

All of these post-Pleistocene deep sands are dominated by species which are deep-sand tolerant or even deep-sand adapted. The absence of sand-adapted species on mesquite dunes attests to the recent origin of the dunes. Table 8.4 is a list of major plants that may comprise the various deep-sand areas of the state. The most ubiquitous shrub of the sand scrub vegetation is Sand Sagebrush (*Armtemisia filifolia*). Note that most (65%) of the deep-sand adapted grasses are widespread in the state (Table 8.4). Various combinations of these plants are found associated with Sand Scrub vegetation. Some of these species are also found in sandy Grassland communities. Examples are Hairy, Blue, and Sideoats Grama, Alkali Sacaton, and Mesa Dropseed. All but Blue and Sideoats Grama are common enough in sand scrub to be included in Table 8.4. Broom Snakeweed (*Gutierrezia sarothrae*) and Estafiata (*Artemisia frigida*) are two shrubs often found in sand scrub but also dominate in other vegetation types.

One of the more unique sand scrub areas is the shinnery scrub found in east-central and southeastern New Mexico. Many habitat studies have been made of the shinnery areas, habitat for the lesser prairie chicken (York and Dick-Peddie, 1969, and Secor et al., 1983). Considerable vegetation information is contained in graduate theses by Ahlborn (1980), Candelaria (1979), Suminski (1977), and Wilson (1982). The vegetation is dominated by the oak shrub Shinoak (*Quercus havardii*), along with a lesser amount of Sand Sagebrush. Other common shrubs are White Ratany (*Krameria grayi*), a Low Yucca (*Yucca campestris*), and Four-Wing Saltbush. The tree-shrub Soapberry (*Sapindus saponaria* var. *drummundii*) is commonly found in small catchments or where the water table is close to the surface. Major grasses include Sand Bluestem (*Andropogon hallii*), Red Lovegrass (*Eragrostis oxylepis*), Purple Three-Awn (*Aristida purpurea*), Sand Paspalum (*Paspalum stramineum*), and

Plains Bristlegrass (*Setaria texana*). The most common forbs are Annual Buckwheat (*Eriogonum annuum*), Sand Verbena (*Abronia angustifolia*), Hierba Del Pollo (*Commelina erecta*), and Texas Croton (*Croton texensis*). Plate 21 is an example of shinnery sand scrub in Eddy County, New Mexico.

Dominant sand scrub plants in New Mexico occur on both gypsum and quartz sands. The dominants found on gypsum, at White Sands, by Shields (1956) and Reid (1979, 1980), were the same as those on quartz sand reported by Dittmer (1959), in the northwest, by Campbell and Campbell (1938), on the Jornada Experimental Range, and by Burgess and Northington (1974), found dominating deep sands in the vicinity of the Guadalupe Mountains (Pl. 22). Except in the east-central and southeastern areas, many of the dominant sand scrub plants are common to all regions of the state. These are:

Shrubs
Artemisia filifolia
Atriplex canescens
Ephedra torreyana
Penstemon ambiguus
Psorothamnus scoparius
Rhus trilobata

Grasses
Aristida purpurea
Eragrostis intermedia
Muhlenbergia pungens
Paspalum stramineum

Setaria levcopila
Sporobolus cryptandrus
Sporobolus airoides

Forbs
Abronia angustifolia
Ambrosia acanthifolia
Caesalpinia jamesii
Eriogonum annuum
Heliotropium convolvulaceum
Maurandya spp.

In the northwestern quarter of the state, the most common sand-scrub vegetation type is *Artemisia filifolia/Muhlenbergia pungens* (Sand Sagebrush/Sandhill Muhly). Other shrubs usually found in these communities are Estafiata (*Artemisia frigida*), Broom Snakeweed, *Chrysothamnus greenii,* Torrey Ephedra, and Fineleaf Yucca (*Yucca angustissima*). Soaptree Yucca (*Yucca elata*) is usually found in sand scrub communities of central and southern New Mexico.

As mentioned earlier, deep-sand areas dominated by coppice dunes of Honey Mesquite (*Prosopis glandulosa*) are often relatively depauperate in sand-adapted species. In most instances, the major plants associated with the mesquite are disturbance types, such as the half-shrub Broom Snakeweed (*Gutierrezia sarothrae*) and forbs such as Tansymustard (*Descurainia pinnata*) and Russian Thistle (*Salsola kali*). Other forbs and grasses are usually those associated with the desert grassland vegetation type (Atwood, 1983, and Hennessy, 1981).

GREAT BASIN DESERT SCRUB

The North American Great Basin Desert has also been called cold desert, sagebrush desert, and saltbush desert. As with all deserts, the Great Basin Desert biome is considered to be of relatively recent origin, that is, 5,000 to 12,000 years ago (Brown, 1982). In New Mexico, this desert is limited to the northwest corner and a tongue in the north-central part (Fig. 8.1D). The Great Basin Desert differs from other North American deserts by receiving most of its moisture in the form of snow, during the winter months. In New Mexico, 50%–60% of the annual precipitation falls during the six-month cool, or low sun-angle, season, October 1–March 31 (Moir, personal communication, 1988). Annual precipitation on the Great Basin Desert in New Mexico is approximately 230 mm (9 in.).

The term desert is quite general, and the Great Basin Desert includes areas where there are vegetation units other than Great Basin Desert Scrub, such as Closed Basin Scrub, desert grassland, and arroyo riparian vegetation. We are only considering here the scrub vegetation unique to the Great Basin Desert. The major shrubs of the Great Basin Desert Scrub are Big Sagebrush (*Artemisia tridentata*), Shadscale (*Atriplex confertifolia*), Greasewood (*Sarcobatus vermiculatus*), and Fourwing Saltbush (*Atriplex canescens*). Shadscale is the best indicator of Great Basin Desert Scrub, because the others can be dominants of other types of vegetation in the state. Big Sagebrush has increased dramatically during the past 125 years, evidently as a result of heavy grazing by domestic livestock. The expansion of Big Sagebrush out of the desert through grassland into woodland has been so extensive through the intermountain West that many assume that sagebrush stands with some grass indicate disturbance conditions. Brown (1982) has pointed out that *Artemisia tridentata* communities with some grass occur on ungrazed sites. However, in New Mexico, most of the vegetation dominated by Big Sagebrush is on sites that were grassland or savanna in the middle of the last century (Gross and Dick-Peddie, 1979).

Sagebrush and saltbush association of the Great Basin Desert Scrub in the state have sparse ground cover, and little of that is grass. This is also the case for this vegetation outside of New Mexico (Brown, 1982). Big Sagebrush communities having a considerable amount of grass cover are considered to be part of the Desert Grassland vegetation and are described in that section (Chapter 7). Classification of Great Basin Desert Scrub communities is included in Table 8.1. Table 8.5 lists the major plants of the Great Basin Desert Scrub in New Mexico.

Sagebrush communities usually occur on higher ground than the saltbush communities (Shreve, 1942). Other sagebrush species often share dominance with Big Sagebrush, as can be seen in Table 8.1: Black Sage (*Artemisia arbuscula*) on some sites and Bigelow Sage (*A. bigelovii* on others.

Fourwing Saltbush is the member of the saltbush genus (*Atriplex*) dominating most of the associations here, while Shadscale is the most important in this vegetation outside New Mexico. Shadscale does share dominance with Fourwing Saltbush on many sites in the state (Pl. 23). The most common association over the largest area is that of *Atriplex canescens–Sarcobatus vermiculatus/Sporobolus airoides*. Cully and Knight (1987) have mapped this type in San Juan, McKinley, Catron, Sandoval, Cibola, Valencia, and Bernalillo counties. Brown (1982) comments that a number of alien annual species are important and increasing in the Great Basin Desert Scrub, probably because native forbs could not withstand the heavy grazing by domestic livestock. These alien species are indicated in Table 8.5. One of the more important members of the saltbush group in Great Basin Desert Scrub is *Atriplex nuttallii* (Goodman and Caldwell, 1971). In New Mexico, this plant usually shares dominance with Fourwing Saltbush and Shadscale (Table 8.1). Winterfat (*Ceratoides lanata*) and rabbitbrush (*Chrysothamnus* spp.) are often members of Great Basin Desert Scrub communities, and in some areas they are dense enough to be considered codominants with Fourwing Saltbush.

CHIHUAHUAN DESERT SCRUB

The Chihuahuan Desert is located primarily in Mexico. There are fingers of the desert extending north onto the mesas bordering the Pecos and Rio Grande floodplains in southern New Mexico and a small area on the mesas east of the San Pedro River floodplain, near Tombstone, in southeastern Arizona. Chihuahuan Desert boundaries in North America are generally based upon the contiguous ranges of Creosotebush (*Larrea tridentata*) and Tarbush (*Flourensia cernua*) on mesas of the central plateau, extending from the end of the Rocky Mountains south to San Luis Potosí, Mexico. The various creosotebush series are considered to be the only major scrubland vegetation types comprising the Chihuahuan Desert, just as sagebrush and saltbush series comprise the Great Basin Desert in New Mexico. Other vegetation types are found within the boundary of the Chihuahuan Desert, such as Closed Basin Scrub, alkali sink scrub, and Desert Grassland, from our classification. Chihuahuan Desert Scrub is a relatively new vegetation type in New Mexico, possibly no older than 4,000 years (Van Devender and Toolin, 1983). The following is from *The Desert Vegetation of North America* (Shreve, 1942):

Also there is trustworthy evidence that the entire elevated region between Benson, Arizona and Deming, New Mexico, formerly supported a fairly heavy stand of perennial grasses, and that *Larrea* and *Prosopis* were much less abundant under the original conditions. It therefore seems best to regard this region as part of the Desert Grassland Transition Region rather than part of the desert.

Our use of Desert Grassland (ecotone) as a subtype in the grassland classification was influenced, in part, by these views of Shreve's. He goes on to state that the Creosotebush-Tarbush codominants of the Chihuahuan Desert are equivalent to the Creosote-Bursage (*Franseria dumosa*) codominants of the Sonoran Desert. Major plants of the Chihuahuan Desert Scrub in New Mexico are listed in Table 8.6, and the distribution of Chihuahuan Desert Scrub in New Mexico is shown in Figure 8.1E.

Shreve (1942) alludes to another feature of the Chihuahuan Desert Scrub vegetation found today in New Mexico. Large areas that supported plains-mesa or desert grassland prior to the second half of the last century, are today occupied by Chihuahuan Desert Scrub vegetation. The cause of this rapid shift in New Mexico appears to be primarily the concentrations in the late 1800s of domestic livestock similar to concentrations on the more mesic grasslands of the Great Plains, to the east. Donart (1984) writes that, "True mismanagement of the western rangelands probably ranged from 70 to 250 years, depending on the location. This period was associated with major change in vegetative structure which may be attributed largely to federal policy, ecological ignorance, land tenure and economic structure of the time" There is a considerable body of literature based upon research into other possible causes for this dramatic shift. The findings indicate that other possible causes, such as climate change, fire suppression, and rodent competition served merely to accelerate the rate of succession. For example, Neilson (1986) indicates that what climate changes there were over the past 150 years were too slight and subtle to cause major shifts in vegetation patterns. However, the changes might have been enough to prevent the reestablishment of previous vegetation patterns.

It is very difficult in the field to tell whether an area is occupied by an original (pre-1850) or a recently established Chihuahuan Desert Scrub vegetation. If there are scattered Soaptree Yuccas (*Yucca elata*) in a relatively pure stand of creosotebush, it is likely that Chihuahuan Desert Scrub has occupied the site no longer than 100–120 years. The absence of Tarbush, the common codominant with Creosotebush, might be considered another indicator of new Chihuahuan Desert Scrub. Desert Holly (*Perezia nana*) plants were found to be extensive, dense, vigorous, and relatively large on a mature, ungrazed area of Desert Grassland in southern New Mexico (Stewart, 1982). This perennial forb is commonly found in low densities and of small size, associated with Creosotebush in Chihuahuan Desert Scrub sites. Desert holly found on a Chihuahuan Desert Scrub site should possibly be considered a relict of Desert Grassland vegetation. Fluff grass (*Erioneuron pulchellum*) displays characteristics of occurrence and appearance similar to those of Desert Holly.

Research is necessary to determine whether the presence of Soaptree Yucca, Desert Holly, and Fluff Grass, and the absence of Tarbush, in Chihuahuan Desert Scrub can be used as indicators of recent succession from a grassland or desert grassland vegetation. It can be assumed from various records, including the Territorial Survey, that the amount of Chihuahuan Desert Scrub in New Mexico today is greater that that found 150 years ago, but due to the difficulties mentioned, no attempts were made on our maps to differentiate "original" from "new" boundaries of this vegetation type.

In New Mexico, most of the Chihuahuan Desert Scrub area has Creosotebush as the sole dominant. There are scattered stands of the Creosotebush-Tarbush combination, and occasionally stands are found where Whitethorn (*Acacia constricta* or *neovernicosa*) codominates (Pl. 24). Other Chihuahuan Desert Scrub shrubs, such as cacti, Squawbush (*Condalia spathulata*), Lechuguilla (*Agave lechuguilla*), and Ocotillo (*Fouquieria splendens*) are found scattered or as local patches through the desert. Some Montane Scrub shrubs can be found in Chihuahuan Desert Scrub vegetation on rocky slopes of mountain ranges in southern New Mexico. Examples, included in Table 8.6, are Spicebush, Mariola, Sacahuista, Resinbush, and Banana Yucca.

Honey Mesquite (*Prosopis glandulosa*), found on mesas mixed with Creosotebush rather than in riparian situations, such as in or along arroyos or bordering playas, is likely due to use and accidental planting by early man or domestic livestock (York and Dick-Peddie, 1969).

The composition and extent of Plains-Mesa Sand Scrub and Montane Scrub have changed little during the past 150 years. Great Basin Desert and Chihuahuan Desert scrublands have increased in extent and will continue to do so as long as succession is initiated on grassland and Desert Grassland areas. Successional-disturbance montane scrub will continue to become established on logged or burned montane coniferous forest sites and will persist until forest is again established.

Table 8.1. Scrubland Vegetation Types in New Mexico.

SCRUBLAND VEGETATION	*Cowania mexicana–Philadelphus microphyllusa*/S
MONTANE SCRUB	*Cercocarpus montanus–Philadelphus microphyllus–Fendlera rupicola–Rhus trilobata*/S
Mountain Mahogany–Mixed Shrub Series	
Cercocarpus montanus–Amelanchier spp.–	

Table 8.1. (continued).

Cercocarpus montanus–Quercus turbinella–Rhus trilobata–Nolina microcarpa–Garrya wrightii/S

Cercocarpus montanus–Juniperus pinchotti–Ceanothus greggii–Quercus turbinella–Quercus undulata/MG

Cercocarpus montanus–Quercus grisea–Quercus turbinella–Rhus trilobata/S

Cercocarpus montanus–Quercus turbinella–Ericameria laricifolia–Nolina microcarpa–Yucca schottii/S

Mixed Evergreen Series

Quercus grisea–Cercocarpus montanus–Yucca baccata/MF

Quercus toumeyi–Arctostaphylos pungens–Ceanothus greggii–Rhus choriophylla/S

Pinus discolor/Quercus toumeyi–Cercocarpus montanus–Arctostaphylos pungens/MG

PLAINS-MESA SAND SCRUB

Shinoak (Shinnery) Series

Quercus havardii/Schizachyrium scoparium–Bothrichloa saccharoides

Quercus havardii–Artemisia filifolia–Yucca campestris/Andropogon hallii–Aristida purpurea

Quercus havardii–Atriplex canescens/Sporobolus airoides–Eriogonum annuum

Quercus havardii–Artemisia filifolia/MG-F

Sand Sagebrush Series

*Artemisia filifolia/Muhlenbergia pungens–*MF

Artemisia filifolia/Andropogon hallii–Stipa comata

Artemisia filifolia/Schizachyrium scoparium–Bothrichloa saccharoides

Artemisia filifolia/Bouteloua eriopoda–Sporobolus spp.

Mixed Shrub Series

Psorothamnus scoparius–Artemisia filifolia–Atriplex canescens/MG-F

Psorothamnus scoparius–Artemisia filifolia–Oryzopsis hymenoides–Sporobolus spp.

*Prosopis glandulosa/Gutierrezia sarothrae/*SMG-F (dunes)

Poliomintha incana–Rhus trilobata/Eriogonum annuum–Abronia angustifolium

GREAT BASIN DESERT SCRUB

Sagebrush Series

Artemisia tridentata/SMG-F

*Artemisia tridentata–Artemisia arbuscula/*SMG-F

*Artemisia tridentata–Artemisia bigelovii/*SMG-F

Saltbush Series

Atriplex canescens–Atriplex confertifolia/Sporobolus airoides

Atriplex canescens–Atriplex nuttallii–Atriplex confertifolia/SMG-F

Atriplex canescens–Sarcobatus vermiculatus/Sporobolus airoides

*Atriplex canescens–Ceratoides lanata/*SMG-F

CHIHUAHUAN DESERT SCRUB

Creosotebush Series

Larrea tridentata/Erioneuron pulchellum

Creosotebush-Mixed Shrub Series

Larrea tridentata–Flourensia cernua/Erioneuron pulchellum

*Larrea tridentata–Acacia constricta/*SMF

*Larrea tridentata–Gutierrezia sarothrae/*SMF

SUCCESSIONAL-DISTURBANCE SCRUB

Oak Successional Series

Quercus gambelii (undulata)/MG-F (toward lower upper montane and lower montane coniferous forest)

Oak-Locust Successional Series

Quercus gambelii (undulata)–Robinia neomexicana/MG-F (toward lower upper montane and lower montane coniferous forest)

Table 8.2. Major Plants Comprising Montane Scrub Vegetation in New Mexico.

Scientific Name	Common Name	Region Found
TREE-SHRUBS		
Acer grandidentatum	Bigtooth Maple	northern, central
Arbutus arizonica	Arizona Madrone	southwestern
Arbutus xalapensis	Texas Madrone	southeastern
Juniperus deppeana	Alligator Juniper	widespread
Juniperus monosperma	One-seed Juniper	central, southern
Prunus virginiana	Chokecherry	widespread
Ptelea angustifolia	Hoptree	widespread
Quercus arizonica	Arizona White Oak	central, southwestern
Quercus grisea	Gray Oak	central, southeastern
Quercus hypoleucoides	Silverleaf Oak	southwestern
Quercus pungens	Sandpaper Oak	widespread
Quercus toumeyi	Toumey Oak	southwestern
SHRUBS AND VINES		
Agave palmeri	Palmer Agave	southwestern
Agave parryi	Parry Agave	southern
Amelanchier oreophila	Mountain Serviceberry	northern
Amelanchier utahensis	Utah Serviceberry	northern
Arctostaphylos pungens	Manzanita	central, southwestern
Artemisia arbuscula	Black Sagebrush	northern
Artemisia tridentata	Big Sagebrush	northern
Berberis fendleri	Colorado Barberry	northern
Berberis fremontii	Fremont Barberry	northern
Berberis haematocarpa	Algerita	central, southern
Ceanothus fendleri	Buckbrush	widespread
Ceanothus greggii	Desert Ceanothus	southern
Ceanothus integerrimus	Deerbrush Ceanothus	southwestern
Cercocarpus montanus	Mountain Mahogany	widespread
Clematis spp.	clematis	widespread
Cowania mexicana	Cliffrose	northwestern, west-central
Dalea formosa	Feather Peabush	widespread
Dasylirion wheeleri	Sotol	southern
Ephedra torreyana	Mormon Tea	northern, central
Ericameria laricifolia	Turpentinebush	southern
Eriogonum wrightii	Wild Buckwheat	widespread
Fallugia paradoxa	Apache Plume	widespread
Fendlera rupicola	Cliff Fendlerbush	north-central, central, southern
Garrya wrightii	Wright Silktassel	southern
Holodiscus dumosus	Oceanspray	widespread
Hedyotis intricata	Shrubby Bluet	southern
Juniperus pinchotii	Redberry Juniper	southeastern
Nolina microcarpa	Sacahuista	southern
Nolina texana	Texas Nolina	southern

Table 8.2.   (continued).

Scientific Name	Common Name	Region Found
Parthenium incanum	Mariola	southern
Philadelphus microphyllus	Mockorange	widespread
Physocarpus monogynus	Ninebark	widespread
Purshia tridentata	Antelope Bitterbrush	northwestern
Quercus turbinella	Shrub Liveoak	widespread
Quercus undulata	Wavyleaf Oak	widespread
Rhamnus californica	Coffeeberry	southwestern
Rhus choriophylla	Evergreen Sumac	south-central, southwestern
Rhus microphylla	Littleleaf Sumac	southern
Rhus trilobata	Skunkbush	widespread
Ribes aureum	Golden Currant	north-central, central
Ribes cereum	Wax Currant	north-central, central
Rosa stellata	Desert Rose	southern
Salvia pinguifolia	Blue Sage	southern
Salvia ramosissima	Sage	south-central, southwest
Symphoricarpos oreophilus	Mountain Snowberry	widespread
Symphoricarpos utahensis	Utah Snowberry	northwest
Viguiera stenoloba	Resinbush	southern
Yucca angustissima	Fineleaf Yucca	northwest, west-central
Yucca baccata	Banana Yucca	central, southern
Yucca schottii	Spanish Lance	southwest

GRASSES
Agropyron smithii	Western Wheatgrass	northern, central
Aristida fendleriana	Fendler Threeawn	widespread
Aristida ternipes	Spidergrass	southern
Bothrochloa barbinodis	Cane Beardgrass	widespread
Bouteloua curtipendula	Sideoats Grama	widespread
Bouteloua gracilis	Blue Grama	widespread
Bouteloua hirsuta	Hairy Grama	widespread
Leptochloa dubia	Green Sprangletop	southern
Leptochloa filiformis	Red Sprangletop	southern
Lycurus phleoides	Wolftail	western
Oryzopsis hymenoides	Indian Ricegrass	northern, central

FORBS
Antennaria parvifolia	Pussytoes	northern, central
Artemisia ludoviciana	Louisiana Wormwood	widespread
Erigeron divergens	Fleabane	northern
Erigeron nudiflorus	Fleabane	northern
Eriogonum spp.	wildbuckwheat	widespread
Eriogonum wrightii		southern
Penstemon fendleri	Fendler Beardtongue	widespread
Penstemon linaroides		widespread

Table 8.2. (continued).

Scientific Name	Common Name	Region Found
Penstemon palmeri	Scented Penstemon	central
Penstemon pseudospectabilis	Canyon Penstemon	widespread
Penstemon thurberi	Thurber Penstemon	southern
Solidago sparsiflorus	Fewflowered Goldenrod	widespread
Verbena wrightii	Desert Verbena	widespread

Table 8.3. Major Tree-Shrubs and Shrubs of the Montane Scrub, by Elevational Zone. Listed by the elevational (available moisture) zones where they dominate in this vegetation type, in New Mexico. * = semiriparian species.

Upper Zone (more moisture)	Middle Zone (mesic)	Lower Zone (less moisture)
Acer glabrum	*Amelanchier oreophila*	*Agave palmeri*
*Acer grandidentatum	*Amelanchier utahensis*	*Agave parryi*
Ceanothus fendleri	*Arbutus arizonica*	*Artemisia arbuscula*
Ceanothus integerrimus	*Arbutus xalapensis*	*Artemisia tridentata*
Fendlera rupicola	*Berberis fremontii*	*Dasylirion wheeleri*
Holodiscus dumosus	*Berberis fendleri*	*Nolina microcarpa*
*Prunus virginiana	*Ceanothus greggii*	*Nolina texana*
Quercus hypoleucoides	*Cercocerpus montanus*	*Parthenium icanum*
Quercus undulata	*Cowania mexicana*	*Quercus turbinella*
*Rhamnus californica	*Ephedra torreyana*	**Rhus microphylla*
*Rhus choriophylla	**Ericameria laricifolia*	*Viguiera stenoloba*
Ribes aureum	*Eriogonum wrightii*	*Yucca angustissima*
Symphoricarpos oreophilus	**Fallugia paradoxa*	*Yucca baccata*
Symphoricarpos utahensis	**Garrya wrightii*	
	Hedyotis intricata	
	Juniperus pinchotii	
	Philadelphus microphyllus	
	Physocarpus monogynus	
	*Ptelea angustifolia	
	Purshia tridentata	
	Quercus arizonica	
	Quercus grisea	
	Quercus pungens	
	Quercus toumeyi	
	Quercus undulata	
	*Rhus trilobata	
	Ribes cereum	
	Salvia pinguifolium	
	Salvia ramosissima	
	Yucca schottii	

Table 8.4. Major Plants Comprising Plains-Mesa Sand Scrub Vegetation in New Mexico.

Scientific Name	Common Name	Region Found
TREE-SHRUBS		
Sapindus saponaria var. *drummundii*	Western Soapberry	central, southern
SHRUBS		
Amorpha canescens	Leadplant	eastern
Artemisia filifolia	Sand Sagebrush	widespread
Artemisia frigida	Estafiata	northern
Atriplex canescens	Fourwing Saltbush	widespread
Ephedra torreyana	Mormon Tea	northern, central
Krameria grayi	White Ratany	eastern
Penstemon ambiguus	Gilia Penstemon	central, southern
Poliomintha incana	Rosemary Mint	northern, central
Prosopis glandulosa	Honey Mesquite	widespread
Psorothamnus scoparius	Broom Indigobush	widespread
Psorothamnus terminalis		widespread
Quercus havardii	Shinoak	east-central, southeastern
Rhus trilobata	Skunkbush	widespread
Yucca angustissima	Fineleaf Yucca	northwestern
Yucca campestris		east-central, southeastern
Yucca glauca	Small Soapweed	eastern
Yucca elata	Soaptree Yucca	central, southern
GRASSES		
Andropogon hallii	Sand Bluestem	widespread
Aristida purpurea	Purple Threeawn	widespread
Bothriochloa barbinodis	Cane Beardgrass	widespread
Bothriochloa saccharoides	Silver Bluestem	southern
Bouteloua hirsuta	Hairy Grama	widespread
Calamovilfa gigantea	Sand Reed	eastern
Eragrostis intermedia	Plains Lovegrass	widespread
Eragrostis oxylepis	Red Lovegrass	eastern
Eragrostis trichodes	Sand Lovegrass	northern, eastern
Leptoloma cognatum	Fall Witchgrass	southern
Muhlenbergia porteri	Bush Muhly	widepread
Muhlenbergia pungens	Sandhill Muhly	widespread
Munroa squarrosa	False Buffalograss	widespread
Oryzopsis hymenoides	Indian Ricegrass	northern, central
Panicum capillare	Witchgrass	widespread
Panicum lindheimeri	Sand Witchgrass	southern
Paspalum ciliatifolium	Fringed Millet	southern
Paspalum stramineum	Sand Paspalum	widespread
Schizachyrium scoparium	Little Bluestem	widespread
Setaria leucopila	Yellow Bristlegrass	southern
Setaria macrostachya	Plains Bristlegrass	widespread
Sporobolus airoides	Alkali Sacaton	widespread

Table 8.4. (continued).

Scientific Name	Common Name	Region Found
Sporobolus contractus	Spike Dropseed	widespread
Sporobolus cryptandrus	Sand Dropseed	widespread
Sporobolus flexuosus	Mesa Dropseed	widespread
Sporobolus giganteus	Giant Dropseed	southern
FORBS		
Abronia angustifolia	Sand Verbena	widespread
Abronia carnea	Sandpuff	widespread
Abronia fragrans	Snowball	eastern, south-central
Ambrosia acanthicarpa	Burweed	widespread
Amsonia arenaria		southern
Arenaria aculeata	Sandwort	northwestern
Asclepias arenaria	Sand Milkweed	eastern
Caesalpinia jamesii	James Rushpea	widespread
Calylophus serrulatus	Yellow Evening Primrose	eastern
Cenchrus insertus	Grassburr	widespread
Commelina erecta	Hierba del Pollo	southern
Corispermum spp.	Bugseed	widespread
Croton texensis	Texas Croton	widespread
Cycloloma atriplicifolium	Winged Pigseed	widespread
Euphorbia glyptosperma	Ridgeseed Spurge	widespread
Euphorbia parryi	Parry Spurge	western, southern
Eriogonum abertianum	Abert Wildbuckwheat	southern
Eriogonum annuum	Annual Wildbuckwheat	widespread
Dalea lanata	Wooly Dalea	eastern
Diodia teres	Buttonweed	eastern, southern
Dithyrea wislezenii	Spectaclepod	widespread
Heliotropium convolvulaceum	Bindweed Heliotrope	widespread
Hymenopappus flavescens	Woolywhite	southern
Ipomoea spp.	Sand Morning-glory	widespread
Linum aristatum		southern
Maurandya spp.	False Snapdragon	southern
Machaeranthera tanacetifolia	Tahoka Daisy	widespread
Mentzelia pumila	Golden Stickleaf	widespread
Palafoxia sphacelata	Sand Palafox	northern, south-central, eastern
Paronychia jamesii	Nailwort	widespread
Parryella filifolia	Dunebroom	northern
Petalostemon candidum	White Prairieclover	eastern
Petalostemon purpureum	Purple Prairieclover	northern
Polygonella americana	Jointweed	central
Psoralea lanceolata	Lemon Scurfpea	northern, western
Psoralea tenuiflora	Slender Scurfpea	widespread
Schrankia occidentalis	Sensitive Brier	northeastern
Sphaeralcea parvifolia		northwestern
Stillingia sylvatica	Common Queensdelight	eastern

Table 8.5. Major Plants Comprising Great Basin Desert Scrub Vegetation in New Mexico. * = alien (introduced) species.

Scientific Name	Common Name	Region Found
SHRUBS		
Artemisia arbuscula	Black Sagebrush	northern
Artemisia bigelovii	Bigelow Sagebrush	northern, south-central, western
Artemisia spinescens	Button Sagebrush	northern
Artemisia tridentata	Big Sagebrush	northern
Atriplex canescens	Fourwing Saltbush	widespread
Atriplex confertifolia	Shadscale	northwestern
Atriplex nuttallii	Nuttall Saltbush	northwestern
Atriplex obovata	Obovateleaf Saltbush	northern
Certatoides lanata	Winterfat	widespread
Chrysothamnus greenei	Green Rabbitbrush	northwestern
Chrysothamnus nauseosus subsp. *bigelovii*	Bigelow Rubber Rabbitbrush	widespread
Echinocactus mesa-verde	Hedgehog Cactus	northwestern
Ephedra nevadensis	Rough Jointfir	widespread
Ephedra torreyana	Mormon Tea	widespread
Gutierrezia microcephala and *sarothrae*	Broom Snakeweed	widespread
Opuntia erinacea	Mohave Pricklypear	northwestern
Opuntia polyacantha	Plains Pricklypear	northern, central
Opuntia whipplei	Whipple Cholla	northwestern
Pediocactus spp.		northern
Sarcobatus vermiculatus	Greasewood	northern, central
Tetradymia canescens	Gray Horsebush	northwestern
GRASSES		
Agropyron smithii	Western Wheatgrass	northern, central
Bouteloua eriopoda	Black Grama	widespread
Bouteloua gracilis	Blue Grama	widespread
Bouteloua hirsuta	Hairy Grama	widespread
Distichlis stricta	Saltgrass	widespread
Festuca octoflora	Sixweeks Fescue	widespread
Muhlenbergia pungens	Sandhill Muhly	widespread
Orysopsis hymenoides	Indian Ricegrass	widespread
Sitanion hystrix	Bottlebrush Squirreltail	widespread
Sporobolus airoides	Alkali Sacaton	widespread
FORBS		
Ambrosia acanthicarpa	Burweed	widespread
Atriplex patula	Fathen Saltbush	northwestern
Atriplex powellii	Powell Saltbush	northwestern, western
Bahia oblongifolia		western
*Bassia hysippifolia**	Smotherweed	widespread
*Descurania pinnata**	Tansymustard	widespread
*Erodium cicutarium**	Filaree	widespread
Kochia americana	Red Sage	northern, central
Salsola spp.*	russian thistle	widespread
*Sisymbrium altissimum**	Tumblemustard	northern
Sphaeralcea parvifolia		northwestern

Table 8.6. Major Plants Comprising Chihuahuan Desert Scrub Vegetation in New Mexico. * = diagnostic (climax) member.

Scientific Name	Common Name
SHRUBS AND HALF-SHRUBS	
Acacia constricta	Whitethorn
Acacia neovernicosa	
Agave lechuguilla	Lechuguilla
Aloysia wrightii	Spicebush
Coldenia spp.	coldenia
Condalia spathulata	Squawbush
Coryphantha micormeris	Pincushion Cactus
Coryphanthan vivipara var. aggregata	Purple Ballcactus
Dalea formosa	Feather Peabush
Dasylirion leiophylla	Smoothleaf Sotol
Dasylirion wheeleri	Wheeler Sotol
Echinocactus horizonthalonius	Hedgehog Cactus
Echinocereus chloranthus	
Echinocereus pectinatus	
Ephedra trifurca	Longleaf Jointfir
Epithelantha micromeris	
Ferocactus wislizenii	Barrel Cactus
Flourensia cernua	Tarbush
Fouquieria splendens	Ocotillo
Gutierrezia sarothrae	Broom Snakeweed
Gutierrezia microcephala	
Koeberlinia spinosa	Allthorn
Krameria glanulosa	Range Ratany
Larrea tridentata	Creosotebush
Mammillaria meiacantha	
Nolina microcarpa	Sacahuista
Opuntia imbricata	Tree Cholla
Opuntia kleiniae	Klein Cholla
Opuntia leptocaulis	Christmas Cactus
Opuntia macrocentra	Purple Pricklypear
Parthenium incanum	Mariola
Prosopis glandulosa	Honey Mesquite
Viguiera stenoloba	Resinbush
Yucca baccata	Banana Yucca
Yucca elata	Soaptree Yucca
Yucca torreyi	Torrey Yucca
Zizyphus obtusifolia	Graythorn
GRASSES	
Bouteloua eriopoda	Black Grama
Distichlis stricta	Saltgrass
Erioneuron pulchellum	Fluffgrass

Sporobolus airoides	Alkali Sacaton
Sporobolus contractus	Spike Dropseed
FORBS	
*Bahia absinthifolia	Field Bahia
Boerhaavia spp.	Spiderling
Cassia bauhunioides	Twinleaf
Dyssodia acerosa	Dogweed
Drymaria pachyphylla	Thickleaf Drymary
Pectis papposa	Lemonweed
Perezia nana	Desert Holly

CHAPTER 8 REFERENCES

SCRUBLAND VEGETATION

Ahlborn, G. G., 1980. Brood-rearing habitat and fall-winter movements of lesser prairie chickens in eastern New Mexico. Thesis, New Mexico State University, Las Cruces.

Atwood, T. L., 1983. Ecology of a sand sage community in southern New Mexico. Thesis, New Mexico State University, Las Cruces.

Atwood, G. G., 1987. Influence of livestock grazing and protection from livestock grazing on vegetational characteristics of *Bouteloua eriopoda* rangelands. Dissertation, New Mexico State University, Las Cruces.

Barbour, M. G., J. A. MacMahon, S. A. Bamberg, and J. A. Ludwig, 1977. The structure and distribution of *Larrea* communities. pp. 227–251. In: T. J. Marby, J. H. Hunziker, and D. R. DiFeo, Jr. (eds.). Creosotebush: biology and chemistry of *Larrea* in new world deserts. Dowden, Hutchinson, and Ross, Stroudsburg, Pennsylvania.

Bedker, E. J., 1966. A study of the flora of the Manzano Mountains. Thesis, University of New Mexico, Albuquerque.

Borden, B. D., 1973. Characteristics of a scaled quail population in southeastern New Mexico. Thesis, New Mexico State University, Las Cruces.

Bowers, J. E., 1982. The plant ecology of dunes in western North America. Journal of Arid Environments, 5: 199–220.

Brown, D. E., 1982. Biotic communities of the American Southwest–United States and Mexico. Desert Plants, 4: 332 pp. Boyce Thompson Southwestern Arboretum, Superior, Arizona.

Burgess, T. L. and D. K. Northington, 1974. Desert vegetation in the Guadalupe Mountains Region. pp. 229–242. In: R. H. Wauer and D. H. Riskind (eds.). Transactions of the symposium of the biological resources of the Chihuahuan Desert region, United States and Mexico. U.S. Department of the Interior, National Park Service Transactions and Proceedings Series No. 3, U.S. Government Printing Office, Washington, D.C.

Campbell, R. S., 1929. Vegetation succession in the *Prosopis* sand dunes of southern New Mexico. Ecology, 10: 392–398.

Campbell, R. S. and I. F. Campbell, 1938. Vegetation on gypsum soils of the Jornada plain, New Mexico. Ecology, 19: 572–577.

Candelaria, M. A., 1979. Movements and habitat-use by lesser prairie chickens in eastern New Mexico. Thesis, New Mexico State University, Las Cruces.

Chavez, A., 1982. Perennial shrubs, soil and

topographic relationships on a basaltic substrate in Chihuahuan Desert *Xanthocephalum-Senecio* community. Thesis, University of Texas, El Paso.

Clements, F. E. and E. S. Clements, 1939. The sagebrush disclimax. Carnegie Institution Washington Yearbook, 38: 139–140.

Cully, A. and P. Knight, 1987. A handbook of vegetation maps of New Mexico counties, 135 pp. New Mexico Natural Resources Department, Santa Fe.

Dick-Peddie, W. A., J. K. Meents, and R. Spellenberg, 1984. Vegetation resource analysis for the Velarde community ditch project, Rio Arriba and Santa Fe Counties, New Mexico. Final Report. U.S. Bureau of Reclamation, Southwest Region, Amarillo, Texas. 251 pp.

Dick-Peddie, W. A. and W. H. Moir, 1970. Vegetation of the Organ Mountains, New Mexico. Range Science Department Science Series No. 4. Colorado State University, Ft. Collins.

Dittmer, H. J., 1959. A study of the root systems of certain sand dune plants in New Mexico. Ecology, 40: 265–273.

Donart, G. B., D. Sylvester, and W. Hickey, 1978. A vegetation classification system for New Mexico, USA. pp. 488–490. In: D. N. Hyder (ed.). Proceedings of 1st International Rangeland Congress, Denver, Colorado.

Donart, G. B., 1984. The history and evolution of western rangelands in relation to woody plant communities. pp. 1235–1258. In: National Research Council/National Academy of Sciences: Developing strategies for rangeland management. Westview Press, Boulder, Colorado. 2022 p.

Elenowitz, A. S., 1983. Habitat use and population dynamics of transplanted desert bighorn sheep in the Peloncillo Mountains, New Mexico. Thesis, New Mexico State University, Las Cruces.

Fletcher, R. A., 1978. A floristic assessment of the Datil Mountains. Thesis, University of New Mexico, Albuquerque.

Gardiner, J. L., 1951. Vegetation of the creosotebush area of the Rio Grande Valley in New Mexico. Ecological Monographs, 21: 379–403.

Gehlbach, F. R., 1967. Vegetation of the Guadalupe Escarpment, New Mexico–Texas. Ecology, 48: 404–419.

Goodman, P. J. and M. M. Caldwell, 1971. Shrub ecotypes in a salt desert. Nature, 232: 571–572.

Griffing, J. P., 1972. Population characteristics and behavior of scaled quail in southeastern New Mexico. Thesis, New Mexico State University, Las Cruces.

Gross, F. A. and W. A. Dick-Peddie, 1979. A map of primeval vegetation in New Mexico. The Southwestern Naturalist, 24: 115–122.

Hendrickson, J., 1974. Saline habitats and halophytic vegetation of the Chihuahuan Desert region. pp. 289–314. In: R. H. Wauer and D. H. Riskind (eds.). Transactions of the symposium of the biological resources of the Chihuahuan Desert region, United States and Mexico. U.S. Department of the Interior, National Park Service Transactions and Proceedings Series No. 3, U.S. Government Printing Office, Washington, D.C.

Hennessy, J. T., 1981. Soil movements and vegetational changes over a forty-five-year period in south-central New Mexico. Thesis, New Mexico State University, Las Cruces.

Johnston, M. C., 1977. Brief resume of botanical, including vegetational features, of the Chihuahuan desert region with special emphasis on their uniqueness. pp. 335–362. In: R. H. Wauer and D. H. Riskind (eds.). Transactions of the symposium of the biological resources of the Chihuahuan Desert region, United States and Mexico. U.S. Department of the Interior, National Park Service Transactions and Proceedings Series No. 3, U.S. Government Printing Office, Washington, D.C.

Kelly, N. E., 1973. Ecology of the Arroyo Hondo Pueblo site. Thesis, University of New Mexico, Albuquerque.

Kittams, W. R., 1972. Effects of fire on the vegetation of the Chihuahuan desert region. Proceedings of the Tall Timbers Fire Ecology Conference, 12: 427–444.

Lebgue, T., 1982. Flora of the Fort Stanton Experimental Ranch, Lincoln County, New Mexico. Thesis, New Mexico State University, Las Cruces.

Little, E. L., Jr. and R. S. Campbell, 1943. Flora of Jornada Experimental Range, New Mexico. The American Midland Naturalist, 626–670.

MacDougal, D. T., 1908. Botanical features of the North American deserts. Carnegie Institute Washington, Publication 99. Washington, D.C.

MacMahon, J. A., 1979. North American deserts: Their floral and faunal components. pp. 21–82. In: D. W. Goodal and R. A. Perry (eds.). Arid-land ecosystems: structure, functioning, and management. Vol. 1, Cambridge University Press, N.Y.

Magoub, E. E., 1982. Growth patterns and chemical composition of mountain mahogany (*Cercocarpus breviflorus* (Gray) Wright) on the Organ Mountains Recreation Land. Thesis, New Mexico State University, Las Cruces.

Merriam, C. H., 1942. The desert vegetation of North America. Botanical Review, 8: 195–246.

Milstead, W. W., 1960. Relict species of the Chihuahuan desert. The Southwestern Naturalist, 5: 53–60.

Moir, W. H., 1963. Vegetational analysis of three southern New Mexico mountain ranges. Thesis, New Mexico State University, Las Cruces.

———, 1979. Soil-vegetation patterns in the central Peloncillo Mountains, New Mexico. The American Midland Naturalist, 102: 317–331.

Muller, C. H., 1940. Plant succession in the *Larrea-Flourensia* climax. Ecology, 21: 206–212.

———, 1953. The association of desert annuals with shrubs. American Journal of Botany, 40: 53–60.

Naylor, J. N., 1964. Plant distributions of the Sandia Mountains area, New Mexico. Thesis, University of New Mexico, Albuquerque.

Neilson, R. P., 1986. High-resolution climatic analysis and southwest biogeography. Science 232: 27–34.

Osborn, N. C., 1962. The flora of Mount Taylor. Thesis, University of New Mexico, Albuquerque.

———, 1966. A comparative floristic study of Mount Taylor and Redondo Peak, New Mexico. Dissertation, University of New Mexico, Albuquerque.

Reid, W. H., 1979. Natural resources inventory and analyses of White Sands National Monument. Laboratory for Environmental Biology: Research Report 11. University of Texas, El Paso.

———, 1980. Final report: White Sands National Monument natural resources and ecosystem analysis. Laboratory for Environmental Biology: Research Report 12. University of Texas, El Paso.

Riffle, N. L., 1973. The flora of Mount Sedgwick and vicinity. Thesis, University of New Mexico, Albuquerque.

Robertson, C. W., 1968. A study of the flora of Cochiti and Bland Canyons of the Jemez Mountains. Thesis, University of New Mexico, Albuquerque.

Robinson, J. L., 1969. Forest survey of the Guadalupe Mountains, Texas. Thesis, University of New Mexico, Albuquerque.

Secor, J. B., S. Shamas, D. Smeal, and A. L. Gennaro, 1983. Soil characteristics of two desert plant community types that occur in the Los Medaños area of southeastern New Mexico. Soil Science 136: 133–144.

Shields, L. M., 1956. Zonation of vegetation within the Tularosa Basin, New Mexico. The Southwestern Naturalist, 1: 49–68.

Shreve, F., 1942. The desert vegetation of North America. Botanical Review, 7: 195–246.

Standley, P. C., 1915. Vegetation of the Brazos Canyon, New Mexico. Plant World, 18: 179–191.

Stewart, L. H., 1982. Desert grassland communities on Otero Mesa, Otero County, New Mexico. Thesis, New Mexico State University, Las Cruces.

Suminski, R. H., 1977. Habitat evaluation for lesser prairie chickens in eastern Chaves County, New Mexico. Thesis, New Mexico State University, Las Cruces.

Tatschl, A. K., 1966. A floristic study of the San

Pedro Parks Wild Area, Rio Arriba County. Thesis, University of New Mexico, Albuquerque.

Thomas, M. G., 1977. A floristic analysis of the Sevilleta Wildlife Refuge and Ladron Mountains. Thesis, University of New Mexico, Albuquerque.

Van Devender, T. R. and L. J. Toolin, 1983. Late Quaternary vegetation of the San Andres Mountains, Sierra County, New Mexico. pp. 33–54. In: P. L. Eidenbach (ed.). The prehistory of Rhodes Canyon. 6585th Test Group, Holloman Air Force Base, Holloman, New Mexico.

Von Loh, J. D., 1977. A flora of the San Andres National Wildlife Refuge, Dona Ana County, New Mexico. Thesis, University of New Mexico, Albuquerque.

Wagner, W. L., 1977. Floristic affinities of the Animas Mountains, southwestern New Mexico. Thesis, University of New Mexico, Albuquerque.

Watson, J. R., 1912. Plant geography of northcentral New Mexico. Botanical Gazette, 54: 194–217.

Wester, D. B., 1984. Ordination and classification of vegetation in the Guadalupe Mountains National Park, Texas. Dissertation, Texas Tech University, Lubbock.

Wester, D. B. and H. A. Wright, 1987. Ordination of vegetation change in Guadalupe Mountains, New Mexico, U.S.A. Vegetatio 72: 27–33.

Whitson, P. D., 1975. Factors contributing to production and distribution of Chihuahuan desert annuals. U.S./International Biological Program, Desert Biome Research Memorandum 75-11. Utah State University, Logan.

Wilson, D. L., 1982. Nesting of lesser prairie chickens in Roosevelt and Lee Counties, New Mexico. Thesis, New Mexico State University, Las Cruces.

Wisdom, M. J., 1980. Nesting habitat of lesser prairie chickens in eastern New Mexico. Thesis, New Mexico State University, Las Cruces.

Yeaton, R., 1978. A cyclical relationship between *Larrea tridentata* and *Opuntia leptocaulis* in the northern Chihuahuan desert. Journal of Ecology, 66: 651–656.

York, J. C. and W. A. Dick-Peddie, 1969. Vegetation changes in southern New Mexico during the past hundred years. pp. 157–166. In: W. G. McGinnies and B. J. Goldman (eds.), Arid lands in perspective. University of Arizona Press, Tucson.

Young, J. A., R. F. Eckert, and R. A. Evans, 1979. Historical perspective regarding the sagebrush ecosystem. pp. 1–13. In: The sagebrush ecosystem: a symposium. Utah State University, Logan.

9

RIPARIAN VEGETATION

CONCEPT OF RIPARIAN VEGETATION

In New Mexico, drainage systems may contain perennial water units, such as streams and rivers, or they may include canyons, arroyos, playas, and swales, which carry water only occasionally. Vegetation associated with these systems is called riparian vegetation. This use of the term is common in the western United States and includes the "floodplain" vegetation of Kuchler (1964). Federal agencies handle the classification of this type of vegetation in various ways. The U.S. Fish and Wildlife Service combines what we are calling riparian, which is terrestrial vegetation with aquatic (emergent and submergent) vegetation and calls it all "wetlands" vegetation. We consider emergent and submergent vegetation systems such as that found in bogs, marshes, and cienegas, as well as that found under water in streams and lakes, as aquatic systems. In this book, "aquatic systems" is synonymous with "wetlands."

The use of "drainage systems" in the definition of riparian vegetation is important for New Mexico and most western states, because riparian vegetation can be recognized and is distinctive in and along arroyos, in closed basins, alkali sinks, swales, and playas, all of which are ephemeral with regard to carrying or holding water. The internally drained features of closed basins, alkali sinks, swales, and playas are typical of lands in the Basin and Range Physiographic Province. The vegetation these areas support is similar throughout the southwestern United States and is unique to these types of habitats.

Brown (1982) makes an interesting observation about montane (river and stream) and floodplain riparian vegetation. He suggests that the deciduous tree species comprising riparian forests are relics of the time when the Southwest was covered by a vegetation similar to the present-day deciduous hardwood forests of the eastern United States. In the Southwest, these species have adapted to riparian habitats and are now what we call obligate riparian species.

Many plant species found in riparian habitats are limited to them and are called obligate (true) riparian species. Examples are Cottonwood and Willow. Some species found in riparian habitats may also be occupying the surrounding open (upland) landscape. These components of riparian vegetation are called facultative riparian species. Sometimes riparian vegetation includes species restricted to the local riparian habitat but typically found in nonriparian habitats at higher (cooler, more moist) locations. These are referred to as restricted riparian species. Ponderosa Pine and Douglas-fir growing in a canyon that runs through pinyon-juniper woodland are examples of restricted riparian species. A few species that appear to be facultative exhibit very different size and density characteristics in riparian and in neighboring nonriparian situations. We refer to these species as semiriparian species. Examples would be Aspen (*Populus tremuloides*) and New Mexican Locust (*Robinia neomexicana*), which are commonly dominant along streams but also may form dense stands on mountain slopes that have been burned or logged. Obligate and semiriparian species are used in this book for the classification of riparian vegetation. Table 9.1 lists the common obligate riparian and semiriparian species found in New Mexico.

Obligate riparian species are often found in habitats that are not part of a true drainage system, such as at the bottom of talus slopes, in roadside ditches, or in erosion gullies. The term "pseudoriparian" is applied to such habitats.

There have been few attempts to classify riparian vegetation. Kuchler (1964) did include a "floodplain" category in his map of the vegetation of the conterminous United States. Brown (1982) included a section of "deciduous woodland" and some other very general categories, but these were not incorporated into his digitized system. One reason why riparian vegetation has not received much classification effort could be that its composition is difficult to predict from the upland (open) surrounding vegetation. Riparian vegetation maintains its continuity in a vertical direction from source to mouth of drainage systems. This is perpendicular to upland vegetation, whose continuity is maintained in a horizontal direction along elevational contours. Composition of riparian vegetation at a site can be better predicted from knowledge of the composition of riparian vegetation above or below the site than from the upland vegetation through which the riparian band is passing. Another reason there have been few attempts to classify riparian vegetation may be that there appear to be few patterns that allow discrete identification; even if units could be delineated, they would be so varied that a given unit would not be repeated on the landscape with sufficient frequency to make a classification feasible. However, if only obligate and semiriparian species are used to characterize riparian communities, the supposed varied and complex patterns are substantially reduced, and relatively discrete units can be segregated. Using this procedure, Dick-Peddie and Hubbard (1977) suggest that similar units are common enough in riparian habitats to support the construction of a useful classification. This assumption was substantiated for riparian vegetation in New Mexico by Browning (1988). A classification of riparian vegetation in New Mexico is presented in Table 9.2

The combinations of obligate species found in riparian vegetation vary considerably around the state. Some of these variations are due to range limitations. For example, Arizona Sycamore (*Platanus wrightii*) is largely restricted in New Mexico to river and stream systems west of the continental divide. Screwbean, or Tornillo (*Prosopis pubescens*), found only in the southern half of the state, is another example of range limitation. The physical features of the riparian zone may also dictate species composition. An old river bed or floodplain may contain terraces (benches) bordering the present channel. Terraces often support different species at varying distances from the current channel bed. These patterns apparently are due to degrees and frequencies of flooding and/or height of the terrace surfaces above the water table. Obligate riparian species occupying the bottom of a narrow canyon may be different species from those occupying a channel of an open drainage at the same elevation, in the same general geographic location.

Plants whose roots are in the water table or its capillary fringe during all or most of the growing season are referred to as phreatophytes. Due to high water tables, riparian habitats often have phreatophytic environments. It is not uncommon to have poor soil-aeration conditions (low oxygen and high carbon dioxide) associated with phreatophytic environments. In addition, the soil solution can be highly saline in these habitats. Some plant species possess adaptations that allow them to occupy such habitats containing poor soil aeration and/or high salinity conditions; this adds to the variations found in riparian vegetation patterns. It should be kept in mind that all plants can thrive under phreatophytic conditions if aeration of the soil solution is adequate and there is little or no salinity. Subsurface irrigation systems take advantage of this fact. Species growing under phreatophytic conditions appear to be similar to so-called "halophytic" species, which have adaptations allowing them to grow in special habitats and thus lessen competition, but they are not obligate halophytes (Barbour, Burk, and Pitts, 1980).

Riparian vegetation functions as a magnet for animals. These corridors of highly diverse habitats are interconnected networks that dissect the upland vegetation through which they run. Except for facultative species, the composition of riparian vegetation tends to be independent of the surrounding upland vegetation. This adds to the habitat diversity as well as to the species richness of an area. Lines of deciduous forest in a coniferous forest; forest stands in a woodland, savanna, or grassland; and shrub-trees and non-sclerophyllous shrubs associated with arroyos running through desert scrubland are examples of added structural diversity supplied by riparian vegetation.

OVERVIEW OF RIPARIAN VEGETATION IN NEW MEXICO

Separating riparian vegetation into distinct types was suggested for New Mexico as early as 1893 by Townsend, in his paper on the vegetation of the Organ Mountains. He describes a tornillo or cottonwood zone (our floodplain riparian) and a mesquite zone (our arroyo riparian). Watson's (1912) classification of the vegetation of north-central New Mexico included River Valley Formations that contained cottonwood forest, *Juncus-Houttuynia,* and *Bigelovia* associations. Under his Douglas Spruce Formation (our montane coniferous forest), he included canyon associations, in which he described vegetation typical of our montane riparian types. Paul Standley (1915), in describing the vegetation of Brazos Canyon, New Mexico, commented separately about plants "along the streams" and "[c]haracteristic of these hydrophilous plants." Whitfield and Anderson (1938) describe sacaton subclimax and saltbush subclimax vegetation. The descriptions of the vegetation in these "Clementsian" categories typify our playa, closed basin, and alkali sink riparian vegetation. Gehlbach (1967) refers to the montane riparian vegetation of the Guadalupe escarpment as deciduous woodland. When added to natural disturbances, grazing by domestic livestock, clearing for agriculture and urbanization, intentional burning, and modifications of the hydrologies of streams and rivers often prove to be too much for the continued existence of natural riparian communities.

ALPINE RIPARIAN

Alpine riparian vegetation is difficult to separate from alpine tundra and subalpine grassland vegetation. Running and standing water often form a complicated network among alpine grasses and forbs. We should be reminded that alpine areas are the source areas for drainage systems at lower elevations, but these areas themselves seldom exhibit pronounced or well-developed riparian sites.

Alpine riparian associations in New Mexico are dominated by sedges, mostly in the genus *Carex,* and grass plants in the *Deschampsia, Agrostis,* and *Glyceria* genera. According to the establishment report for one of the Forest Service's incipient research natural areas, McCrystal Meadow, it is primarily composed of vegetation typical of alpine riparian vegetation: "Sedges dominate the vegetation and doubtless comprise the bulk of the recently accumulated peat. Common species include *Carex aquatilis* and *C. nebraskensis*. Where mineral soils (Cumulis cryaquolls) occur, the wet meadow vegetation becomes more diverse. Wet meadow grasses include *Calamagrostis inexpansa, Hordeum brachyantherum, Deschampsia caespitosa,* and species of *Agrostis*. Other herbs include *Juncus arcticus* spp. *ater, J. austromontana,* and such broad-leafed plants as *Swertia perennis, Gentiana thermaiis, Parnassia parviflora, Pedicularis groenlandica,* and *Polygonum viviparum.*" Alpine riparian vegetation is found in New Mexico only on the highest peaks and ridges, which have snow most of the year, such as the sedge series found on Wheeler Peak.

MONTANE RIPARIAN

The vegetation associated with permanent and intermittent streams and rivers that drain mountain masses is the most extensive and varied riparian vegetation in the state. It is resilient because it has evolved under conditions where an occasional flood may completely wipe out riparian vegetation, including mature trees. Most of the obligate riparian species of these montane habitats are well adapted to catastrophic flooding events and respond with rapid reproduction and colonization. As might be expected, the adaptive mechanisms are highly varied, from prolific seed production, to efficient dispersal, to efficient and rapid vegetative reproduction. These features are often coupled with fast growth and relatively short life cycles, or at least the rapid achievement of the reproductive stage. Some species regenerate from massive root networks, following denudation.

Blue Spruce can dominate riparian associations at higher elevations. Boxelder, Arizona Alder, and Arizona Sycamore-dominated associations are common at middle elevations. Netleaf Hackberry and Velvet Ash associations comprise most of the lower montane riparian vegetation. Brown, Lowe, and Hausler (1977), Pase and Layser (1977), and Brown (1982) are some of the workers who have recognized the elevational gradients of species dominants in montane riparian vegetation. Boles and Dick-Peddie (1983), Freeman and Dick-Peddie (1970), Henry (1981), Hardesty (1986), and Browning (1988) have all looked at elevational trends of montane riparian vegetation in New Mexico. The relative elevational positions of montane riparian association dominants in New Mexico is illustrated in Figure 9.1. Virtually all of the plants in Figure 9.1 have elevational amplitudes of occurrence well beyond those shown in the figure, but the graphic depicts the elevational zones where the species have their greatest influence on montane riparian vegetation.

Mature montane riparian vegetation often takes the form of a forest with a closed canopy; these stands are called "gallery forests." Gallery forests are more common in floodplain riparian vegetation than in montane, but they are still important in both. Galleries are very influential ecosystems. Pioneer species, which flourished when the sites were more open, are replaced by other, more shade-tolerant species in the newly created understory. Soil temperatures under the canopy are greatly reduced during the growing season, resulting in more available moisture. The mature stands of trees tend to resist all but the most extreme flooding conditions and thereby reduce erosion, contain channel banks, and retard nutrient loss.

Montane riparian vegetation running through high alpine meadows is commonly dominated by willows, such as *Salix scouleriana, S. bebbiana, S. amygdaloides,* or *S. irrorata,* combined with Mountain Alder (*Alnus tenuifolia*), in the highest meadows, and Red-osier Dogwood (*Cornus sto-*

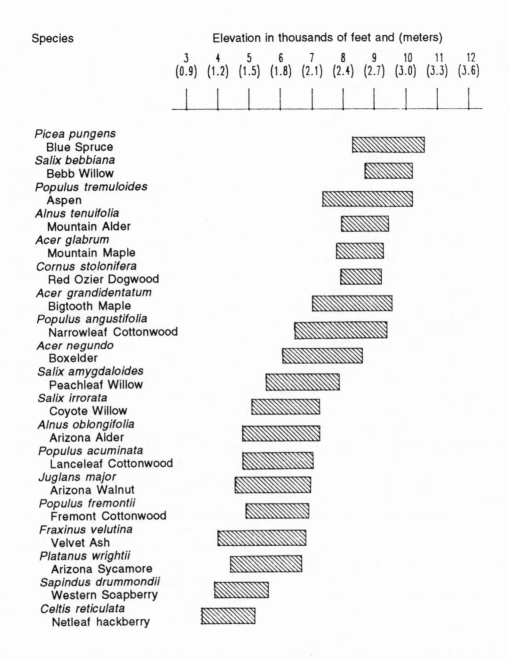

Figure 9.1 Elevation zones of Greatest Dominance of Montane Riparian Trees and Shrubs in New Mexico.

lonifera), in meadows lower down. Rocky Mountain Maple (*Acer glabrum*) and Blueberry Elder (*Sambucus glauca*) may be scattered in these high, treeless riparian sites. Blue Spruce (*Picea pungens*) and Aspen (*Populus tremuloides*) trees may occur as scattered individuals in this type of vegetation. Occasionally, small stands of Water Birch (*Betula occidentalis*) are found. This kind of high, open riparian vegetation was called riparian scrubland by Brown (1982) (Pl. 25).

Some of the highest montane riparian associations dominated by trees are those dominated by Blue Spruce; they have been accepted as habitat types. One is the blue spruce/mixed shrub/sedge (*Carex foenea*) type. Another similar habitat type has sedge replaced by Kentucky Bluegrass (*Poa pratensis*). Sometimes mountain alder dominates the shrub layer under the spruce (Table 9.2). Occasionally, these blue-spruce riparian associations are at relatively low elevations and have the understory layer dominated by oaks (*Quercus gambelii* or *Q. arizonica*). These associations are found mostly in west-central New Mexico. A few high-mountain riparian communities dominated by aspen can be found in the northern part of the state. It is interesting that most of the aspen in New Mexico is found as postdisturbance successional vegetation in nonriparian or pseudoriparian habitats, not in riparian habitats.

As might be inferred from the elevational overlap of species seen in Figure 9.1, the montane riparian tree-dominated communities are formed by combinations of dominant trees. Starting at high elevations there are some areas codominated by Blue Spruce and Narrowleaf Cottonwood (*Populus angustifolia*) or Aspen and Bigtooth Maple (*Acer grandidentatum*). Farther down, Boxelder (*Acer negundo*) may share dominance with Arizona Alder (*Alnus oblongifolia*), with an understory dominated by *Prunus serotina* (Black Cherry) and *Salix exigua* (Sandbar Willow). Narrowleaf Cottonwood by itself dominates many communities and is also commonly found as a codom-inant with each of the major montane riparian trees (Table 9.2).

It is not surprising that *Populus acuminata* (Lanceleaf Cottonwood), considered to be a hybrid of *P. angustifolia x P. fremontii*, dominates or codominates associations at transitional elevations between Narrowleaf and Fremont Cottonwood dominated communities. California Brickelbush (*Brickellia californica*) is often found as a shrub member of montane riparian vegetation. False Indigo (*Amorpha fruticosa*) is an occasional codominant with Lanceleaf Cottonwood. Associations dominated or codominated by Arizona Sycamore (*Platanus wrightii*) are mid-elevational, but are found only in west-central New Mexico.

There are more stands comprised of associations in the Narrowleaf Cottonwood and Narrowleaf Cottonwood–mixed deciduous series of montane riparian vegetation than of any others in New Mexico (Table 9.2). Browning (1988) obtained data and literature information on thirty-five different areas dominated by associations in these two series. The areas ranged from Township-4-South to Township-31-North (more than 200 miles) with an east-west span of 275 miles, from Range-25-East to Range-21-West. Associations in the Fremont Cottonwood and Fremont Cottonwood–mixed deciduous series are the next most prevalent in the state. Browning encountered twenty-seven areas, ranging from Township-31-South to Township-25-North (over 390 miles) and from Range-25-East, to Range-21-West, 275 miles.

The montane riparian vegetation of the southern mountains in New Mexico is commonly composed of associations codominated by Fremont Cottonwood (*Populus fremontii*) and Netleaf Hackberry (*Celtis reticulata*), or Velvet Ash (*Fraxinus velutina*) and Goodding Willow (*Salix gooddingii*). Codominants of some of these stands are Western Soapberry (*Sapindus drummundi*) and Little Walnut (*Juglans microcarpa*) in the east, replaced by Arizona Walnut (*Juglans major*), west of the Rio Grande. The semiriparian shrub California Brickelbush (*Brickellia californica*), which is often a member of montane scrub, may be found in lower montane and higher arroyo riparian sites.

FLOODPLAIN-PLAINS RIPARIAN

In New Mexico, floodplain-plains riparian vegetation has probably suffered more from man's activities than any other type of riparian vegetation. The floodplain portion of this vegetation type is typically found on older, meandering river systems where, as the name implies, there are extensive floodplains, such as on the Rio Grande. The plains portion is that which grows along intermittent streams of the eastern river systems, such as the Pecos system. Plains riparian drainages run through the plains grasslands of eastern New Mexico. Most of the riparian trees were harvested long ago, and few stands have been allowed to regenerate and mature.

One manifestation of human activities in or near drainage systems has been the virtual explosion of the alien shrub-tree species Russian Olive (*Elaeagnus angustifolia*), in north-central New Mexico and Salt Cedar (*Tamarix* spp.) in southern and northwestern New Mexico. Most people realize that it has been human activities that have allowed Russian Olive and Salt Cedar to make such spectacular gains that it appears to the casual observer that they are out-competing native vegetation. In the case of Salt Cedar, the changes have been extensive, rapid, and recent. The gallery forests that covered the floodplain river systems were composed of large trees and were often close to early settlements. Trees were cut initially for fuel and shelter purposes, such as ceiling beams, or "vigas," for adobe buildings. Riparian trees were also cut to clear land for agriculture and urbanization. Much of the heavy utilization of gallery-forest trees had subsided by the beginning of the twentieth century. Natural resiliency of riparian vegetation might have eventually resulted in the restoration of the former forest, if it had not been for the initiation of a new and devastating human-initiated disruption. This was the impoundment of stream and river surface waters. The hydrologies of dammed river systems were altered in ways that thwarted some of the reproductive mechanisms of native riparian species. Water tables dropped, reducing the extent and number of phreatophytic habitats. Not only were flow rates below dams reduced, but the annual flooding of benches and terraces bordering channels was eliminated. This drastically curtailed the major reproductive mechanism of cottonwoods, which is vegetative growth from buried twig and branch nodes (layering) and sprouting (suckering) from root systems. The absence of annual bench flooding also virtually stopped cottonwood seed germination. If bench flooding does occur, it is quickly followed by dense, almost hairlike stands of cottonwood sprouts and seedlings. Ironically, these young plants are highly relished by livestock and deer, and often not a single survivor can be found. One such event was documented by Boles (1978), while conducting his thesis research on the Mimbres River. Continued destruction of trees and the altered available-water regime on rivers and streams, coupled with livestock grazing, set a perfect stage for the establishment and expansion of salt cedar on most of the floodplain and plains drainage systems. If gallery forests were allowed to regenerate and mature, the resulting heavy shade would likely curtail the reproductive success of salt cedar, and eventually only a few old individuals would be left in the understory. This has been observed to be the case for the other alien riparian species, Russian Olive (Dick-Peddie, Meents, and Spellenberg, 1984). The human-induced changes in channel hydrology are probably permanent, and if it were deemed desirable to reestablish cottonwood gallery forests, it might be necessary for people to assist in the process.

The primary type of natural vegetation on the floodplain portion of the floodplain-plains riparian is that dominated by Fremont Cottonwood. In the southern part of New Mexico, Fremont Cottonwood commonly shares dominance with the tree Goodding Willow (*Salix gooddingii*). Farther north, Peachleaf Willow (*Salix amygdaloides*) occasionally shares dominance with Fremont Cottonwood. The understory layers in Fremont Cottonwood associations may be dominated by New Mexican Olive (*Forestiera neomexicana*), Skunkbush, Rabbitbrush (*Chrysothamnus* spp.), or Sandbar Willow (Table 9.2). Plains riparian is dominated singly or in combination by Western Soapberry and Little Walnut. The plains riparian stands are widely scattered and most of the channels have little vegetation. Saltcedar associations representing the successional-disburbance vegetation type are common on water courses formerly supporting floodplain and plains riparian vegetation. From Albuquerque north, in the central part of the state, Russian Olive tends to replace Saltcedar as the dominant of successional-disturbance riparian communities (Campbell and Dick-Peddie, 1964, and Dick-Peddie, Meents, and Spellenberg, 1984). Riparian thickets, or bosques, on the Rio Grande, in the southern part of the state, are often composed of Tornillo (*Prosopis pubescens*), with Skunkbush, Seep Willow (*Baccharis glutinosa*), Wolfberry (*Lycium* spp.), and Arrowweed (*Pluchea sericea*). Mesquite bosques may have seep willow and four-wing saltbush in the understory (Pl. 26 and 27).

Like aspen in montane riparian, communities dominated by Mesquite (*Prosopis glandulosa*) are not as common in riparian vegetation as in other nonriparian communities, such as sand scrub communities. These dual occurrences in riparian and nonriparian vegetation are why aspen and mesquite are considered semiriparian species.

ARROYO RIPARIAN

This type of riparian vegetation occupies drainages that dissect bajadas and mesas of the state. The upland vegetation surrounding this type may be plains-mesa grassland, desert grassland, Great Basin Desert Scrub, Chihuahuan Desert Scrub, or Plains-Mesa Sand Scrub (Table 9.1). In the northwest quarter of New Mexico, the arroyo riparian associations are usually dominated by Greasewood (*Sarcobatus vermiculatus*). A common arroyo dominant of the northern two-thirds of the state is a variety of Rabbitbrush, *Chrysothamnus nauseosus* var. *graveolens*. *C. nauseosus* var. *graveolens* tends to be replaced in the northwest by *C. nauseosus* var. *bigelovii*. In the southern third of the state lower portions of arroyos, where the beds widen, are dominated singly or in combination by Burrobrush (*Hymenoclea monogyra*), Apache Plume (*Fallugia paradoxa*), Littleleaf Sumac (*Rhus microphylla*), and Brickellia (*Brickellia laciniata*). Obviously, arroyo riparian vegetation of bajadas grades into montane riparian vegetation, and some sites may contain plants of both types. Seepwillow (*Baccharis glutinosa*) is a shrub commonly found in both types and is also common in floodplain riparian communities. (Pl. 28).

CLOSED BASIN–PLAYA–ALKALI SINK RIPARIAN

These riparian habitats are internally drained depressions having varied sizes of watersheds. Closed basins are common and large enough in New Mexico to be mapped (Fig. 9.2) and vegetation map insert). These are usually broad, flat or gently sloping areas, where water tends to spread rather than form gullies. However, the water moves slow enough after light rains to evaporate and cause an increase in salinity. Closed basin vegetation could have been included under scrubland vegetation. West (1983) called these vegetation types intermountain salt-desert shrubland, but the

nature of these areas in New Mexico is such that they are usually part of a drainage system. Large areas with dense stands of Fourwing Saltbush (*Atriplex canescens*) are typical of closed basin riparian vegetation (Pl. 29). Occasionally, Pale Wolfberry (*Lycium pallidum*) is a codominant. Ground cover is sparse, often with clumps of Burrowgrass (*Scleropogon brevifolius*) and occasional forbs, usually from the surrounding desert grassland vegetation type. In the northern and northwestern areas, Fourwing Saltbush often codominates with Shadscale, and ground cover may be considerable. Most of the forbs belong to the goosefoot family, Chenopodiaceae. Common forbs are Red Sage (*Kochia americana*), and *Atriplex patula* and *A. corrugata*. Greasewood (*Sarcobatus vermiculatus*) can be found dominating upper portions of the basins. On the closed basins near White Sands, Reid (1980) found large areas dominated by *Coldenia hispidissimus,* with Scattered Allthorn (*Koeberlinia spinosa*) and Squawbush. Dominant forbs were Gyp Moonpod (*Selinocarpus lanceolatus*) and the mustard *Nerisyrenia camporum*. In southeastern New Mexico, Burgess and Northington (1974) report large areas of Closed Basin vegetation dominated by Fourwing Saltbush, *Coldenia hispidissimus,* and Scattered Alkali Sacaton. Closed basin soils are often highly gypsiferous and occupied by grasses such as Gyp Grama (*Bouteloua breviseta*) and Gyp Dropseed (*Sporobolus nealleyi*).

Playas often have water that stands long enough to prevent the establishment of perennials in their centers. Most of the time, sandy playa boundaries will be occupied by mesquite and sometimes Soaptree Yucca (*Yucca elata*). Annuals such as Bitterweed (*Hymenoxys odorata*) often form pure stands on the dried playa beds. If the boundary is not sandy, it is not uncommon to find a zone of Tarbush (*Flourensia cernua*) on the outside, with Burrograss extending toward the center. Many areas near playas and closed basins have transi-

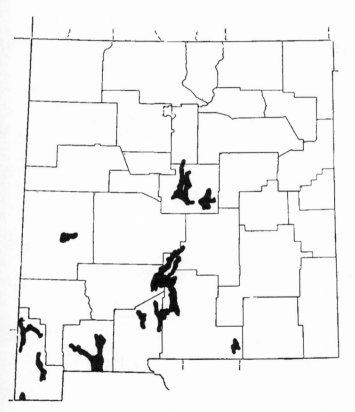

Figure 9.2 Distribution of Closed Basin Riparian Vegetation in New Mexico. See New Mexico County Map on page xiv for reference, if needed, on county names.

tion zones where sacaton associations are found. These communities are dominated by Alkali Sacaton, Sacaton (*S. wrightii*), or Sand Dropseed (*S. cryptandrus*).

Alkali sink riparian vegetation is very distinctive, because the habitats are extremely saline; plant species associated with sinks have evolved a tolerance for this high salinity. These saline-tolerant plants are sometimes called halophytes. They are not obligate halophytes, but while adapting to saline conditions, many species appear to have lost the ability to compete well with other plants in nonsaline habitats. Hendrickson (1974) lists species that dominate alkali sink vegetation:

Shrubs
Allenrolfia occidentalis
Pseudoclappia arenaria
Suaeda depressa
Suaeda suffrutescens
Suaeda torreyana

Grasses
Distichlis stricta
Eragrostis obtusiflora

Forbs
Cleomella longipes
Frankenia jamesii
Helliotropium curassavicum
Limonium limbatum
Salicornia bigelovii
Sesuvium verrucosum

Bitter Lake, in Chaves County, has extensive areas covered by alkali sink vegetation. The alien shrub-tree, Saltcedar (*Tamarix* spp.), is a common component of alkali sink vegetation. The major shrub of alkali sink associations in New Mexico is Iodinebush (*Allenrolfea occidentalis*). Occasionally, Fourwing Saltbush is a codominant, and in northwestern New Mexico, Greasewood may be common. Ground cover of alkali sinks is dominated by Seepweed (*Suaeda* spp.) and Saltgrass. On the less saline borders of sinks, alkali sacaton may be found. There are a number of plants that can tolerate the salinity at the margins of alkali sinks and are also important in other vegetation types. Examples are Shadscale, Fourwing Saltbush, greasewood, *Atriplex acanthocarpa,* Sacaton, and Alkali Sacaton. A very common dominant of alkali sink vegetation in New Mexico is the introduced tree-shrub, Saltcedar (*Tamarix* spp.).

Table 9.1. Major Obligate Riparian and Semiriparian Plants in New Mexico. 1 = Alpine-Montane, 2 = Floodplain-Plains, 3 = Arroyos, 4 = Alkali Sinks, Closed Basins, 5 = Playas. * = alien (introduced).

Scientific Name	Common Name	Occurrence
TREES (obligate riparian)		
Acer negundo	Boxelder	1
Alnus oblongifolia	Arizona Alder	1
Betula occidentalis	Water Birch	1
Celtis reticulata	Netleaf Hackberry	1, 2, 3
Fraxinus pennsylvanica	Velvet Ash	1, 2
Juglans major	Arizona Walnut	1, 2
Juglans microcarpa	Little Walnut	2
Morus microphylla	Texas Mulberry	1
Platanus wrightii	Arizona Sycamore	2

Table 9.1. (continued).

Scientific Name	Common Name	Occurrence
Populus acuminata	Lanceleaf Cottonwood	1
Populus angustifolia	Narrowleaf Cottonwood	1
Populus fremontii	Fremont Cottonwood	1, 2
Salix gooddingii	Goodding Willow	1, 2
TREES (semiriparian)		
Picea pungens	Blue Spruce	1
Populus tremuloides	Aspen	1
Quercus gambelii	Gambel Oak	1
Prunus serotina	Black Cherry	1, 3
Prunus virginiana	Chokecherry	1
Sapindus saponaria	Western Soapberry	2, 3
SHRUB-TREES (obligate riparian)		
Alnus tenuifolia	Mountain Alder	1
Amorpha fruticosa	False Indigo	1, 2
Cercis occidentalis	Redbud	1
Chilopsis linearis	Desertwillow	2, 3
Crataegus spp.	hawthorn	1
Elaeagnus angustifolia	*Russian Olive	2
Prosopis pubescens	Screwbean or Tornillo	2, 3
SHRUB-TREES (semiriparian)		
Acer grandidentatum	Bigtooth Maple	1
Fraxinus cuspidatus	Flowering Ash	1, 3
Robinia neomexicana	New Mexican Locust	1
Prosopis glanulosa	Honey Mesquite	2, 3, 4, 5
Ptelea angustifolia	Hoptree	1
Salix amygdaloides	Peachleaf Willow	1
Salix scoueriana	Scouler Willow	1
Tamarix spp.	*saltcedar	2, 3, 4, 5
SHRUBS (obligate riparian)		
Acacia greggii	Catclaw Acacia	3
Allenrolfea occidentalis	Iodinebush	4
Baccharis glutinosa	Seepwillow	1, 2, 3
Brickellia californica	California Brickelbush	1
Brickellia floribunda	Bigleaf Brickelbush	1, 2, 3
Brickellia laciniata	Cutleaf Brickelbush	2, 3
Chrysothamnus nauseosus var. *graveolens*	Rubber Rabbitbrush	2, 3
Chrysothamnus pulchellus	Southwest Rabbitbrush	1
Cornus stolonifera	Redosier Dogwood	1
Forestiera neomexicana	New Mexican Olive	1, 2
Hymenoclea monogyra	Burroweed	3
Lonicera involucrata	Inkberry	1
Lycium torreyi	Tomatillo	2, 4, 5

Table 9.1. (continued).

Scientific Name	Common Name	Occurrence
Pluchea sericea	Arrowweed	2
Potentilla fruticosa	Shrubby Cinquefoil	1
Rhamnus betulaefolia	Birchleaf Buckthorn	2
Salix exigua	Coyote Willow	1, 2
Salix irrorata	Bluestem Willow	1
Salix lutea	Yellow Willow	1
Salix bebbiana	Bebb Willow	1
Sambucus spp.	elderberry	1
Sarcobatus vermiculatus	Greasewood	3, 4
Shepherdia argentea	Buffaloberry	1
Suaeda depressa	Pursh Seepweed	1
Suaeda suffrutescens	Desert Seepweed	4
Suaeda torreyana	Torrey Seepweed	4

SHRUB (semiriparian)

Acacia constricta	Whitethorn	3
Acacia neovernicosa		3
Anisacanthus thurberi	Birdshade	3
Atriplex canescens	Fourwing Saltbush	2, 3, 4
Atriplex confertifolia	Shadscale	3
Bumelia lanuginosa	Woolly Buckthorn	1, 3
Chrysothamnus nauseosus var. *bigelovii*	Bigelow Rubber Rabbitbrush	3
Ericameria laricifolia	Turpentinebush	1
Fallugia paradoxa	Apacheplume	2, 3
Frankenia jamesii	Frankenia	4
Garrya wrightii	Wright Silktassel	1
Jamesii americana	Cliffbush	1
Lippia wrightii	Spicebush	1, 3
Lycium pallidum	Pale Wolfberry	2, 3, 4, 5
Mimosa biuncifera	Catclaw Mimosa	3
Philadelphus microphyllus	Mock Orange	1
Pseudoclappia arenaria		4
Rhus choriophylla	Mearns Sumac	3
Rhus glabra	Smooth Sumac	1
Rhus microphylla	Littleleaf Sumac	3, 4, 5
Rhus trilobata	Skunkbush	1, 2, 4
Yucca elata	Soaptree Yucca	4, 5

VINES

Clematis spp.	clematis	1, 3
Parthenocissus inserta	Virginia Creeper	1
Vitis arizonica	Arizona Grape	1, 2, 3

GRASSES AND GRASSLIKE PLANTS (obligate riparian)

Alopecurus spp.	*foxtail	

Table 9.1.   (continued).

Scientific Name	Common Name	Occurrence
Arundo donax	*Giant Reedgrass	2, 3
Carex foenea	Wind Sedge	1, 2
Carex rupestris	Rock Sedge	1, 2
Catabrosa aquatica	*Brookgrass	1
Cyperus spp.	flatsedge	1, 2
Deschampsia caespitosa	Tufted Hairgrass	1
Distichlis stricta	Saltgrass	2, 3, 4
Eleocharis spp.	spikerush	1, 2
Elymus canadensis	Canada Wildrye	1
Equisetum spp.	horsetail	1, 2
Eragrostis obtusiflora	Alkali Lovegrass	4
Eriochloa gracilis	Southwest Cutgrass	1
Glyceria borealis	Mannagrass	1
Hordeum hystrix	*Mediterranean Barley	1
Hordeum jubatum	*Foxtail Barley	1
Juncus spp.	rush	1, 2
Leersia oryzoides	*Cutgrass	
Luzua spp.	woodrush	1, 2
Paspalum distichum	Knotgrass	1
Phragmites communis	*Common Reed	1, 2
Polypogon spp.	rabbitfoot	1
Scirpus spp.	bulrush	1
Sporobolus wrightii	Sacaton	3, 4, 5
Typha latifolia	Common Cattail	1, 2

GRASSES AND GRASSLIKE PLANTS (semiriparian)

Agrostis scabra	Rough Bent	1
Agrostis semiverticillata	Water Bent	1
Bouteloua breviseta	Gyp Grama	4, 5
Poa pratensis	Kentucky Bluegrass	1, 2
Schizachyrium scoparium	Little Bluestem	4, 5
Scleropogon brevifolius	Burrograss	4, 5
Sporobolus airoides	Alkali Sacaton	2, 3, 4
Sporobolus cryptandrus	Sand Dropseed	2, 3, 4
Sporobolus gigantea	Giant Dropseed	4, 5
Sporobolus nealleyi	Gypgrass	4, 5
Sporobolus texanus	Texas Dropseed	3, 4

FORBS (obligate riparian)

Anemopsis californica	Yerbamanza	1, 2
Angelica pinnata	Small-leaf Angelica	1
Apocynum spp.	dogbane	1, 2
Aquilegia chrysantha	Yellow Columbine	1
Caltha leptosepala	Marshmarigold	1
Cardamine cordifolia	Bittercress	1, 2

Table 9.1. (continued).

Scientific Name	Common Name	Occurrence
Cicuta douglasii	Waterhemlock	1
Dodecatheon radicatum	Southern Shootingstar	1
Eustoma exaltum	Catchfly Gentian	2, 4
Galium aparine	Bedstraw	1
Glycyrrhiza epidota	Licorice	2
Habenaria sparciflora	Bog Orchid	1
Heliotropium curassavicum	Quailplant	4
Heracleum lanatum	Cowparsnip	1
Hydrophyllus fendleri	Squawlettuce	1
Hypericum formosum	Southwestern St. Johnswort	1
Iris missouriensis	Rockymountain Iris	1
Kochia americana	Redsage	3, 4, 5
Limonium limbatum	Marsh Rosemary	4
Lobelia cardinalis	Cardinalflower	1
Mertensia franciscana	Franciscan Bluebells	1
Mimulus spp.	monkeyflower	1
Phyla (Lippia) spp.	fogfruit	1, 2
Polygonum spp.	knotweed	1, 2
Prunella vulgaris	Selfheal	1
Ranunculus spp.	buttercup	1
Rorippa spp.	watercress	1
Rudbeckia laciniata	Cutleaf Coneflower	1
Rumex crispus	Curly Dock	1, 2
Salicornia bigelovii	Bigelow Glasswort	4
Sesbania macrocarpa	Colorado Riverhemp	1, 2
Sidalcea candida	Marsh Mallow	1
Suckleya suckleyana	Poison Suckleya	4, 5
Veronica spp.	speedwell	1
Viola nephrophylla	Northern Bog Violet	1
FORBS (semiriparian)		
Actaea arguta	Western Baneberry	1
Atriplex acanthocarpa	Tubercled Saltbush	4
Atriplex corrugata	Mat Saltbush	4
Atriplex patula	Fathen	4
Bassia hyssopifolia	Smotherweed	2, 3
Coldenia hispidissima	Coldenia	4
Epilobium angustifolium	Willowherb Fireweed	1
Frankenia jamesii	James Frankenia	4
Geranium richardsonii	Richardson Geranium	1
Helenium spp.	sneezeweed	1
Hymenoxys odorata	Bitterweed	3, 4
Polemonium spp.	Jacob's Ladder	1
Urtica spp.	nettle	1

Table 9.2. Riparian Vegetation Types in New Mexico.

RIPARIAN VEGETATION

ALPINE RIPARIAN
  Sedge Series
    Carex rupestris var. *drummondiana*
    Carex spp.–*Deschampsia caespitosa*
    Carex spp.–*Agrostis scabra–Glyceria*
      borealis

MONTANE RIPARIAN
  Willow Series
    Salix amygdaloides/MS/MG-F

  Willow–Mountain Alder Series
    Salix bebbiana–Alnus tenufolia/S

  Willow-Dogwood Series
    Salix irrorata–Cornus stolonifera/S

  Blue Spruce Series
    Picea pungens/Alnus tenuifolia/MG-F
    Picea pungens/Cornus stolonifera/MG-F

  Aspen Series
    Populus tremuloides/MS/MG-F

  Aspen-Maple Series
    Populus tremuloides–Acer Grandidentatum/
      MS/MG

  Boxelder Series
    Acer negundo/Alnus tenuifolia/MG-F
    Acer negundo/MS/*Poa pratensis*

  Boxelder-Alder Series
    Acer negundo–Alnus oblongifolia/MS/MG-F

  Alder Series
    Alnus oblongifolia/MS/MG-F

  Narrowleaf Cottonwood Series
    Populus angustifolia/MS/MG-F
    Populus angustifolia/Salix irrorata/MG-F

  Narrowleaf Cottonwood-Mixed Deciduous
  Series
    Populus angustifolia–Alnus oblongifolia–
      Acer negundo/MS/MG-F
    Populus angustifolia–Alnus oblongifolia/MS/
      MG-F
    Populus angustifolia–Acer negundo/MS/
      MG-F

  Populus angustifolia–Juglans major/MS/
    MG-F
  Populus angustifolia–Populus fremontii/MS/
    MG-F

  Broadleaf Cottonwood Series
    Populus acuminata–Populus fremontii/MS/
      MG-F
    Populus acuminata/MS/MG-F
    Populus fremontii/MS/MG-F

  Broadleaf Cottonwood-Mixed Deciduous Series
    Populus fremontii–Platanus wrightii–Juglans
      major/MS/MG-F
    Populus fremontii–Platanus wrightii–Celtis
      reticulata/Baccharis glutinosa/MG-F
    Populus fremontii–Platanus wrightii–Salix
      gooddingii/MS/MG-F
    Populus fremontii–Platanus wrightii/S/MG-F
    Populus fremontii–Fraxinus pennsylvanica–
      Salix gooddingii/MS/S
    Populus fremontii–Celtis reticulata/MS/MG-F
    Populus fremontii–Salix gooddingii–Juglans
      major/MS/MG-F

  Sycamore Series
    Platanus wrightii/MS/MG-F

  Ash Series
    Fraxinus pennsylvanica/MS/MG-F

  Ash–Arizona Walnut Series
    Fraxinus pennsylvanica–Juglans major/MS/
      MG-F

  Arizona Walnut Series
    Juglans major/Brickellia californica/MG-F
    Juglans major/MS/MG-F

  Hackberry Series
    Celtis reticulata/MS/S

FLOODPLAIN-PLAINS RIPARIAN
  Cottonwood Series
    Populus fremontii/Forestiera neomexicana/
      MG-F
    Populus fremontii/Rhus trilobata/MG-F
    Populus fremontii/Salix exigua/MG-F
    Populus fremontii/Chrysothamnus nauseosus/
      MG-F

Table 9.2. (continued).

Populus fremontii–Prosopis pubescens/MS/MG-F

Cottonwood-Willow Series
Populus fremontii–Salix gooddingii/Prosopis pubescens/MG-F
Populus fremontii–Salix gooddingii/MS/MG-F
Populus fremontii–Salix amygdaloides/MS/MG-F

Little Walnut Series
Juglans microcarpa/MS/MG-F

Little Walnut–Soapberry Series
Juglans microcarpa–Sapindus drummondii/MS/MG-F

Soapberry Series
Sapindus drummondii/S/MG-F

Mesquite Series
Prosopis glandulosa/Atriplex canescens/MF
Prosopis glandulosa/MS/MG-F

ARROYO RIPARIAN
Greasewood–Desert Shrub Series
Sarcobatus vermiculatus/MF
Sarcobatus vermiculatus–Atriplex canescens–Chrysothamnus nauseosus var. *graveolens*/S
Sarcobatus vermiculatus–Chrysothamnus nauseosus var. *graveolens*/S
Sarcobatus vermiculatus–Chrysothamnus nauseosus var. *bigelovii*/S

Saltbush Series
Atriplex confertifolia/Sporobolus airoides
Atriplex canescens/MF

Rabbitbrush Series
Chrysothamnus nauseosus var. *graveolens*/S

Apacheplume Series
Fallugia paradoxa/MF
Fallugia paradoxa–Chrysothamnus nauseosus/MF

Burrobrush Series
Hymenoclea monogyra/S

Brickelbush Series
Brickellia laciniata/S

Mesquite Series
Prosopis glandulosa/MF

CLOSED BASIN–PLAYA–ALKALI SINK RIPARIAN
Iodineweed Series
Allenrolfea occidentalis/Suaeda spp.–*Distichlis stricta*
Allenrolfea occidentalis–Atriplex canescens/S

Fourwing Saltbush Series
Atriplex canescens/SMF
Atriplex canescens–Lycium pallidum/Sporobolus airoides–SMF
Atriplex canescens/Coldenia hispidissima–Scleropogon brevifolius–SMF
Atriplex canescens–Atriplex confertifolia/Sporobolus airoides–Atriplex spp.

Mesquite Series
Prosopis gladulosa/MF

SUCCESSIONAL-DISTURBANCE RIPARIAN
Russian Olive Series
Elaeagnus angustifolia/MS/S
(toward Cottonwood Associations)

Saltcedar Series
Tamarix spp.
(toward Cottonwood Associations)
Some saltcedar stands appear to be stable and not successional

CHAPTER 9 REFERENCES

RIPARIAN VEGETATION

Barbour, M. G., J. H. Burk, and W. D. Pitts, 1980. Terrestrial Plant Ecology. The Benjamin Cummings Publishing Company, Inc., Menlo Park, California.

Boles, P. H., 1978. Analysis of five plant communities along a fifty-six kilometer section of the Mimbres River in southwestern New Mexico. Thesis, New Mexico State University, Las Cruces.

Boles, P. H. and W. A. Dick-Peddie, 1983. Woody riparian vegetation patterns on a segment of the Mimbres River in southwestern New Mexico. The Southwestern Naturalist, 28: 81–87.

Brown, D. E. (ed.), 1982. Biotic communities of the American Southwest–United States and Mexico. Desert Plants, 4: 1–342.

Brown, D. E., N. B. Carmony, and R. M. Turner, 1977. Inventory of riparian habitats. pp. 10–13. In: R. R. Johnson and D. A. Jones (eds.). Importance, preservation and management of riparian habitats: a symposium. U.S. Department of Agriculture, Forest Service General Technical Report RM-43, Ft. Collins, Colorado.

Brown, D. E., C. H. Lowe, and J. F. Hausler, 1977. Southwestern riparian communities: their biotic importance and management in Arizona. pp. 201–211. In: R. R. Johnson and D. A. Jones (eds.). Importance, preservation and management of riparian habitat: a symposium. U.S. Department of Agriculture, Forest Service General Technical Report RM-43. Ft. Collins, Colorado.

Browning, J., 1988. Classification of riparian vegetation in New Mexico. Thesis, New Mexico State University, Las Cruces.

Bryan, K. M., 1928. Change in plant associations by change in ground water level. Ecology, 9: 474–478.

Burgess, T. L. and D. K. Northington, 1974. Desert vegetation in the Guadalupe Mountains Region. pp. 229–242. In: R. H. Wauer and D. H. Riskind (eds.). Transactions of the symposium of the biological resources of the Chihuahuan Desert region, United States and Mexico. U.S. Department of the Interior, National Park Service Transactions and Proceedings Series No. 3, U.S. Government Printing Office, Washington, D.C.

Campbell, C. J., 1970. Ecological implications of riparian vegetation management. Journal of Soil and Water Conservation, 25: 45–52.

Campbell, C. J. and W. A. Dick-Peddie, 1964. Comparison of phreatophyte communities on the Rio Grande in New Mexico. Ecology, 45: 492–502.

Campbell, C. J. and W. Green, 1968. Perpetual succession of stream channel vegetation in a semiarid region. Journal of Arizona Academy of Science, 5: 86–98.

Corley, E., 1964. The Gila River of the Southwest. University of Nebraska Press, Lincoln.

Crumpacker, D. W., 1984. Regional riparian research and a multi-university approach to the special problem of livestock grazing in the Rocky Mountains and Great Plains. In: R. E. Warner and K. M. Hendrix (eds.). California riparian systems: ecology, conservation, and productive management. University of California Press, Berkeley.

Depree, E. and J. A. Ludwig, 1978. Vegetative and reproductive growth patterns in desert willow, *Chilopsis linearis*. The Southwestern Naturalist, 23: 239–246.

Dick-Peddie, W. A., J. V. Hardesty, E. Muldavin, and B. Sallach, 1987. Soil-vegetation correlations on the riparian zones of the Gila and San Francisco rivers in New Mexico. U.S. Department of Interior, Fish and Wildlife Service, Biological Report 87(9), Washington, D.C.

Dick-Peddie, W. A. and J. P. Hubbard, 1977. Classification of riparian vegetation. pp. 85–90. In: R. R. Johnson and D. A. Jones (eds.). Importance, preservation and management of riparian habitat: a symposium. U.S. Department of Agriculture, Forest Service General Technical Report RM-43, Ft. Collins, Colorado.

Dick-Peddie, W. A., J. K. Meents, and R. Spellenberg, 1984. Vegetation resource analysis for the Velarde community ditch project, Rio Arriba and Santa Fe Counties, New Mexico. Final Report. U.S. Bureau of Reclamation, Southwest Region, Amarillo, Texas. 251 pp.

Egbert, J. C., 1981. Biological inventory of the Gila Riparian Preserve. The Nature Conservancy Report. Western Regional Office, San Francisco, California.

Freehling, M. D., 1982. Riparian woodlands of the Middle Rio Grande Valley, New Mexico: a study of bird populations and vegetation with special reference to russian olive (*Elaeagnus angustifolia*). U.S. Department of Interior, Fish and Wildlife Service, Albuquerque, New Mexico.

Freeman, C. E. and W. A. Dick-Peddie, 1970. Woody riparian vegetation in the Black and Sacramento mountain ranges, southern New Mexico. The Southwestern Naturalist, 15: 145–164.

Gehlbach, F. R., 1967. Vegetation of the Guadalupe escarpment, New Mexico–Texas. Ecology, 48: 403–419.

Glinsky, R. L., 1977. Regeneration and distribution of sycamore and cottonwood trees along Sonoita Creek, Santa Cruz County, Arizona. pp. 116–123. In: R. R. Johnson and D. A. Jones (eds.). Importance, preservation and management of riparian habitat: a symposium. U.S. Department of Agriculture, Forest Service General Technical Report RM-43, Ft. Collins, Colorado.

Hardesty, J. V., 1986. Riparian vegetation at three sites along the Gila River in southwestern New Mexico. Thesis, New Mexico State University, Las Cruces.

Haase, E. F., 1974. Survey of floodplain vegetation along the lower Gila River in southwest Arizona. Journal of Arizona Academy of Science, 7: 66–81.

Hendrickson, J., 1974. Saline habitats and halophytic vegetation in the Chihuahuan Desert region. pp. 289–314. In: R. H. Wauer and D. H. Riskind (eds.). Transactions of the symposium of the biological resources of the Chihuahuan Desert region, United States and Mexico. U.S. Department of the Interior, National Park Service Transactions and Proceedings Series No. 3, U.S. Government Printing Office, Washington, D.C.

Henry, R. J., 1981. Riparian vegetation of two mountain ranges in southwestern New Mexico. Thesis, New Mexico State University, Las Cruces.

Horton, J. S., 1977. The development and perpetuation of the permanent tamarisk type in the phreatophyte zone of the Southwest. pp. 124–127. In: R. R. Johnson and D. A. Jones (eds.). Importance, preservation and management of riparian habitat: a symposium. U.S. Department of Agriculture, Forest Service General Technical Report RM-43, Ft. Collins, Colorado.

Hubbard, J. P., 1977. Importance of riparian ecosystems: biotic considerations. pp. 14–18. In: R. R. Johnson and D. A. Jones (eds.). Importance, preservation and management of riparian habitat: a symposium. U.S. Department of Agriculture, Forest Service General Technical Report RM-43, Ft. Collins, Colorado.

Johnson, R. R. and D. A. Jones, 1977. Importance, preservation and management of riparian habitat: a symposium. U.S. Department of Agriculture, Forest Service General Technical Report RM-43, Ft. Collins, Colorado.

Jojola, J. R., 1977. Bird populations and habitat in a riparian woodland, Isleta Indian Reservation, central New Mexico. Thesis, New Mexico State University, Las Cruces.

King, H. T., 1976. Bird abundance and habitat in a southern New Mexico bosque. Thesis, New Mexico State University, Las Cruces.

Kuchler, A. W., 1964. The potential natural vegetation of the conterminous United States (manual and map). American Geographical

Society Special Publication: 361. New York.

Medina, A. L., 1984. Riparian plant communities and soils of the Ft. Bayard watershed in southwestern New Mexico. Ms submitted to The Southwestern Naturalist.

Pase, C. P. and E. F. Layser, 1977. Classification of riparian habitat in the southwest. pp. 5–9. In: R. R. Johnson and D. A. Jones (eds.). Importance, preservation and management of riparian habitat: a symposium. U.S. Department of Agriculture, Forest Service General Technical Report RM-43, Ft. Collins, Colorado.

Reichenbacher, F. W., 1984. Ecology and evolution of Southwestern riparian plant communities. Desert Plants, 6: 14–22.

Reid, W. H., 1980. White Sands National Monument natural resource inventory and analysis. Field Report CX 702900001. National Park Service. pp. 104.

Rucks, M. G., 1984. Composition and trend of riparian vegetation on five perennial streams in southeastern Arizona. In: R. E. Warner and K. M. Hendrix (eds.). California riparian systems: ecology, conservation, and productive management. University of California Press, Berkeley.

Standley, P. C., 1915. Vegetation of the Brazos Canyon, New Mexico. Plant World, 18: 179–191.

Townsend, C. H. T., 1893. On life zones of the Organ Mountains and adjacent region in southern New Mexico, with notes on the fauna of the range. Science, 22: 313–315.

Van Cleave, M., 1936. Vegetation changes in the Middle Rio Grande Conservancy District. Thesis, University of New Mexico, Albuquerque.

Warner, R. E. and K. M. Hendrix (eds.), 1983. California riparian systems: ecology, conservation, and productive management. University of California Press, Berkeley.

Watson, J. R., 1912. Plant geography of north central New Mexico. Botanical Gazette, 54: 194–217.

Wauer, R. H., 1977. Significance of Rio Grande riparian systems upon the avifauna. pp. 165–174. In: R. R. Johnson and D. A. Jones (eds.). Importance, preservation and management of riparian habitats: a symposium. U.S. Department of Agriculture, Forest Service General Technical Report RM-43, Ft. Collins, Colorado.

West, N. E., 1983. Intermountain salt-desert shrubland. pp. 375–397. In: N. E. West (ed.). Ecosystems of the World, vol. 5. Temperate Deserts and Semi-Deserts. Elsevier Science Publishers, Amsterdam.

Whitfield, C. J. and H. L. Anderson, 1938. Secondary succession in the desert plains grassland. Ecology, 19: 171–180.

Williams, J. S., 1969. West slope arroyo vegetation of the Franklin Mountains, El Paso, Texas. Texas Journal of Science, 20: 299–300.

10

VEGETATION OF
SPECIAL HABITATS

This book deals primarily with major terrestrial vegetation found in New Mexico. This chapter examines aquatic vegetation and certain rela- tively unique terrestrial vegetation found in the state.

AQUATIC VEGETATION

A hierarchical code-system for water bodies of New Mexico, developed by Hubbard (1988), includes running water (both permanent and intermittent), such as rivers, creeks, arroyos, draws, washes, runs, forks, and canyons, and embraces lakes and ponds (both natural and man-made), major springs, and closed basins. The scheme arranges the water bodies by major drainages and their subsystems in the state and is similar to that used for the Hydrological Unit Map of New Mexico (1986 reprint of 1974 edition) of the U.S. Geological Survey. This water bodies system includes more than 400 separate water body entries. Even though 80% of the entries are running or intermittent waters, the list still includes more than 65 lakes. It is possible that 20% of the listed water bodies do not support true aquatic vegetation. These include many of the closed basins, arroyos, draws, washes, and canyons. However, over 300 water bodies in New Mexico can be expected to support emergent and/or submersed vegetation. In a publication on New Mexico fishing waters, the New Mexico Department of Game and Fish identified and located more than 1,600 fishing waters in the state and it is assumed that most of these areas support some type of aquatic vegetation.

In a system adapted from Correll and Correll (1972), water bodies in New Mexico can be categorized as follows:

Open water surface the most conspicuous feature (lakes, ponds, and reservoirs). Vegetation more conspicuous than water surface (marshes, cienegas, and bogs).

Most of the lakes in New Mexico are manmade. The establishment of vegetation in manmade lakes tends to follow similar patterns. Once stabilized, artificial lakes often develop vegetation zones similar to northern natural lakes (Correll and Correll, 1972).

Shallow margins have emergent vegetation (plants rooted under water with photosynthetic parts mostly above water). Emergent vegetation consists of grasslike plants such as sedges, rushes, and cattails along with nongrasslike forms such as water plantain (*Alisma*) and arrowhead (*Sagittaria*). Usually the first plant to become established is cattail (*Typha*). Next bulrushes (*Scirpus*) become established on shallow sites and then expand outward to deeper water as dead organic matter builds up from the bottom.

The next zone toward the center is that of floating-leaf plants. These may be rooted or free-floating and may include water shield (*Brasenia*), pondweed (*Potamogeton*), water fern (*Azolla*), water lettuce (*Pistia stratiotes*), duckweed (*Lemna*), and floating algae (phytoplankton). Not limited to lakes alone, these first two zones are also typical of marshes linked with stream systems. Browning (1988) found over 45 cattail marshes associated with major areas of riparian vegetation throughout the state.

Farther out in the lake are found the submersed plants dominated by milfoil (*Myriophyllum*), hornwort (*Ceratophyllum*), waterweed (*Elodea*), stoneworts (*Chara* and *Nitella*), and fixed algae. Some lakes naturally do not contain all zones, and the zones themselves do not always have discrete, recognizable boundaries. These generalized zones of aquatic vegetation in the southwestern United States have been extracted from Correll and Correll (1972). Although the Water Fern (*Ceratopteris*) and Water Lettuce (*Pistia stratiotes*) do not occur in New Mexico, the zonation pattern is typical of aquatic vegetation in the state.

Most of the man-made lakes in the state (over 55) maintain relatively stable water levels and support two or more zones of aquatic vegetation. Many of these lakes have shallow areas that support marsh vegetation. Some man-made lakes are subject to large and rapid changes in water levels and rarely have stable plant communities. The vegetation that does occur in these unstable lakes contains fewer plant species than that found in relatively stable ones.

Simple nonseed plants such as green algae, blue-green algae, diatoms and so forth are seldom recorded or quantified as components of vegetation. However, Lindsey (1951), Flowers (1961), Cole, Weigmann, and Hayes (1984), and Turner (1987) included nonseed aquatic plants in their discussions of aquatic vegetation. The algae dominating wet situations in the Navajo Reservoir Basin (Flowers, 1961) included 9 species of blue-green algae (Cyanophyta), 9 species of diatoms and 3 species of golden algae (Chrysophyta), and 17 species of green algae (Chlorophyta). A saline reservoir in southern New Mexico was dominated in early summer by the blue-green algae *Anabaena* and *Microcystis* and shared dominance in late summer with green algae and diatoms (Cole, Weigmann, and Hayes,

1984). Lindsey (1951) described rock-crustal algal and ooze communities in the lava sink-hole ponds of the Grants lava flow. The rock-crust was formed of half-inch-thick, nonslippery crusts, composed primarily of four species of blue-green algae. The ooze communities were dominated by four species of diatoms and several species of green algae. Duckweed (*Lemna minor*) was common on the surfaces of ponds supporting ooze communities. Lindsey found some ponds with a pure growth of a green algae (*Scenedesmus bujuga*) virtually covering the surface of the bottom muck. Tables 10.1 and 10.2 list major plants comprising aquatic vegetation in New Mexico.

In *Cienegas of the American Southwest*, Hendrickson and Minckley (1984) concentrated on middle-elevation marshy habitats in southeastern Arizona. The authors chose to use the term *cienega* to refer to marshy habitats found between elevations of 1,000 and 2,000 meters (3,300 and 6,600 feet). Their definition of *cienega* included both riparian and aquatic vegetation as they are used in this book. They separated southwestern wetlands into alpine meadowlands over 2,000 m in elevation, cienegas between 1,000 and 2,000 m in elevation, and riverine marshes lower than 1,000 m in elevation and exposed to no freezing temperatures. The first two categories suit the wetlands of New Mexico, but the last category, riverine marshes, does not. New Mexico has many sites where the vegetation fits the characteristics of the riverine marsh type, but most are higher than 1,000 m and all are subject to some freezing. The taxa listed by Hendrickson and Minckley are separated into "aquatic," "semiaquatic," and "riparian." The aquatic and many of the semiaquatic taxa are the same as those found in the marshy sites of New Mexico.

Due to lack of sufficient data, any classification of aquatic vegetation in New Mexico must be both general and tentative at this time. However, some groupings appear to be warranted; indeed they seem most reliable at the generic level and might be considered as "Series." The following are groups of aquatic plants, most of which have been reported often from various parts of the state:

Emergent (rooted and free floating)

shallower	Sedge (*Carex*, etc.)—Horsetail (*Equisetum*) Series
(wetland)	Cattail (*Typha*)—Water Parsnip (*Berula erecta*) Series
↓	Cattail—Rush (*Juncus*)—Bulrush (*Scirpus*) Series
↓	Sedge—Rush—Spikerush (*Eleocharis*) Series
↓	Bulrush—Arrowhead (*Sagittaria*)—Water Plantain (*Alisma*) Series
↓	Arrowhead—Speedwell (*Veronica*)—Watercress (*Rorippa*) Series
↓	Smartweed (*Polygonum*)—Crowfoot (*Ranunculus*) Series
deeper	Pondweed (*Potamogeton*)—Water Fern (*Azolla*)—Duckweed (*Lemna*)
(water bed)	Algae (various genera) Series

Submersed

Milfoil (*Myriophyllum*)—Hornwort (*Ceratophyllum*)—Stonewort (*Chara*)—Fixed algae (various genera)
Milfoil—Bladderwort (*Utricularia*)

Considerable variation exists in dominant species from water body to water body. The most common cattail reported is *Typha latifolia*. Fennel-Leaved Pondweed (*Potamogeton pectinatus*) has been reported more often and from a wider area than any other pondweed. Needle-Beaked Spikerush (*Eleocharis rostellata*) appears to be the most ubiquitous spikerush in the state. One of the most common arrowheads is Duck Potato (*Sagittaria latifolia*).

Aquatic vegetation may be composed of only a few species on sites where vegetation is in an early stage of succession. Aspect (seasonal) dominance by one or a few species of phytoplankton, referred to as algal bloom, is common during mid- to late summer in many small New Mexico lakes.

Goerndt, Schemnitz, and Zeedyk (1985) found that Watercress (*Rorippa nasturtium-aquaticum*), growing in "spring/seep" areas of the Sacramento Mountains of New Mexico, was common and plentiful enough to constitute a sizable portion of the diets of wild turkeys. According to definitions generated by Healy (1977), a "spring" discharges water from a fixed point and usually forms a small stream, while a "seep" percolates the ground water to the surface and forms a saturated area. The authors noted that most of the spring/seep areas were located in openings in stands of Douglas-Fir–White-Fir vegetation, with Blueberry Elder (*Sambucus neomexicana*) and Rock Spiraea (*Holodiscus dumosus*) dominating the understory.

In lava sink-hole ponds of the Grants Malpais, Lindsey (1954) found aquatic vegetation in various stages of succession. Ponds in early stages had Stonewort (*Chara contraria*) out in the water with Pondweed (*Potamogeton pectinatus*) and Watercress (*Rorippa nasturtium-aquaticum*) near the banks. Lindsey referred to an advanced sere as the "reed marsh" stage, during which the Great Bulrush (*Scirpus validus*), Cattail (*Typha latifolia*), and Water Parsnip (*Berula erecta*) shared dominance. He noted that common Reedgrass (*Phragmites communis*) would establish itself in very late stages when dead organic matter had accumulated to sufficient levels.

While investigating habitats of the Bitter Lake National Wildlife Refuge in east-central New Mexico, Peterson and Rasmussen (1986) found salt-marsh vegetation containing Widgeongrass (*Ruppia maritima*) and stonewort (*Chara* spp.). Some marsh areas contained Prairie Cordgrass (*Spartina pectinata*), Saltmarsh Bulrush (*Scirpus maritimus*), and Scratchgrass Muhly (*Muhlenbergia asperfolia*). They also reported that alkali sink vegetation contained Iodineweed (*Allenrolfia occidentalis*) and species of salicornia (*Salicornia* sp.) and seepweed (*Suaeda* sp.).

The following are examples of New Mexico vegetation that is difficult to categorize either as aquatic or terrestrial (riparian):

1. Vegetation found on sites, such as wet meadows and marsh margins, that are too dry to support true emergent plants. See Chapters 7 and 9.

2. Vegetation located on closed basins having alkaline or saline soils. See the discussion on the Alkali Sink type of riparian vegetation in Chapter 9.

3. Vegetation inhabiting the low areas of floodplains, near marshes or in cottonwood bosques where the water is not permanent enough to support emergent species. The soils of these gentle depressions are usually alkaline or saline. This type of vegetation is often composed almost entirely of stands of Yerba Mansa (*Anemopsis californica*) surrounded by the Riparian Saltgrass (*Distichlis stricta*). Yerba Mansa is considered an aquatic plant and is listed in Table 10.2.

VEGETATION OF ROCKY UNWEATHERED (SHALLOW OR NO SOIL) SURFACES

In numerous areas of New Mexico, the landscape forms outcrops of rocky surfaces, such as lava flows, escarpments (including cliffs), rock fields (talus slopes and scree), gypsum fields and others. These formations either have geologically young surfaces whose soils have not yet formed or position surfaces so that small rock particles, sand, and dust from weathering erode off the surfaces, instead of accumulating to initiate soil formation.

Plants differing widely in their moisture requirements compose the vegetation of these rocky surfaces. The relatively consolidated smooth surfaces permit only plants with low available moisture tolerance and/or a tolerance for extremely shallow soils. The plants on these dry (xeric) sites are mostly lichens, mosses, liverworts, and ferns. On the other hand, water running off the smooth rock surfaces tends to collect in any fissure or crevice. These catchment areas become relatively moist microsites that support species having moisture requirements even higher than the plants growing in soil off the rocky areas.

LAVA FLOW VEGETATION

There are a number of rather extensive lava flows in New Mexico. The largest are located on the vegetation map insert. In the southwestern United States these flows are usually referred to as "malpais." The word *malpais* is a modification of the Spanish words *mal* and *pais* meaning bad place or bad land (Dice, 1940 and Shields and Crispin, 1956). According to Shields and Crispin, the Carrizozo malpais is one of the youngest in the United States, having been extruded within the past 1,000 years. Because most malpaises in New Mexico are geologically young, they support few if any endemic plant species.

In lava flows, water runs off surfaces into cracks, crevices, and pockets. Consequently, moisture for plants is more available in these catchments than in surrounding nonlava soil. Lindsey (1951) documented this phenomenon by observing that many species flourishing in shallow crevices on some sites were found only in deep crevices on sites at lower elevations. Concerning vegetation of the Grants malpais, Lindsey stated: "The environmental resultant, in each of the two higher belts, supports a more mesophytic vegetation type than occurs on adjacent non-lava land." Shields (1956) made similar observations about the Carrizozo malpais: "As a result of the sinkholes and the crevices which efficiently collect and retain moisture, the lava flow in places provides a much more favorable environment for vegetation than surrounding areas receiving the same precipitation." The 18 species that Shields and Crispin found common to the Carrizozo and Grants malpaises grow throughout the state on nonlava soils, but on more mesic sites farther north, sites at higher elevations, or in sites in canyons. Table 10.3 is a list of major plants at the Grants and Carrizozo malpaises. Plants not found nearby on nonlava sites are indicated.

The Carrizozo malpais is surrounded by desert grassland or Chihuahuan desert scrub vegetation. Although found on the Carrizozo flow, the following plants are not found on sites bordering the flow but still are commonly seen in more mesic situations in that part of the state: the shrub tree *Juniperus monosperma;* the montane shrubs *Cercocarpus montanus* and *Berberis haematocarpa;* the forb *Mirabilis oxybaphoides;* the mustards *Sisymbrium linearifolium* and *Thelypodium wrightii;* the sages *Salvia henryi* and *Salvia vinacea;* the Indian paintbrush *Castilleja integra;* and Horehound (*Marrubium vulgare*). Lindsey described three major vegetation types on the Grants malpais. These were Douglas Fir Belt, Ponderosa Pine Belt, and Apache Plume Belt. The vegetation of all of these types was more mesic than that of the nonlava lands bordering them.

It is interesting to note that the major understory shrubs of Lindsey's Douglas Fir and Ponderosa Pine types and the major shrubs of his Apache Plume type, which were Apache Plume (*Fallugia paradoxa*), New Mexican Olive (*Forestiera neomexicana*), California Brickellia (*Brickellia californica*), and Skunkbush (*Rhus trilobata*), are all semi- or obligate riparian shrubs. Likewise on the Carrizozo malpais, Shields (1956) found riparian trees such as Netleaf Hackberry (*Celtis reticulata*), Arizona Walnut (*Juglans major*), and Littleleaf Mulberry (*Morus microphylla*) along with the riparian shrubs Wright's Silktassel (*Garrya wrightii*) and California Brickellia (*Brickellia californica*). The presence of these plants further attests to the water accumulation in the malpais.

In some areas of the Grants malpais, Ponderosa Pines are stunted and distorted and resemble a small pygmy forest (Spellenberg, 1979). This striking stand should be of considerable interest to the general public.

VEGETATION OF EXPOSED ROCK

Most of the vegetation of these sites is successional, and the species composition on a site tends to indicate the stage. On steep slopes (cliff faces), the vegetation generally remains in a perpetual early successional stage because there is no opportunity for the accumulation of "fines" (rock particles, sand grains, and dust) or dead organic matter.

In early successional stages on level or moderately sloping sites, the composition of the sparse vegetation is combinations of lichens, liverworts, mosses, and ferns. Lichens typically dominate the early stages and ferns, the later stages. Soil formation facilitates subsequent stages. Lindsey (1951) documented various stages of plant succession on the Grants lava flow. Fosberg (1940) reported successional stages on the western face of the Organ Mountains and on the northern face of the eastern spur of the Organ Mountains. Table 10.4 is a list of major plants comprising the vegetation of early successional stages on rocky surfaces in New Mexico.

VEGETATION OF GYPSUM SURFACES

Gypsum outcrops are common in New Mexico. Only a limited number of species can tolerate gypsum or strongly gypseous soils. Table 10.5 lists major plants comprising the vegetation of gypsum soils in New Mexico. Some species appear to have lost the ability to compete in nongypsum situations and are considered to be endemic to gypsum sites. These plants may be referred to as "gypsophiles" (Powell and Turner, 1974; Parsons, 1976; and Meyer, 1986). Most of the plants listed in Table 10.5 are categorized as gypsophiles by one or more of these authors. Some gypsum-tolerant plants also have calcareous and/or alkaline tolerances and may be found as members of closed basin riparian vegetation (see Chapter 9) or of sand scrub vegetation (see Chapter 8).

Table 10.1. Major Floating and Submersed Plants Comprising Aquatic Vegetation in New Mexico.

NONSEED PLANTS

CHLOROPHYTA (green algae)

Chara contraria
Chlamydomonas spp.
Chlorella vulgaris
Cladophora glomerata
Cladophora kuetzingiana
Closterium acerosum
Cosmarium galeritum
Cosmarium menegnenii
Cosmarium undulatum crenulatum
Mougeotia capucine paruula
Nitella spp.
Oedogonium spp.
Pediastrum tetras
Scenedesmus bujuga
Scenedesmus obliquus
Spirogyra crassa
Stigioclonium tenue
Zygnema insigne cruciatum

CHRYSOPHYTA

Achnanthes spp.
Amphora ovalis
Botrydium granulatum
Cocconeis spp.
Cymbella spp.
Diatoma hiemale var. *mesodon*
Gomphonema spp.
Gyrosigma spp.
Melosira spp.
Navicula spp.
Rhopalodia gibba
Surirella spp.
Synedra ulana
Tabellaria floccosa
Tribonema bombycinum
Vaucheria spp.

CYANOPHYTA (blue-green algae)

Amphithrix janthina
Amphithrix parietina
Anabaena inaequalis
Anabaena oscillaroides
Calothrix parietina
Chroococcus turgidus
Coeloshaerium collinsii
Eutophysalis coruuana
Lyngbya aestuarii
Microcystis sp.
Nostoc ellipsosporum
Nostoc sphaeroides
Oscillatoria animalis
Oscillatoria chatybea
Oscillatoria prolifica
Oscillatoria sancta
Phormindium incrustatum
Phormindium tenue
Plectonema nostocorum
Spirulina major

PYROPHYTA and CRYPTOPHYTA

Ceratium spp.
Cryptomonas spp.
Rhodomonas spp.

FERN

Azolla mexicana   Waterfern

SEED PLANTS

Ceratophyllum demersum	Common Hornwort	*Ranunculus aquaticus*	Water Crowfoot
Elodea canadensis	Waterweed	var. *capillaceus*	
Lemna minor	Duckweed	*Ranuncus longirostris*	Whitewater Crowfoot
Myriophyllum exalbescens	American Milfoil	*Utricularia vulgaris*	Common Bladderwort
Myriophyllum spicatum	Spiked Watermilfoil		

Table 10.2. Major Attached Plants (leaves and/or flowers emergent) Comprising Aquatic Vegetation in New Mexico.

Scientific Name	Common Name	Scientific Name	Common Name
FERN		*Carex simulata*	Shortbeaked Sedge
Marsilea mucronata	Water Clover	*Carex stipata*	
		Carex vulpinoidea	Foxtail Sedge
CATTAILS AND HORSETAILS		*Cladium jamaicense*	Sawgrass
Equisetum arvense	Bottlebrush	*Cyperus esculentus*	Yellow Nutgrass
Equisetum hyemale	Canuela	*Eleocharis acicularis*	Needle Spikerush
Equisetum kansanum	Summer Scouringrush	*Eleocharis macrostachya*	Creeping Spikerush
Equisetum laevigatum	Cola de Caballe	*Eleocharis montana*	
Typha angustifolia	Narrowleaved Cattail	*Eleocharis rostellata*	Beaked Spikerush
Typha domingensis	Tule	*Eriophorum angustifolium*	Cottongrass
Typha latifolia	Broadleaved Cattail	*Scirpus acutus*	Hardstem Bulrush
		Scirpus americanus	Threesquare
ARROW-WEEDS, DITCHGRASSES, AND PONDWEEDS		*Scirpus californica*	Giant Bulrush
Potamogeton crispus	Curled Pondweed	*Scirpus lacustris* var. *occidentalis*	
Potamogeton foliosus		*Scirpus maritimus* var. *paludosus*	Saltmarsh Bulrush
Potamogeton natans	Broadleaved Pondweed	*Scirpus olneyi*	
Potamogeton nodosus		*Scirpus validus*	Great Bulrush
Potamogeton pectinatus	Fennelleaved Pondweed		
Potamogeton pusillus	Dwarf Pondweed	**RUSHES**	
Ruppia maritima	Widgeongrass	*Juncus balticus* var. *montanus*	Wire Rush
Triglochin maritimum	Seaside Arrowgrass	*Juncus bufonius*	Toad Rush
Zannichellia palustris	Horned Pondweed	*Juncus mexicanus*	
		Juncus tenuis	Slender Rush
GRASSES		*Juncus torreyi*	Torrey Rush
Alopecurus aequalis	Meadow Foxtail		
Glyceria borealis	Northern Mannagrass	**FORBS**	
Glyceria grandis	American Mannagrass	*Alisma subcordatum*	Water Plantain
Glyceria striata	Fowl Mannagrass	*Alisma trivale*	Mud Plantain
Muhlenbergia asperifolia	Scratchgrass Muhly	*Anemopsis californica*	Yerba Mansa
		Berula erecta	Water Parsnip
Phragmites communis	Common Reedgrass	*Caltha leptosepala*	Elk's Lip
Polypogon monspeliensis	Rabbitfoot Grass	*Elatine brachysperma*	Waterwort
		Fagopyrum sagittatum	Buckwheat
Spartina pectinata	Prairie Cordgrass	*Flaveria campestris*	
		Flaveria chloraefolia	
SEDGES		*Heteranthera limosa*	Mud Plantain
Bulbostylis capillaris	Threadleaf Sedge	*Limonium limbatum*	Marsh Rosemary
Carex aquatilis	Water Sedge	*Lythrum californicum*	Hierba del Cancer
Carex hystericina	Porcupine Caricwolly Sedge		
Carex lanuginosa			
Carex scoparia	Broom Sedge		

Table 10.2.   (continued).

Scientific Name	Common Name	Scientific Name	Common Name
Mimulus glabratus		*Ranunculus macounii*	
Mimulus guttatus	Spotted Monkeyflower	*Rorippa nasturtium-aquaticum*	Watercress
Myosurus aristatus	Mousetail		
Myosurus minimus		*Rorippa obtusa*	Fieldcress
Polygonum amphibium	Floating Knotweed	*Rorippa sinuata*	
Polygonum bicorne	Pink Smartweed	*Sagittaria cuneata*	Wapato
Polygonum convolvulus	Black Bindweed	*Sagittaria latifolia*	Duck Potato
Polygonum lapathifolium	Willow Smartweed	*Sagittaria montevidensis*	Arrowhead
Polygonum pensylvanicum	Pinkweed	*Samolus cuneatus*	Water Pimpernel
		Veronica americana	American Brooklime
Polygonum persicaria	Lady's Thumb	*Veronica anagalis-aquatica*	Water Speedwell
Polygonum punctatum	Water Smartweed		
Ranunculus aquatilis var. *capillaceus*		*Veronica nephrophylla*	
Ranunculus cardiophyllus		*Veronica peregrina* var. *xalapensis*	Mexican Speedwell
Ranunculus cymbalaria		*Veronica serphylli-folia*	Thymeleaved Speedwell
		Veronica wormskjoldii	
Ranunculus hydrocharoides		*Veronica nephrophylla*	
Ranunculus inamoenus		*Viola nephrophylla*	Northern Bog Violet

Table 10.3.   Major Plants Comprising Vegetation of the Grants and Carrizozo Malpaises in New Mexico.
* = not found on areas bordering malpais.

TREES
  *Juniperus deppeana**
  Juniperus monosperma
  *Juniperus scopulorum**
  *Pinus edulis**
  *Pinus ponderosa**
  *Populus tremuloides**
  *Pseudotsuga menziesii**

SHRUBS
  *Berberis repens**
  *Berberis haematocarpa**
  Brickellia californica
  *Cercocarpus montanus**
  Fallugia paradoxa
  Forestiera neomexicana
  Opuntia arborescens
  *Physocarpus monogynus**
  *Quercus gambelii**
  Quercus grisea
  *Rhus glabra**
  Rhus trilobata
  *Ribes cereum**
  Ribes inebrians

FORBS
  *Agastache neomexicana**
  Artemisia carrythii
  *Arabis fendleri**
  *Castilleja integra**
  Chrysopsis hispida
  Erigeron divergens
  *Geranium caespitosum**
  *Ipomopsis aggregatum**
  Lappula redowskii
  *Lithospermum multiflorum**
  Lotus wrightii
  Marrubium vulgare
  *Mirabilis oxybaphoides**
  *Pericome caudata**
  *Salvia henryi**
  *Salvia vinacea**
  *Silene liciniata**
  *Sisymbrium linearifolium**
  *Swertia radiata**
  *Thalictrum fendleri**
  *Thelypodium wrightii**
  Verbena wrightii

GRASSES
  Bouteloua curtipendula
  Bouteloua gracilis
  *Festuca arizonica**

Table 10.4. Major Plants Comprising Early Successional Vegetation of Exposed Rock Sites in New Mexico.

SEEDLESS PLANTS

LICHENS AND ALGAE

Acarospora evoluta
Acarospora schleicheri
Acarospora texana
Anaptychia mexicana
Aspicilia sp.
Buellia retrovertens
Caloplaca bracteata
Candelina submexicana
Cladonia fimbriata
Dermatocarpon miniatum
Gasparrinia elegans var. *brachyloba*
Parmelia conspersa
Parmelia dendritica
Parmelia neoconspersa
Peltigera canina
Physcia pulverulenta
Protococcus viridis
Ramalina pollinaria
Xanthoparmeria mexicana
Xanthoparmeria psoromifera

MOSSES AND LIVERWORTS

Amblystegiella sprucei
Bryum capillare
Ceratodon purpureus
Cystopteris fragilis
Dicranum rhabdocarpum
Drepanocladus uncinatus
Grimmia arizonae
Homomallium mexicanum
Hypuum revolutum
Leskea tectorum
Marchantia polymorpha
Orthotrichum anomalum
Paraleucobryum enerve
Plagiochila asplenioides
Reboulia hemisphaerica
Selaginella rupincola
Tortula burtramii

FERNS

Asplenium trichomanes
Aythrium filix-femina
Bommeria hispida
Cheilanthes feei
Cheilanthes lindheimeri
Cheilanthes villosa
Notholaena sinuata
Notholaena standleya
Pellaea longimucronata
Pteridium aquilinum
Woodsia plumerae

Table 10.5. Major Plants of Gypsum or Strongly Gypseous Soils.

Scientific Name	Common Name
HALFSHRUBS	
Coldenia hispidissima	
Pseudoclappia arenaria	
GRASSES	
Enneapogon desvauxii	Spiked Pappusgrass
Sporobolus nealleyi	Nealley Dropseed
FORBS	
Anulocaulis gypsogenus	Gyp Ringstem
Anulocaulis leiosolenus	Waterfall
Frankenia jamesii	Frankenia
Mentzelia perennis	
Mentzelia pumila var. procera	
Nama carnosum	
Nerisyrenia linearifolia	
Oenothera hartwegii	
Oenothera pallida	Pale Evening Primrose
Phacelia integrifolia	
Sartwellia flaveriae	Gyspumweed
Selinocarpus lanceolata	Gyp Moonpod

CHAPTER 10 REFERENCES

VEGETATION OF SPECIAL HABITATS

Baad, M. and J. P. Hubbard, 1964. List of plants collected in the Mogollon Mountains, New Mexico. Thesis, University of Michigan, Ann Arbor.

Bailey, V., 1913. Life zones and crop zones of New Mexico. North American Fauna, 35: 1–100.

Bridges, C. D., 1976. San Simon Cienega wildlife habitat development analysis record. U.S. Department of Interior, BLM, Las Cruces, New Mexico.

Campbell, A. D., 1973. Vegetation type report of the Bitter Lake Migratory Waterfowl Refuge, USFWS, Chaves County, New Mexico.

Campbell, R. S. and I. F. Campbell, 1938. Vegetation on the gypsum soils of the Jornada Plain, New Mexico. Ecology, 19: 572–577.

Cole, R. A., D. L. Weigmann, and M. C. Hayes, 1984. Limnology of a shallow, brackish, hypereutrophic reservoir in southern New Mexico. Agriculture Experiment Station Bulletin 709, New Mexico State University, Las Cruces.

Correll, D. S. and H. B. Correll, 1972. Aquatic and wetland plants of southwestern United States. Environmental Protection Agency, Washington, D.C., 1777 pp.

Cully, A. and P. J. Knight, 1987. A handbook of vegetation maps of New Mexico counties. New Mexico Department of Natural Resources, Santa Fe. 135 pp.

Cully, J. F., Jr., 1983. Tucumcari Lake wildlife

inventory. Castetter Laboratory Technical Series No. 97, University of New Mexico, Albuquerque.

Dice, L. R., 1940. The Tularosa malpais. Scientific Monthly, 50: 419–424.

Egbert, J., 1987. Trip report on plans of Centerfire Bog. Heritage Office, New Mexico Department of Natural Resources, Santa Fe.

Flowers, S., 1961. Vegetation of the Navajo Reservoir Basin in Colorado and New Mexico. pp. 15–87. In: D. M. Pendergast and C. C. Stout (eds.). Ecological studies of the flora and fauna of Navajo Reservoir Basin, Colorado and New Mexico. Anthropological Papers, No. 55, Department of Anthropology, University of Utah, Salt Lake City.

Fosberg, F. R., 1940. The aestival flora of the Mesilla Valley region, New Mexico. American Midland Naturalist, 23: 573–593.

Goerndt, D. L., S. D. Schemnitz, and W. D. Zeedyk, 1985. Managing common watercress and spring/seeps for Merriam's turkey in New Mexico. Wildlife Society Bulletin, 13: 297–301.

Hayward, B., T. Heiner, and R. Miller, 1978. Inventory of the Alamo Hueco–Big Hatchet–Sierra Rica Mountain complex. U.S. Department of Interior, Bureau of Land Management, Las Cruces, New Mexico.

Hendrickson, D. A. and W. L. Minckley, 1984. Cienegas—vanishing climax communities of the American southwest. Desert Plants, 6: 130–175.

Hubbard, J. P., 1971. The summer birds of the Gila Valley, New Mexico. Nemouria Number 2. 35 pp.

————, 1988. A hierarchical code-system for the water bodies of New Mexico. Unpublished Ms. New Mexico Department of Fish and Game, Santa Fe.

Huey, W. S. and J. R. Travis, 1961. Burford Lake, New Mexico, revisited. Auk, 78: 607–626.

Limerick, S., 1984. Gypsum wild buckwheat (*Eriogonum gypsophilum*) recovery plan. U.S. Department of Interior, Fish and Wildlife Service, Albuquerque, New Mexico. 34 pp.

Lindsey, A. A., 1951. Vegetation and habitats in a southwestern volcanic area. Ecological Monographs, 21: 227–253.

Martin, W., R. Fletcher, and P. Knight, 1981. An analysis of the flora of the Canadian River Canyon–Mills Canyon section. Report for Range Management Division, U.S. Forest Service, Region 3, Albuquerque, New Mexico.

Meyer, S. E., 1986. The ecology of gypsophile endemism in the eastern Mojave Desert. Ecology, 67: 1303–1313.

Osborn, N. L., 1966. A comparative floristic study of Mount Taylor and Redondo Peak, New Mexico. Ph.D. Dissertation, University of New Mexico, Albuquerque.

Parsons, R. F., 1976. Gypsophyly in plants—a review. American Midland Naturalist, 96: 1–19.

Peterson, R. S., 1988. Aquatic plants of the Bitter Lake National Wildlife Refuge. Personal communication.

Peterson, R. S. and E. Rasmussen, 1986. Research Natural Areas in New Mexico. U.S. Department of Agriculture, Rocky Mountain Forest and Range Experiment Station, Albuquerque, New Mexico.

Potter, L. D., 1975. Mescalero Sands Natural Landmark evaluation. Great Plains Natural Region Ecological Site, New Mexico. Report of U.S. Department of Interior, National Park Service, Washington, D.C. 31 pp.

Powell, A. M. and B. L. Turner, 1977. Aspects of the plant biology of the gypsum outcrops of the Chihuahuan Desert. pp. 315–325. In: R. H. Wauer and D. H. Riskind (eds.). Transactions of the symposium of the biological resources of the Chihuahuan Desert region, United States and Mexico. U.S. Department of the Interior, National Park Service Transactions and Proceedings Series No. 3, U.S. Government Printing Office, Washington, D.C.

Riffle, N. L., 1959. The flora of Mount Sedgwick and vicinity. M.S. Thesis, University of New Mexico, Albuquerque.

Shields, L. M., 1956. Zonation of vegetation within the Tularosa Basin, New Mexico. The Southwestern Naturalist, 1: 49–68.

Shields, L. M. and J. Crispin, 1956. Vascular vegetation of a recent volcanic area in New Mexico. Ecology, 37: 341–351.

Smith, G. M., 1950. Fresh-water algae of the United States. McGraw-Hill Book Company, Inc., New York. 719 pp.

Spellenberg, R., 1979. A report on the survey for threatened, endangered, or rare plant species on the Grants Malpais, Valencia County, New Mexico, with general comments on the vegetation. For: Heritage Section, New Mexico Department of Natural Resources, Santa Fe.

Stahlecker, D. W., 1987. Summer birds of Stinking (Buford) Lake, a third visit. Eagle Ecological Services, Albuquerque, New Mexico.

Turner, P. R., 1987. Ecology and management needs of the White Sands pupfish in the Tularosa Basin of New Mexico. Final report on Contract No. DAAD 07-84-M-2242, Environmental Division, Wildlife Branch, U.S. Department of the Army, White Sands Missile Range, New Mexico.

Waterfall, U. T., 1946. Observations on the desert gypsum flora of southwest Texas and adjacent New Mexico. American Midland Naturalist, 36: 456–466.

Wetmore, A., 1920. Observations on the habits of birds at Lake Burford, New Mexico. Auk, 37: 221–412.

11

SPECIES OF SPECIAL CONCERN

RICHARD SPELLENBERG

Previous chapters have indicated that certain plant species often occur together in response to variables of geography, climate, and geology. The interaction of the genotype of a species and its environment ultimately limits the geographical boundaries of any one species. The environment of a species has physical (climate, geology, and so forth) and biological (competition, pollinator, human, and so forth) components. Over time, the environment of a plant species often changes in extent, expanding or shrinking as geological and climatic forces shape the biological world. These forces still operate on plant ranges, but in the past century a new and powerful biological component of the environment—the human component—has made severe inroads on the ranges and populations of many species.

A species becomes rarer as its habitat shrinks. In other instances, a peculiar set of environmental circumstances may have combined to produce a habitat that is, and always has been, very spatially limited. A plant residing in a highly restricted area may be in danger of extinction simply because its small population cannot withstand chance events, many of which are the result of the strong human component of the environment at the present time. Sometimes built without the consideration of either the plant or animal life in the area, a dam produces a reservoir that completely inundates the ranges of rare and local plant species. In other cases, water may be diverted, ranges covered with subdivisions, and habitats may be altered for human food production. When such man-caused chance events occur, competition among species for resources and living room also may follow. Worldwide, the human impact is now so severe that it threatens broadly distributed species with large numbers of individuals. Human encroachment no longer threatens only isolated species but the ecosystems that support them as well.

THE NATURE OF THE PROBLEM

In the context of this chapter, *species of special concern* refers to those plant species whose continued existence on Earth is imperiled or appears to be easily jeopardized either by natural events or by human acts, or a combination of both. Most such species are now rare, and many have always been so. In developed countries at least, most people are aware of the increasing attention given to rare species. At the same time, however, few individuals are informed adequately about the accelerating extinction of plant and animal species and the rapidly diminishing diversity of the world's biota.

Population growth for any species generally follows a J-shaped curve. When available resources limit the increase of a population, the carrying capacity of the environment has been reached and population growth is forced to cease. The population may even collapse in a nature-mediated decline. Mankind has now rounded the curve of population growth and is progressing up the stem of the "J." Human population growth, the use of resources, and the production of waste materials is overwhelming the world environment. If population growth is left unchecked, it will threaten the environmental life-support systems that presently sustain humankind (Miller, 1985, p. 461).

The problem is now most critical in the tropical underdeveloped countries where burgeoning human populations and perhaps some rise in standard of living have combined to remove tropical forests at the rate of 50 acres per minute in the period between 1964 and 1972 (Whitmore, 1980), a rate that probably continues or accelerates. This translates into the loss of $2/3$ of the tropical forests by the year 2000 and a loss of 10 to 20% of the species that now inhabit the earth (Lovejoy, 1980). In absolute numbers, hundreds of thousands or even a million species are predicted to disappear by the end of this century (Lovejoy, 1986), perhaps even 750,000 in Latin America alone (Ehrlich, 1980), most of which will have been unknown to science. Although abnormal, massive extinctions have occurred and mark the end of the Ordovician, Permian, Triassic, and Cretaceous periods, at these geologically abrupt transitions the extinct were replaced by the newly evolving. Today's rates of extinction, 40 to 400 times the rates during most of the past, are unprecedented, given the brief geological time in which they are now occurring. These massive extinctions are accompanied by man-induced homogenization of the environment. New species are not replacing old, and the loss of species and biotic diversity is permanently altering the course of evolution. Lovejoy (1979) has suggested that the unprecedented rate of loss of diversity is so extreme that it marks the end of the Recent Epoch of the geological time scale and signals the beginning of the Epoch of Biotic Impoverishment.

The impact of human numbers and wealth may seem less severe in the developed temperate countries where there simply is neither the tremendous biological diversity to begin with nor the impoverished human millions eking a sub-sistence life by felling forests and establishing short-lived farms. In the developed countries societal controls are more effective. Alteration of the environment is made in increments that appear to be less threatening, but when taken as an aggregate, considerable change has occurred nonetheless in the biota of these countries. Only 200 to 400 years have passed since the initial settlement of the Western Hemisphere by European immigrants, and in reality, most of the environmental change has occurred within the last century. Many vegetation types that were once widespread are now fragmented or completely eradicated. In the United States alone, of the estimated 22,200 species of flowering plants that are native to the country, 90 (0.4%) are now extinct, 850 (3.8%) might be endangered by extinction, and 1,200 (5.4%) may be threatened (Ayensu and DeFilipps, 1975). Although the loss of a single species may seem insignificant, threatened and endangered species as a group indicate a stressed environment (Lovejoy, 1979), and herald a number of incrementally developing ecological problems that are likely to pose a threat to the continued comfortable existence of humankind (Lovejoy, 1986).

Once pressures resulting in the decline of a species develop, the process of population extinctions is as a series of continuous stages proceeding from rare, to threatened, to endangered, and finally to extinction unless reversal is effected. The terms *rare, threatened, endangered,* and even *species* are prone to subjective definitions. They will be reviewed briefly from the perspective of the conservation of endangered species.

According to the International Union for the Conservation of Nature and Natural Resources, a rare species has a total population of less than 20,000 individuals (Koopowitz and Kaye, 1983, p. 8), whereas Reveal (1981) ignores absolute population size and defines rarity on a much more relative basis. In Reveal's opinion, an organism is rare if either its numbers or its area stand at a level demonstrably less than that of the majority of its relatives at any comparable taxonomic rank.

Because agencies responsible for managing rare species usually have limited funds and manpower, the causes of rarity must be clearly understood. An organism may both be rare and not be threated with extinction. A newly emerging species is initially rare but may increase its population with time. Due to the short time that humankind has been monitoring species, together with the accelerating rate at which it has been altering the environment, and consequently the selective pressures, such a rare species would be difficult to identify. More common are constitutively rare species that have evolved in response to peculiar and local selective pressures and that are restricted to, but are persisting well in, a spatially limited, unique habitat. Although these vulnerable species certainly require periodic monitoring, they usually are not in decline.

Once decline in a rare species is detected, or a major threat to a limited habitat is evident, a species then may become threatened or endangered. International definitions of *threatened* and *endangered* are similar to those developed in the United States for the Endangered Species Act of 1973. A species that is in danger of extinction throughout all or a significant portion of its range is "endangered." One that is likely to become an endangered species within the foreseeable future is "threatened" (Reveal, 1981). The definition and application of the term *species* have been argued in the biological literature for decades. In the case of the Endangered Species Act, *species* refers to taxa named at the specific or infraspecific levels. In the case of vertebrates only, individual populations may be considered endangered, and they need not bear a name unique from that of other populations of the species.

Extinction results when, over time, the addition of individuals to a population by any means fails to offset the loss from the population. For plants, addition to a population results from the establishment of individuals from seed or by vegetative means. Loss usually results from mortality, but in some groups of plants, such as cacti, loss may result primarily from persistent removal of adult individuals from a population. The extinction process may be initiated by the introduction of a novel element to the environment or by the development of environmentally stressful conditions. The rate of population decline is affected by a number of factors (Slobodkin, 1986; Terborgh and Winter, 1980; Vermeij, 1986). For plants, paramount among these variables are habitat destruction, small population size in terms of numbers and/or area, introduction of exotic herbivores or competing species, and the direct removal of individuals for human consumption or enjoyment.

WHY SAVE SPECIES?

This question is often asked by individuals who are unaware of the human impact on the environment and particularly of the potential for irreparable damage and a lower quality of life. The question becomes particularly pertinent when the extinction of the 90 or so plant species in the United States appears to have had no negative effect on society, in economic terms at least (Moran, Morgan, and Wiersma, 1986). The argument for the preservation of species usually condenses into four main areas: those of economic impact; aesthetic–recreational potential; ethical obligation; and ecological impact (Miller, 1985; Smithsonian Institution, 1975).

The economic argument is most easily accepted in much of the Western World where the growth mentality is dominant and where most sectors of society resist the use of tax revenue for the protection of organisms of no direct material benefit. Although legislative bodies are rarely swayed with ease in this respect, the economic argument for species preservation is perhaps most easily accepted in highly developed countries with huge, heavily capitalized corporations that "develop" the landscape, but are willing and financially able to mitigate their impact. Understandably, before poverty-stricken developing countries that desire more than a subsistence-level economy can accept the argument, they must provide for their citizens a reasonable life that does not require the immediate exploitation of the most readily attainable resources. According to the economic argument, the maintenance of the diversity of plants has the potential for direct economic benefit. Once lost, a species can no longer contribute to human welfare. If species are conserved, they provide humankind the opportunity to discover new foods, new fibers, fuels, drugs, pest controls, new horticultural plants, new plant species for the reclamation of abandoned land, and so on. The continual evolution of agricultural pests makes imperative the maintenance of a wild, diverse gene pool for the development of resistance in crop plants, especially now that scientists can move genes between very different organisms by methods other than crossbreeding. The study of species may generate information of economic value or may simply increase humanity's body of knowledge, both of which may be of direct or indirect use in the solution of numerous kinds of biological problems. As Lovejoy (1979) points out, one of the most important contributions of biologists today may be the preservation of species so that future biologists may have them simply for their information content.

In the face of economic development, and certainly when confronted with the prospect of poverty and starvation, the aesthetic–recreational argument for species conservation is among the weakest. Nevertheless it is important in contending that diversity in nature, untamed organisms, and wild, unspoiled places are crucial to the well-being of individuals and provide recreational relief from the fast pace of life.

Also difficult to defend in the face of economic interests, the ethical argument originates in a conviction that species have some sort of intrinsic, not only instrumental, value (Callicot, 1986), that species have a fundamental right to exist and that the destruction of a species is ethically and morally wrong. As Callicot discusses, any doubt about the ethical position probably derives from the Judeo-Christian tradition, which on the one hand, gives humanity dominion over the rest of Creation, but which on the other charges Adam to "dress the garden and keep it."

Associated with the first two, the ecological argument is by far the most important but is the least understood by nonbiologists. To be effective, the resource potential of complex and diverse ecosystems must be defined in terms simple enough to be understood as a political issue. Local extinctions serve as barometers for judging the health of biological communities and their potential to support humankind. As an example, the loss of species and genetic diversity in some of our southwestern grassland ecosystems has resulted in forage-productivity losses of as much as 90% (Fletcher, pers. comm.).

Once such arguments are explained and understood, then the public interest demands that governments provide protection for such an important but diffuse natural resource, the complexity of world biota. Societies simply do not realize that most of the energy ultimately used by humanity flows through a poorly understood, giant, complex system of natural communities, not through technology and agriculture (Woodwell, 1977). The complete loss of even one organism, especially if it occupies a high place in the food chain or constitutes a major part of the canopy vegetation, will affect the interactions and selection pressures that control the population levels in other organisms (Vermeij, 1986).

This system incorporates considerable redundancy and overlap, and obviously can sustain (as it has) considerable damage, but the complexity and redundancy are its source of stability. Thriving, healthy ecosystems provide a myriad of free services that help to regulate climate, to maintain quality air and water, to process and dispose of human-generated wastes, to produce and maintain soils and run nutrient cycles, and to contribute to crop pollination and to disease and pest control (Ehrlich, 1980). Within the next few decades, humankind must make the choice between continuing on as usual or changing direction and working effectively to maintain this system. A biologically simpler world will become ever more economically expensive and will become ever more vulnerable to disasters that directly impinge on man. There is a strong potential for the human population to overshoot ecologically the capacity of the world to support comfortably the burgeoning population, giving rise to new societal problems that will lead to new environmental pressures. This may lead to another danger, one where human societies will attempt to solve new problems with deeply entrenched past methods that were only partially effective at lower population levels rather than face squarely the environmental question (Lovejoy, 1986). As Lovejoy writes, the choice at this time is either for a continued rich or an increasingly impoverished existence for humankind (Lovejoy, 1980).

PRESENT ACTION

In the face of an expanding human population, the establishment of nature reserves has provided a staying action against the extinction of some species. Once again, the problem of maintaining species is poorly comprehended by legislators and is only somewhat better understood by biologists. Scientific evidence suggests that the size and number of reserves is inversely correlated with the rate of decay of diversity (Soule and Wilcox, 1980). In addition, studies indicate that the size of a reserve necessary to ensure the perpetuation of diversity is greatly underestimated (Ehrlich, 1980). The smaller the reserve and the fewer reserves there are, the more rapid the loss of species is. Reserves should be established with the largest distribution type in mind, and should be related to density of individuals of a species so that the debilitating effects of inbreeding are reduced. Several large reserves would reduce the potential of loss through disaster and genetic decline, provide a balance of successional stages in the environment, aid in providing a heterogeneous habitat, and maximize diversity (Soule and Wilcox, 1980).

An alternative often suggested is to maintain species in botanical gardens (zoos in the case of animals) or in seed banks, with the intent of someday reintroducing them into a reconstituted wild. Although this is practiced, and is a stopgap against extinction, it falls sorely short as a satisfactory alternative. It is difficult to cultivate more than a few individuals of a species, probably far fewer than is necessary to prevent the loss of alleles, and hence variability, through inbreeding. More genotypes may be preserved in a seed bank, but the bank is vulnerable to vagaries in funding, catastrophes of mechanical failure, and varying levels of commitment of the staff through the years. Most populations show some ecotypic differentiation, and a reintroduction is likely to be in an environment not quite the same as that giving rise to the original differentiation. Many plants are edaphic specialists and fare very poorly, if at all, under cultivation. Finally, the elimination of one species from a community allows competing species to enter its place (spatially, and more broadly, ecologically). Later, the community may be closed to successful reintroduction without intensive management (Slobodkin, 1986).

At the international level, coordination is intensifying to preserve habitat. The International Union for the Conservation of Nature and Natural Resources (IUCN) has published a number of Red Data Books that elucidate the most critically endangered species. They have also published a World Conservation Strategy that is based upon three majors goals: (1) to maintain essential ecological processes and life support systems; (2) to preserve genetic diversity; and (3) to utilize species and ecosystems and species sustainably (Ehrlich and Ehrlich, 1981). Several areas are targeted for immediate attention by the Strategy: (1) in agricultural systems, to halt erosion, desertification, and the loss of genetic resources; (2) to slow or stop destruction of forests, especially where watersheds are affected; (3) to protect oceans, estuaries, and nearshore wetlands; and (4) to give priority to protection of endangered species, especially those that are very different genetically, that are culturally or economically important, or that live in a species-rich area where the intended preservation of one or a few species also includes many others and therefore protects diversity. Critical geographic areas are being identified and, where possible, are being incorporated into parks or reserves. Where governments have not been able to act for a variety of reasons, the International Nature Conservancy attempts, when possible, to purchase tracts of land for preservation.

In the United States a mass movement for environmental conservation has developed from a tradition of wildlife conservation continuous from at least the turn of the century. At the federal level it culminated in two very important pieces of legislation in 1970 and 1973—the National Environmental Policy Act (NEPA) and the Endangered Species Act, respectively. These acts, along with other environmentally oriented legislation, have strongly influenced projects involving federal funding, and have proven to be powerful forces encouraging environmental protection.

The Endangered Species Act is a tough, effective, uncompromising law, and was the first to recognize the necessity of protecting habitat that supports a species. It states that in the case of terrestrial species, federal agencies would consult with the Secretary of the Interior to ensure that actions taken would not diminish the continued potential for survival of a listed threatened or endangered species. The law also prohibited killing, capturing, importing, exporting, or selling any endangered or threatened species. The U.S. Fish and Wildlife Service was given responsibility of implementing the law. Many environmentally destructive activities have federal involvement, and the law was soon learned to have power, probably much more than envisioned by the legislators who, through its implementation, perhaps learned just how serious environmental problems had become. However, as uncompromising as the law seemed, thousands of consultations between interested parties and the U.S. Fish and Wildlife Service proceeded without difficulty. Yet the law was shown to be so powerful through a few highly publicized cases that it was amended in 1978 to make it much more difficult to list a species as threatened or endangered. Simultaneously, an "Endangered Species Committee" was created, composed of high ranking officials in the U.S. government who could mediate "irresolvable conflicts" and grant exemption to the Act when benefits of a project clearly outweighed those of alternative actions, even when an exemption could cause the extinction of a species (Ehrlich and Ehrlich, 1981).

As elsewhere, in New Mexico the Endangered Species Act provided managers of public lands the impetus to begin protecting potentially endangered and threatened plant species that occurred within their jurisdiction, even though legally the species would not require such protection until listed as "threatened" or "endangered." At the time in the state, plant identification and knowledge of distribution of rare species lay primarily with specialists. Knowledgeable members of the Bureau of Land Management, the National Park Service, the Soil Conservation Service, the U.S. Forest Service, the N.M. Department of Natural Resources, and the universities formed a loosely knit body that assembled and supplied information regarding species initially listed by the U.S. government. This body, in so far as possible, also assessed the nature and level of the threats that faced these and other plant species of concern in New Mexico. This information was summarized about ten years after the passage of the Endangered Species Act (NMNPAC, 1984). In conjunction with the U.S. Fish and Wildlife Service, this group continues to provide information and recommendations to persons or agencies who require it for New Mexico plant species.

THE PROCESS OF LISTING A PLANT SPECIES

Plant species may be considered for protection either by individual states, the federal government, or both. The rationale for listing a species may or may not differ between the two levels of government. A state often affords protection to species that are unique to the state or that barely extend into the state from other regions as peripheral populations and are therefore of interest at the local level. States also protect some of the sought after common plants. The Endangered Species Act, however, gives protection to globally threatened or endangered species. States often give added protection to these species, but the process described below is for protecting plant species by their addition to the federal endangered species list. In the sense of the Act, "species" is used broadly and includes subspecies and variety, but excludes various kinds of forms, races, and hybrids, even though they may have been formally named. For animals, *populations* also qualify as "species" under the definition.

The procedure of listing a plant species as endangered or threatened involves a series of steps, numerous individuals, and several levels of government. A species of concern proceeds through three stages, (1) a candidate species, (2) a proposed species, and (3) a listed species. The process takes two or three years except under the most dire circumstances, when emergency listing may be achieved. Information regarding the progress of final stages of the listing process for a species appears in the *Federal Register* and in the *Endangered Species Technical Bulletin*, the latter a publication of the U.S. Fish and Wildlife Service Endangered Species Program.

Listed and candidate plants for threatened or endangered status now number about 4,200 (*Federal Register* 2/21/90). Even though many of these names are not under active consideration at this time, the magnitude of this list illustrates man's impact on his environment in the United States. This list stems from the original Smithsonian Institution list (1975), prepared after polling professional botanists throughout the nation. Updates were published in 1980, 1983, 1985, 1990 (FR 2/21/90). Each update involves review of many taxa by professional botanists and advisory committees of botanically trained people in many states. Concerned citizens may also petition the U.S. Fish and Wildlife Service to add a species to the list, providing the petition contains adequate documentation demonstrating that the species may qualify for listing.

The periodic review of the list results in altered status for many of the candidate species. New ones may be added and the status level of others may be changed. In the list, species names are preceded by numbers that indicate the overall understanding of the level of threat against the taxon. Although the ranking of 1, 2, and 3 seems at first glance to indicate a hierarchy of threat levels, the ranking actually indicates two paths most names take. Most new additions to the list are entered as "category 2," that is, the Service has information indicating that the listing of the species may be appropriate, but insufficient information exists to support the preparation of listing documents.

Because of the large number of species needing investigation, many remain at this status for some time. They are a signal to conservationists that there is a potential for resource management conflicts and that these species need periodic monitoring. As information accumulates, or threats increase, a species may be moved to category 1, a level signifying considerable concern. For a category 1 plant the U.S. Fish and Wildlife Service has substantial information on hand supporting vulnerability of the species and threats to its survival. These species are the most commonly proposed for full legal protection, that is, moved forward in the process of listing as either "threatened" or "endangered."

For other names originally included in category 2, additional information may indicate that a species (used very broadly and including synonyms) does not qualify for listing, and the name is moved to category 3. At this level there are three subcategories: "3A," the taxon is presumed extinct; "3B," on the basis of current understanding, the name listed no longer meets the Act's criteria of "species," and "3C," the taxon has been found to be more widespread and abundant than previously believed, or is not faced with any identifiable threat. The taxon may be quickly moved to category 1 or 2 if new information indicates that category 3 is incorrect. Names are never removed from the candidate list, but are held at category 3 to forestall an endless cycle of attempts to add names thought to have been overlooked when actions regarding those names are long forgotten.

Periodic reviews of the candidate species identify those believed to be in most peril and a status report is prepared for each. This report reviews all relevant information that can be gleaned from herbaria and literature as well as that generated by field work designed to estimate threats, population size, reproductive potential, and other aspects of the biology. These data influence the decision either to proceed further with listing, assign the taxon to category 3, or to make no change until more information is acquired. Once the decision is made by the U.S. Fish and Wildlife Service to proceed with listing, a formal proposal is prepared.

The proposal to list the species is published in the *Federal Register* for public review. The species is now a proposed threatened or endangered taxon, as the case may be. At this point, all background work documenting the proposed listing must be completed, for the Act requires that within the following 12 months the species be either listed as threatened or endangered or it must be withdrawn from consideration. A proposal is withdrawn when new data indicate the taxon does not qualify for listing. Once the proposed rule is published in the *Federal Register,* the public has 60 days to respond in writing, and 45 days to request a public meeting to be held within the geographic region affected by the listing. If comments are minor, the proposal is modified. If they are substantial, the U.S. Fish and Wildlife Service may request a six-month extension to gather further information, often derived from another season of field work. If the comments show that the proposed ruling is incorrect, the proposal is withdrawn.

If the proposed listing is not withdrawn, a final ruling, which becomes part of law, is published in the *Federal Register*. This final ruling again summarizes information about the taxon, the threats that affect it, and the comments that were received during the public review period. After 30 days from the date of publication the species is categorized as either a "threatened" or an "endangered" species, whichever was proposed, and is afforded the protection of the Endangered Species Act.

The Endangered Species Act specifies that critical habitat of a species be designated at the time the species is formally listed as threatened or endangered. Critical habitat is those specific areas within the geographic range of the species where those physical or biological features essential to the conservation of the species are found, and which may require special management considerations or protection. It also may include areas outside the geographical area occupied by the species that are essential for its conservation. For many plants, critical habitat has not been designated. This is allowed by a provision of the Act specifying that critical habitat need not be designated if such designation would increase threats to the species or would not otherwise be beneficial.

Because plants are sedentary, the designation of critical habitat in many cases is superfluous, for once the species is listed, protecting the pop-

ulations where they grow also protects their habitat. Legal designation of critical habitat requires the public disclosure of precise locations of populations and maps of critical habitat areas. Public knowledge of the precise locations of rare species, especially once they are indicated to be of special importance, often works against conservation. Cacti and orchids may quickly be removed by unscrupulous collectors, vandals may visit and destroy the population, and disgruntled users of the land may simply eradicate the population. However, the significance of the taxon is made clear to landowners and land managers when known populations of threatened or endangered species occur within their areas of interest. On public lands, populations of the listed taxon are managed according to the stipulations of the law. On private land, cooperation is sought for management that will help to preserve the species.

FEATURES OF NEW MEXICO LEADING TO ENDEMISM

New Mexico is essentially a large rectangle of 121,666 sq. mi. (315,114 km^2) situated at the southern end of the Rocky Mountains, with intervening ranges connecting loosely to the Sierra Madre Occidental of Mexico. As such, it is a mountainous state with an average elevation of 5,700 ft. (1,740 m), with an elevational range from 2,867 ft. (677 m) in the extreme southeastern corner to 13,160 ft. (4,011 m) on Mt. Wheeler in the north-central region. The size of the state and its topographical diversity provide numerous environments that are home to about 3,900 vascular plant taxa (Martin and Hutchins, 1980, 1981; Spellenberg et al., 1986). The composition of the flora is complex and derives from several floristic regions. In the eastern portion of the state, the floristic influence is primarily from the Great Plains. In the south, at lower elevations, the Chihuahuan Desert has its northernmost extension. In the north, and at higher elevations in the south, the taxa that make up the vegetation are mostly Rocky Mountain in origin. At the lower elevations in the northwestern corner, the vegetation is related to that of the Great Basin, and at middle elevations in the southwestern corner the vegetation is basically Sierra Madrean. These vegetation types contribute to the species richness of the state. However, New Mexico has no floristic association that is uniquely its own and few extensive areas with very strong selective pressures markedly different from those of surrounding regions. Thus, there is a rather low level of endemism.

ENDEMISM IN NEW MEXICO

The threat of extinction to a species is often aggravated by its narrow geographic distribution, and a tally of endemic species in New Mexico quickly illustrates portions of the state where attention is most needed (Table 11.1, Fig. 11.1). Even though the state is large and the vegetation complex, less than 150 taxa are endemic to New Mexico (or nearly endemic) (NMNPPAC, 1984), which is less than 4% of the total flora. Endemic taxa can be found in most parts of the state but, as would be expected by isolation of mountain ranges in certain areas and the presence of strongly selective soils in others, these taxa are not evenly distributed throughout. Even though the method of tabulation for Table 11.1 emphasizes county size and obscures regional endemism (county boundaries do not follow ecological gradients), it is apparent that some portions of the state have greater endemism than others.

The least endemism occurs on High Plains in the climatically and topographically less diverse northeastern counties. Most of the endemics listed for this region occur in the western portion where the land is broken and rises as the front ranges

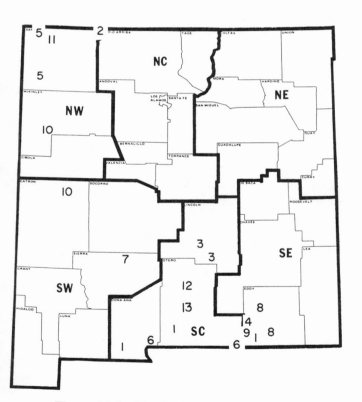

1—*Echinocereus Lloydii* (E) 10/16/79
2—*Pediocactus knowltonii* (E) 10/19/79
3—*Echinocereus fendleri* var. *kuenzleri* (E) 10/26/79
4—*Coryphantha sneedii* var. *Leei* (T) 10/26/79
5—*Sclerocactus mesa-verdae* (T) 10/30/79
6—*Coryphantha sneedii* var. *sneedii* (E) 11/7/79
7—*Hedeoma todsenii* (E) 1/9/81
8—*Eriogonum gypsophilum* (T) 1/19/81
9—*Hedeoma apiculatum* (T) 7/13/82
10—*Erigeron rhizomatous* (T) 4/24/84
11—*Astragalus humillimus* (E) 6/27/85
12—*Cirsium vinaceum* (T) 6/16/87
13—*Argemone pleiacantha* subsp. *pinnatisecta* (E) 8/24/89

Figure 11.1 Distribution of federal threatened (T) and endangered (E) plant species in New Mexico, including dates listed. Sectors of the state are northwest (NW), north-central (NC), northeast (NE), southwest (SW), south-central (SC), and southeast (SE).

of the southern Rocky Mountains. Endemism in the mountainous north-central portion of the state, at the southern tip of the Rocky Mountains, is considerably higher. Most of these endemics are middle elevation plants, isolated from other such taxa by broad valleys and plains. The higher elevation taxa are not endemics for the most part, but are the same as species to the north. It is noteworthy that no federally threatened or endangered taxa occur in these two regions of the state (Table 11.1).

Endemic taxa in the northwestern part of the state occur primarily on the strongly selective saline clay soils of the Mancos and the Fruitland formations. Edaphic endemics, species restricted by local selective pressures of peculiar soils, also form a large component of the endemics occurring in the southeastern corner of New Mexico, but there they occur on gypsum. The Guadalupe Mountains harbor most of the remainder in the southeastern corner: their limestone substrate, their canyons isolated from those of other mountains, and their elevation provide a unique habitat island that encourages endemism.

It is readily apparent from Table 11.1, however, that the south-central and southwestern parts of the state have the most endemics. As divided, the south-central portion encompasses a portion of the Guadalupe Mountains, the geologically different and isolated White and Sacramento mountains, the low Tularosa Basin, and the isolated, dissected Organ Mountains. Endemics occur in response to selective pressures of limestone versus igneous rock and the isolation of microhabitats provided by the tremendous topographical diversity. In southwestern New Mexico, most endemism is due to the topographical diversity of the isolated Mogollon Mountains and the isolation provided by lesser mountain ranges to the south and east, especially the Black Range and the Animas, Big Hatchet, Caballo, and Mimbres mountains.

The species endemic in the state are representative of the composition of the state's flora. The conspicuous vegetation of New Mexico is often large woody perennials, but most of the species are perennial forbs. Of those species reviewed in a recent publication (NMNPPAC, 1984) 94 are perennial forbs (excluding cacti), 24 are cacti, 10 are annuals, 9 are woody shrubs, 3 are biennials, and 2 (*Brickellia chenopodina* and *Helianthus praetermissus*) are known only from fragmentary material collected more than a century ago that do not allow the determination of habit.

A misconception may exist among the public that, in the Southwest, most plant species of concern are highly desirable cacti or, after the Furbish Lousewort controversy (Scrophulariaceae) in the northeastern states, that threatened and endangered species of plants are inconspicuous and of little intrinsic interest. In New Mexico 32 plant families contribute species that are of concern to land managers and conservationists (Table 11.2). Most of those species are in the sunflower family (Asteraceae), reflecting the size of this family, largest in New Mexico and, worldwide, second to the Orchidaceae. The pea family (Fabaceae) contributes the next greatest number of species primarily because of the contribution of the huge western genus *Astragalus*, a genus that tends to speciate on a very local basis. The third family with a large number of species of special concern is the Cactaceae, the cactus family, the species mostly threatened by cactus collectors. The snapdragon family (Scrophulariaceae) has some showy species that are favorites of *Penstemon* fanciers, but the threat to this family in New Mexico comes from mostly narrow endemism combined, in some species, with being green in the spring and therefore providing forage during this characteristically dry time of year. These four families contribute more than one-half of the species of special concern and illustrate the varying causes of vulnerability to extinction; the remaining families each make a small contribution to the total number.

TAXONOMY AND PLANT SPECIES IN NEW MEXICO

As a large, sparsely populated state, the knowledge of the flora of the state at the time of passage of the Endangered Species Act was, and remains, very uneven. Many areas are accessible only by foot or by horseback. Numerous entities were known from only one or two sites, from very few collections, and were poorly integrated into modern taxonomic classification. In the last decade much has been learned about New Mexico plants through the work of a number of individuals. Exploration for plants, and studies of critical populations, were often part of surveys associated with environmental impact statements for various projects, as mandated by the National Environmental Protection Act. Even with this extensive work, a number of entities that have been named are of questionable taxonomic significance (Table 11.2). Twenty-four of these, 16% of the 153 in Table 11.3, appear in the listing of rare plants for New Mexico (NMNPPAC, 1984). Although this figure may seem large, it is down considerably from the proportion of taxonomically questionable entities that made up the original draft lists of sensitive species in 1975. This reduction is due to the field work of numerous individuals and the interaction of the New Mexico Native Plant Protection Advisory Board with field workers and professional botanists throughout the nation.

The spirit of the Endangered Species Act was to protect genetically unique entities. For plants, however, if the population is not named at least at the varietal level, which then constitutes a "species" as defined by the Act, no protection will be afforded. Credibility is lost if names published under an earlier typological philosophy, but that are now understood not to signify a biologically significant taxon in nature, are given strong, federal, legal protection, thereby increasing the potential for the weakening of the Act by justifiably frustrated legislators. The biological significance of a name may be fraught with judgments based on personal preferences, but such determinations are important and must be made with care. Even then, the result is often highly arguable.

Three species illustrate the occasional difficulty in determining that a name applies to a valid taxon, or that a taxon is in need of listing as threatened or endangered. Setting aside for a moment the issue of validity of the taxon, it is not certain that these entities persist in the wild any longer, and taxonomic interpretation is involved with assessing the need for extensive research regarding them.

The Gila Brickelbush, *Brickellia chenopodina*, is known from a single collection made in 1903 at "the crossing of the Gila." Repeated searches along the Gila River have failed to find any plant matching the old collection. Over the past decades, extensive flooding has scoured the river, grazing has changed the nature of the riparian vegetation, and flood control projects have altered the margin of the river. Little, if any, original habitat remains. However, the plant may have been a waif in the first place, possibly established briefly from upstream seed sources hidden along some small tributary of the Gila River. Perhaps the real home of the species is yet to be found. Another plausible explanation of the mystery surrounding *B. chenopodina,* other than the extinction of a valid taxon, involves a relative. All along the Gila River grows the widespread and weedy *B. floribunda.* Leaves and inflorescence structure of these two species in this taxonomically difficult genus are different, but the immediate environment can easily alter these features. Less easily modified features, flower-head structure and kind of the pubescence, are similar. Early collectors often sought and named peculiar forms of common species as new, distinct species, an archaic typological philosophy divorced from evolutionary principles and the concepts of population biology. *Brickellia chenopodina* may simply represent an odd *B. floribunda,* perhaps a shade form occurring in newly flood-deposited silt beneath the cottonwoods that once formed a more extensive, and ungrazed, gallery along the river. The problem cannot be resolved with information at hand.

The problem of taxonomic validity is illustrated further by Woolly Buckwheat (*Eriogonum densum*), a distinct entity named as a species from a single collection of perhaps two plants made in the 1880s near Silver City. Another collection made in 1903 included only about a dozen plants. Though intensely sought, the species has not been seen since, and remains known from only 15 or fewer individuals. Some references (e.g., Martin and Hutchins, 1980) regard *Eriogonum densum* to be a "good" species. A number of botanists have noted its similarity to the common *E. polycladon,* and Spellenberg et al. (1988) argue that it is only a rare phase of *E. polycladon* and is not worthy of taxonomic recognition. Once again, the problem cannot be solved unequivocally until *E. densum* can be rediscovered, its range of variation studied, and its association with *E. polycladon* determined.

The final example is based upon a more re-cently described species, but once again it has proven impossible to determine the nature of the organism to which the name applies. Porter's Globemallow, *Sphaeralcea procera,* was collected and named in 1943 from perhaps only a single plant growing northeast of Deming, its location reported precisely. Searches of the exact arroyo where it grew, and nearby ones, have failed to find it. There is no particularly unique habitat anywhere in the area where it might be growing (Knight, 1985), from which the plant was established as a waif. In this case, the expert in *Sphaeralcea* has considered the application of the name *S. procera* to be unresolvable, and has named a well-documented previously unrecognized distinct species common in the region as *S. polychroma* (LaDuke, 1985). *Sphaeralcea procera* may have been an aberrant phase of this new species, but now the name *S. procera* is relegated to the musty closets of taxonomic ambiguity. In all three examples, sufficient consideration has been given the biological significance of the names that none is now under active consideration by the federal government for threatened or endangered status.

THE NATURE OF THREATS TO SPECIES OF CONCERN

The list of species of special concern (Table 11.3) represents 153 entities identified in 1984 (NMNPPAC, 1984), with later additions. Most of the entries on the list have very narrow geographic ranges and occur in only one or a few counties. Thus, in any area where human activity is involved, rarity and narrow endemism, in and of themselves, are important reasons for monitoring plant species, for small populations can quickly to obliterated by single, intensive projects, or less quickly but deceptively fast by environmental change that occurs from more subtle influences controlled by humans.

In New Mexico, 132 taxa (86% of the 153 listed in Table 11.3) have geographic ranges whose total extent is estimated at roughly 5% or less of the area of New Mexico. This includes intervening areas not occupied by the species, not simply habitat occupied, a much smaller figure. Fifty-nine species (39%) may be considered very narrow endemics, the entire geographic range of the species being less than 1% (usually much less) of the total area of the state. Although all narrow endemics probably are limited by rather precise adaptation to a restricted environment, the primary feature of the environment affecting the restriction can be identified in only 51 (33% of the 153) of the species whose geographic range occupies less than 5% of the area of New Mexico (Table 11.4). Many others are limited to canyons and mountain tops that provide a microhabitat different from the general environment, or are in a microhabitat well isolated from similar ones by intervening expanses inhospitable to the genotype in question. The most easily identified selective features of the environment that most often result in narrow endemism in New Mexico are: (1) edaphic features, that is, soils such as clay and gypsum outcrops; (2) cliff faces, the plants on them called chasmophiles; (3) the presence of free standing or flowing water, a limited and often developed resource in the state; and (4) old growth coniferous forest on isolated mountain ranges.

Only 80 (52%) species in the state seem to be declining in numbers because of human-induced impacts (Table 11.3). These impacts may be grouped into ten categories, and a few New Mexico species are apparently threatened from sources in up to six of these categories (Table 11.3). These threats, in turn, can be ordered into two general *non-exclusive groups,* one of highly focused threats, the other of more diffuse threats. Included in the former category are loss of habitat due to urban development, tilled agriculture, mineral extraction, recreational use of land, road right-of-way maintenance and development, encroachment by exotics, and the direct predation upon populations by collectors. They are easily identified, and often they are the most easily controlled. Habitat alteration through grazing, logging, or water development may be viewed as more diffuse threats, although all may have very direct components.

In descending order, the number of times each threat listed in Table 11.3 affects a species is as follows. Road development and management impacts at least 44 species, twice as many as the next most common threat, that of plant collectors, who cause serious depredations of many rare cacti populations. Grazing by domestic livestock or wildlife, and recreational use of land, each affect at least 19 species. Water development and management influences the populations of about 11 species, and urban development and logging each affect 7 or 8. Various mining activities affect 5 species, and encroachment by exotics may be important in assessing the status of 3 or 4 species. Tilled agriculture also affects only about 3.

This order, however, may misrepresent the importance of threats because it is oriented toward individual species. The primary menace to all natural communities in New Mexico and the species that occupy them is the alteration of habitat that results from the combined action of many of these man-induced forces. The effects of the diffuse threats are often subtle, slow to develop, difficult to document, often very widespread, and have a broad economic impact. Consequently, for those charged with identifying and protecting they are commonly the most difficult to manage for any one species.

EXAMPLES OF SPECIES OF SPECIAL CONCERN IN NEW MEXICO

Even though New Mexico is mostly a rural state with a sparse population, there are now 13 species of plants whose continued existence is so precarious that they have been listed federally as threatened or endangered (USFWS, 1987). These "endangered" plants are four cacti, *Coryphantha* *sneedii* var. *sneedii, Echinocereus fendleri* var. *kuenzleri, E. lloydii, Pediocactus knowltonii,* a locoweed relative, *Astragalus humillimus,* a prickly-poppy, *Argemone pleiacantha* subsp. *pinnatisecta,* and the recently discovered mint *Hedeoma todsenii.* The six threatened species are two cacti, *Coryphantha sneedii* var. *leei* and *Sclerocactus mesa-verdae,* two members of the sunflower family, *Cirsium vinaceum* and *Erigeron rhizomatous,* a mint, *Hedeoma apiculatum,* and a wild buckwheat, *Eriogonum gypsophilum.* These listed species are included in Table 11.3, and their general locations are plotted in Figure 11.1

Figure 11.1 illustrates that many of the listed species occur in the northwestern corner of the state, where they occur on clays and sandstone peculiar to the region. Many also occur in the south-central and southeastern portions of New Mexico on gypsum outcrops and on limestone in the isolation of the Guadalupe Mountains. These, and several other species of special concern, are discussed from the standpoints of their unique nature and the kinds of threats that impinge upon the potential for their existence. The threats affecting them are presented in Table 11.3.

REVIEW OF SOURCES OF THREATS

URBAN DEVELOPMENT
(Table 11.3, column 3)

Compared to coastal California and states along the Eastern Seaboard, urban development does not pose a major threat to plant species in New Mexico. It is ranked sixth among the ten categories of threats listed in Table 11.3. There are four areas in the state where urbanization is threatening plants: Albuquerque; Santa Fe; the Las Cruces–El Paso region; and the Farmington–Shiprock area.

Urban development affects rare plant species in at least two ways. The most obvious is the direct loss of habitat to building sites and pavement. Less obvious at first glance, but fully as serious because of its slow-to-develop but persistent and virtually uncontrollable nature, is the disturbance to the natural vegetation that occurs around the periphery of the urban area, decreasing with distance from the city, as the land is tracked, trashed, and intensively used for recreation. Of the eight rare species in New Mexico impacted by urban development, three cacti are discussed as examples. Each also illustrates other problems associated with rare, threatened, or endangered species.

Mostly isolated on a hillside within a park, *Opuntia viridiflora,* the Santa Fe Cholla, is completely confined by the city of Santa Fe (Ferguson and Brack, 1986). The species has been known for more than 50 years. Surrounding it is considerable taxonomic and biological controversy. Chollas hybridize readily. Several botanists in New Mexico believe the Santa Fe Cholla to be a self-perpetuating hybrid between *O. imbricata,* the Tree Cholla, which grows with it in profusion, and possibly the Whipple Cholla, *O. whipplei.* The last is now not in the immediate vicinity, but may have been there in the past. Other botanists question the origin of this species by hybridization. In either case the population is in danger of extinction, but if it is a hybrid the need for legal protection is not so clear. On the one hand, hybrids are excluded from consideration by the Endangered Species Act. On the other, *O. viridiflora* vigorously reproduces by detached, fallen joints, and is therefore a "good" apomictic species, deserving protection, for the Act does not specify how species must reproduce to be protected. However, if *O. viridiflora* is a hybrid, to protect might not save particularly unique genotypes from extinction, which is the spirit of the Act, for they occur in the putative parents. Its protection would save only a peculiar combination of common genotypes that might be easily reassembled by intentional hybridization.

If the interpretation is taken that *O. viridiflora*

meets the criteria defining "species," then this taxon is one of the most critically endangered in the state. It is threatened by high land values, construction and road improvement, and constant disturbance to its habitat. The probability of its survival could be quickly enhanced by the establishment of new populations, easily accomplished by transplanting of joints in suitable habitat. However responsible this may appear on the surface, in a state where chollas are noxious on heavily grazed lands to release this potential weed from the confines of Santa Fe might be met, understandably, with less than enthusiasm by the ranching industry.

At the opposite end of the state, on the limestone mountain ranges near El Paso and Las Cruces, *Coryphantha sneedii* var. *sneedii* was among the first group of plant species to be formally listed as "endangered." The pressures on the species were the expansion upslope in the Franklin Mountains of the residential areas of El Paso and the highly focused activities of cactus collectors who removed plants for gardens and for sale (Heil and Brack, 1986). A decade has passed since this cactus was listed, and its situation has changed for several reasons, illustrative of the interaction of concepts and activities that often surround a species of special concern. First, habitat for this species is now protected from further urbanization through establishment of a state park in the Franklin Mountains. Second, there is a tendency among cactus fanciers to name as distinct entities those that appear to other botanists to be rather minor variants. Once named, the plant qualifies as a "species" under the Endangered Species Act. A careful scientific study has reinterpreted *C. sneedii* var. *sneedii* to be a more inclusive taxon that occurs from mountains in Luna County, New Mexico, eastward to the Guadalupe Mountains of Eddy County and in mountains in Texas to the immediate south. The cactus is now being considered for downlisting.

The final example is *Toumeya papyracantha* (synonym *Pediocactus papyracanthus*), the grama grass cactus, a species found sporadically across central New Mexico, barely extending into extreme western Texas and eastern Arizona. It once seems to have been common on the grassy outwash fans at the western edge of the Sandia Mountains, but now virtually that entire area is consumed by the eastward expansion of Albuquerque. The cactus is unbelievably cryptic, its flat, twisted, flexible gray spines perfectly mimicking a small clump of grama grass. It is easily seen only when in flower. Populations, once discovered, are quickly depleted by cactus collectors, a fate accelerated by proximity to an urban area. Destruction of plants and habitat by off-road vehicles is often most severe near cities also. As discussed in the final portion of the section on grazing effects, this cactus is among those species whose survival is of great concern to New Mexican botanists.

TILLED AGRICULTURE
(Table 11.3, column 4)

The effects of tilled agriculture are difficult to assess. Most arable lands within the state have been in production for many years. Interspersed remnants of native vegetation are not representative of presettlement situation on arable land. In most instances, no records exist to document species of special concern from areas now farmed. Since the rare species considered here occupy, for the most part, nonagricultural land, only two examples are provided, and these are, at best, marginal.

Opuntia arenaria, the Sand Prickly Pear, is a sand-adapted species that has comparatively large seeds that may successfully germinate where moisture is more reliable well below the surface of the sand. Buried stem-joints, attached to the mother plant or disjoined, readily produce new stems that quickly reach the surface, an important adaptation in shifting sand. In New Mexico, the species is now known from a few sites along the eastern edge of the Rio Grande Valley near the southern border of the state, where it occurs on marginally agricultural land. Some areas where the cactus grew two or more decades ago are now irrigated farms. Near the southern border of the state, it is also found along the upper edges of the valley too sandy for agricultural use.

Opuntia arenaria illustrates the interplay of various factors that influence decisions on the need to protect a species. Because the species is rare it is coveted by some cactus collectors, but it is seldom taken because it is not particularly attractive. However, when it is, the collector may dig the entire plant rather than remove a few joints for rooting later in the garden or the greenhouse. The species is threated in Texas by the growth of El Paso where virtually all land is privately owned. However, immediately across the border, in New Mexico, significant populations occur on land administered by the Bureau of Land Management, and thereby receive some protection from urban development. The extent of its range in less developed northern Chihuahua is unknown.

The ability of this cactus to readily reproduce vegetatively contributes to a sense that protection is not urgently needed. Also, there seems to be ample suitable habitat, although at this time the cactus is not found in most of it. In New Mexico, the species occurs in areas used for grazing and for recreation by owners of off road vehicles. Cattle and ORVs simultaneously may have negative and positive effects. They certainly disturb the sand and smash clumps of the species, but they also may break off pads and scatter them to new locations where they can root. In addition, in southern New Mexico considerable "new" habitat can be identified for this species because of changes in the last 100 years. Much of the area from the east base of the Franklin Mountains westward to the vicinity of Columbus is now mesquite coppice dunes. This community developed in large part due to grazing practices on desert grassland and subsequent wind erosion

(Chapters 2, 7, 8). In following decades the Sand Prickly Pear may colonize this new habitat, perhaps assisted in its dispersal by ORVs and cattle.

A very different kind of situation with regard to agriculture is provided by *Atriplex pleiantha*, the Mancos Saltbush (Cully et al., 1985). This rare annual is found on badlands in the Four Corners region, never occurring on soils with agricultural potential (the species also is discussed in the section on mining impact). Vegetation of badlands is xerically adapted, the plants active during periods of adequate moisture, perennials often becoming dormant during dry seasons, annuals surviving dry periods as dormant seeds. *Atriplex pleiantha* follows the latter pattern, appearing only in those years, at those sites, where there has been adequate winter precipitation. Near Farmington, the badlands where *A. pleiantha* grows interfinger as eroded gullies with the higher, sandy plain to the east, now developed for agriculture. Water from irrigation may reach the badland strata or drain into the gullies. Effects on badland vegetation are unknown, but it can be expected that supplemental water during the dry season may create a hostile environment for plants precisely adapted to survive seasonal or prolonged drought.

GRAZING BY WILDLIFE AND DOMESTIC LIVESTOCK
(Table 11.3, column 5)

Much of the land in New Mexico, and a significant portion of the state's economy, is involved with the ranching industry. Although this category is ranked fourth when threats are tallied in a species-by-species manner (Table 11.3), the use of land for grazing purposes may be the single most serious man-induced threat to many plant species of special concern and, perhaps more significantly, to the maintenance of species diversity in general. The impact on native species combines of the direct effect of animals on specific plants and soil surface, and the diffuse, arguable, long-term changes in the natural vegetation that have occurred in the last century, and continue to occur, as a result of grazing pressures (Chapters 2, 7, 8). Five very different species are used to illustrate the problems for rare plants resulting from direct or indirect effects of grazing.

Allium gooddingi, Goodding's Onion, occurs here and there in high, isolated mountain ranges from south-central New Mexico to south-central Arizona, requiring mesic sites in canyons along streams or around springs under at least partial cover of mature coniferous forest (Fletcher, 1984; Spellenberg, 1982). It is, therefore, also discussed under the impacts of logging. Most populations were threatened in the past or were in decline, but recent management decisions by the U.S. Forest Service may reduce the trend.

Perennial sites of water, such as those which support *Allium gooddingii,* are important for wildlife and livestock alike in the Southwest. When such sites are few and grazing pressures are heavy these wet places become congregating sites and the natural condition is quickly destroyed. If not occluded by fallen timber, canyon bottoms are regularly used as paths of easy access by livestock and trampling of the vegetation and compacting of the soil results. *Allium gooddingii* is also eaten by livestock, in some areas nearly to the surface of the soil. As the photosynthetic tops are repeatedly removed, the subterranean food-storing bulbs may die. Remnants of once more extensive populations may persist only in sites of difficult access as reduced cover exposes plants and the soil is compacted.

The Night-Blooming Cereus, *Cereus greggii* (synonym *Peniocereus greggii*), is widespread but never abundant from southern Arizona to western Texas, southward into Mexico. It is a very popular cactus among cactus growers and is, therefore, also discussed in the section regarding threats by collectors. The species occurs in desert grassland and among desert scrub vegetation, often concealed among the stems of shrubs. It is eaten by livestock only under very poor range conditions. The effect of habitat deterioration and the compacting of the soil on this cactus is not well known, but it is probable that sites for seed germination and seedling establishment are reduced or lost when grazing is heavy. The Night-Blooming Cereus is presently uncommon or absent in many areas that were once rocky, desert grassland (Chapter 7) where it was once known. It is still present, though uncommon and inconspicuous, in some areas within its original geographic range where rangeland still has substantial grass.

Decline of the Night-Blooming Cereus in heavily grazed areas might also result from its tall, spindly, brittle nature. It can be expected to survive in open areas if not repeatedly broken by passing animals. Where the plants grow at the base of shrubs in deteriorated rangeland, stems of the *Cereus greggii* may be broken as livestock nose into the bush to reach grass. Although not specifically studied, removal of the photosynthetic stem can be predicted to deplete the food stored in the tuberous root, resulting in reduced vigor of the regenerated stem and reduction of reproductive potential as fewer flowers and seeds are produced. As populations become sparser the distance between plants bearing the handsome, moth-attracting flowers increases. If cross-breeding is important, reproduction and the maintenance of genetic variability then may be less successful. Repeated breakage ultimately may completely deplete the root, resulting in death, the species finally disappearing from the area.

The Mescalero Thistle, *Cirsium vinaceum,* is a tall, spectacular thistle with maroon stems and nodding purple flower heads. It is restricted to a few wet areas in the Sacramento Mountains above 8,000 ft. (2,470 m) elevation (Fletcher, 1978; USFWS, 1987). There are at least six easily identified sources of threat impinging upon this species (Table 11.3), more than for any other species in New Mexico. Because it grows in springs and seeps it is often directly affected by trampling where grazing is heavy. Springs near roads are affected by road maintenance and increased visitation and trampling by people. Recreational development of the mountains affects the species in a few areas, especially through development of water sources for human use. In at least one instance since the species has been listed as threatened, however, there has been successful compromise between between the need to protect the species and the development of a spring as a water source.

Logging and grazing are also economic activities that indirectly impinge on the existence of *Cirsium vinaceum*. Both activities remove cover from the mountain sides. The reduction of the ability of the vegetation and ground litter to retard run-off following spring snowmelt or intense summer convectional storms can result in increased erosion of some of the valley bottoms, loss of habitat where the species grows, and less recharge of ground water that later surfaces at springs.

Penstemon alamosensis, the Alamo Beardtongue, is a brilliantly red-flowered limestone endemic in the San Andres and Sacramento mountains (Spellenberg, 1981). It is closely related to the western Texas *P. havardii,* and may represent a subspecies of that species. Because of its limited range and its attractiveness, it is believed to have sustained some collecting in more accessible places by *Penstemon* growers, but the impact may have been judged too serious. The species illustrates, however, problems faced by "cool season" plants in the Southwest. Unlike so many of the species at the lower elevations in the region, which become vegetatively active at the onset of summer rains, *P. alamosensis* is green in the spring and dormant during the warm season. It is one of the relatively few species that provide green, spring forage for livestock and wildlife. Plants that grow on open slopes where animals can readily graze or browse are often eaten to stalks only a few inches tall. Seed production is then low or nonexistent for the season. If reduced seed production persists over a long-

period of time, even in perennials such as this, population decline ultimately occurs as seed reserve in the soil is depleted, and adult individuals die without replacement by juveniles.

Finally, one of the special concerns in New Mexico most repeatedly discussed by endangered species specialists is the condition of *Toumeya papyracantha* (synonym *Pediocactus papyracantha*), the Grama Grass Cactus (also reviewed previously in the section "urban development"). Most agree that many populations are probably in decline, some severely so. However, the wide geographic range of the species and the ubiquity of its potential habitat, combined with the extremely inconspicuous nature of this cactus, make it difficult to study and estimate the seriousness of decline. *Toumeya papyracantha* grows in sandy loam, often on highly erodable sites, usually among species of grama grass (*Bouteloua*) and dropseeds (*Sporobolus*). In ungrazed areas where the species occurs, tall plants are fairly common; where grazing is moderate, they are far less frequent and trampled plants are often seen. Trampling, probably the major direct effect of livestock with regard to this cactus, also compacts the soil. Seedlings, always cryptic and difficult to see, are most frequently found in rather loose soil.

The indirect effects of grazing on this *Toumeya* are far more pervasive but are subtle, interrelated, and poorly understood. Light grazing may open grass cover slightly and facilitate seedling establishment. More intensive grazing, in addition to the effects of trampling and soil compaction, is believed to have two other major negative effects. First, grass cover is reduced, potentially exposing plants to increased predation by small herbivores, which would seek alternate food items as other forbs are eaten by livestock. Second, the loss of grass cover results in erosion of topsoil, thereby accelerating loss of sites for seedling establishment and hastening population decline. Because of the legal and economic impacts that would be generated by listing this species as "threatened," as it very well may be, considerable caution is being exercised by reviewers in making recommendations to various government agencies responsible for land management and species protection. The relationship between condition of range, grazing by domestic livestock, and the vigor of populations of the grama grass cactus is now being tested using exclosure experiments in north-central New Mexico.

LOGGING
(Table 11.3, column 6)

When the changes in dominant vegetation due to logging are considered, and all the disturbance produced by bulldozers and highleads recalled, it may seem surprising that logging ranks only seventh in the twelve kinds of threats to plant species in New Mexico (Table 11.3). Although very disruptive at the local level, the planned harvest of forests—when done in small patches on a rotation cycle of long duration—apparently does not threaten the existence of most species of plants. Efforts by U.S. Forest Service botanists to find where rare plants occur have resulted in the discovery of many hitherto unknown populations, providing much new information about the distribution of plants in the Southwest. Most sites now receive consideration prior to timber harvest. Others, harboring especially critical species, have been placed in reserves or at least receive a modicum of protection. The less direct and often more pervasive effects of logging—increased water run-off and erosion, and decreased recharging of the water table—have been mentioned under *Cirsium vinaceum* in the previous section on effects of grazing.

A few plants of southwestern forests have limited geographic ranges and depend upon mature, old growth coniferous forest for survival. Their local populations cannot persist through periods of openness and ecological succession as new forest replaces the one removed. Three such species are briefly mentioned as illustrating this point.

All are from southern mountains of New Mexico and Arizona, are restricted to especially moist and shady sites, and are endemics on small habitat islands in otherwise warmer and drier coniferous forests, which themselves are larger habitat islands on the isolated mountain ranges that rise from surrounding arid regions.

Allium gooddingii, Goodding's Onion (discussed also under effects of grazing), *Senecio cardamine*, Heartleaf Senecio, and *S. quarens*, Gila Groundsel, are examples of plants that seem dependent on the shade and moisture provided by mature coniferous forests. All quickly decline or disappear when the cover of forest is lost. They may burn by long exposure to direct sunlight and heat, and the soil in which they grow becomes, for them, intolerably warm and dry.

A taxonomic problem surrounds *Senecio quarens,* for it may represent a variant of the more widely ranging *S. hartianus* (Fletcher, 1980). Both are in the difficult *Senecio* group *Aurei,* where species are poorly differentiated. As narrowly defined, *S. quarens* seems very vulnerable, for it is found in a strictly limited area; if determined to be taxonomically distinct from *S. hartianus* it merits close attention, if determined to be a synonym it will receive little or no protection. Pressures on *Allium gooddingii* are only slightly less because it is more widespread, but it still occurs in populations so small that a single logging operation could cause the decline or local extinction of any one. *Senecio cardamine,* found in high, cool spruce/fir forests, is very limited in New Mexico. After it was placed on the federal candidate list it was discovered to be fairly common in the White Mountains of southeastern Arizona. It is, therefore, less severely threatened by population decline than are the others.

We often envision western forests as great expanses of cool, nearly continuous shade. So they were. However, prior to the arrival of settlers, temporary openings in forests developed from such natural causes as treefall, land slides, flooding streams, and fire. A number of species found within coniferous forests are adapted to these fleeting disturbed sites and depend for their existence on the repeated, regular formation of them within dispersal distance. Logging also opens the forest, and species that are successional on burned areas and other naturally disturbed sites in old growth forests may also be found as successional species in logged areas. The relationship between periodic opening of the forest and vigor of populations of several plant species of special concern is poorly understood, but such species may be part of the natural cycles of the regeneration in a coniferous forest. The nature of the rarity, therefore, must be determined before management for rare species is undertaken. A species requiring a shaded, undisturbed, mature forest would require very different management than would one that persists for a few years on an open site, only to decline if no disturbance were allowed.

Two species isolated in the Sacramento Mountains represent the few forest species that are rare and that seem to require disturbed sites. After their initial discoveries in the late 1890s they remained virtually unknown until the 1970s. One, *Astragalus altus,* the Tall Milkvetch, is a perennial pea relative, the other *Lesquerella aurea,* the Golden Bladderpod, is a small, annual mustard. Both may have been quite rare in the pristine situation. Although they are highly restricted in area and limited in distribution to the lower reaches of the coniferous forest, both are frequent in a few places, usually on road banks or gully sides a few years after disturbance. They are very scattered or nonexistent in mature, shaded, undisturbed forest.

For such species, the annual fluctuation in climate or the stage in ecological succession can affect the accuracy of estimation of vigor of the species, estimations important for effective management. At some sites where the perennial *Astragalus* was common it may be rare a few years later, or it may not appear until well after summer rains (Fletcher, 1979a). New disturbance may result in a vigorous population, but only after several years pass. At one site the annual *Lesquierella* may fluctuate markedly (Soreng, 1981), being so common as to cause one to wonder why it was ever considered for listing in the first place. In a following year there may be no plants at a site, to be followed in a subsequent year by a large, vigorous population. For species locally common in the early stages of succession in coniferous forests, it is possible that they may benefit from logging if the original local source of seed is not eradicated.

MINERAL EXTRACTION
(Table 11.3, column 7)

Mineral extraction in New Mexico involves the petroleum and mining industries. Effects on vegetation around operations depend on the method of recovery of the resource. In drilling for oil, placement of roads and drilling rigs should be planned in advance to ensure minimal environmental damage. Roads and well-pad sites are the primary threat to the highly local *Astragalus humillimus* the recently rediscovered Mancos Milkvetch. "Hardrock mining," where most of the disturbance is subsurface, may have less effect on population decline or loss of a rare species than widespread alteration of the landscape that results from strip mining. As with other human activities, access roads for mineral exploration and extraction, the construction of supply lines for energy and water, and other peripheral activities of mines or well-fields may disturb populations of rare plants if preliminary surveys are not made. If they are made conscientiously, impacts on rare species usually can be mitigated.

Because of its potential for widespread environmental damage, the mining industry lately has been closely monitored by federal and state government. In New Mexico it is through the cooperation of the mining industry, the Bureau of Land Management, and the U.S. Forest Service, and specifically their efforts to provide detailed environmental impact statements, that many discoveries have been made of previously unknown populations of rare plants. Botanists, surveying proposed mine sites and access corridors for roads, water lines, and powerlines, have examined many areas of the state previously unexplored botanically.

Minerals are concentrated in soils that deviate from "normal" soils where widespread species and common vegetation types occur. Many plants that grow on proposed mine sites, or adjacent to active mines, are edaphic endemics, that is they are adapted to the peculiar chemical and physical characteristics of the soil or rock. Their geographic range is often highly restricted, and therefore they are unusually vulnerable to extinction by a single event. Because of their extreme specialization to soils inhospitable to other plants, they commonly do not withstand transplating to "better" soils or new sites. Some may require the peculiar chemicals of the mineralized soil for specific physiological needs. Others do not, but instead have lost the ability to compete with plants that grow crowded on better soils. For these reasons, edaphic endemics usually do not survive in reclaimed areas, most of which are planned to be rangelands more productive than the original land surface.

In New Mexico individual populations of plants of special concern are threatened by the petroleum industry, and by the mining of copper, uranium, coal, and gypsum. Gypsum, a strongly selective substrate of calcium sulfate, commonly has a suite of plants found on no other soil type. These gypsum-restricted species are called "gypsophiles." *Abronia bigelovii, Astragalus gypsodes,* and *Eriogonum gypsophilum* are examples of gypsophiles whose populations are so restricted that they are of concern to management agencies and conservationists (see discussion on *Eriogonum gypsophilum* in the following section). Gypsum outcrops occur from north-central New Mexico to the southeastern corner of the state, extending from there extensively into west-ern Texas and adjacent Mexico. Some gypsophiles occur nearly throughout, but most are either more northern (as the *Abronia*) or southern (the *Astragalus* and the *Erigonum*). Many are regional endemics on the extensive Permian gypsum beds of southeastern New Mexico and adjacent Texas, but because they are not endemic to the state, and are not particularly rare or threatened, they are not listed here or in the review of rare plants of New Mexico (NMNPPAC, 1984).

For the most part, gypsophilic plants are not seriously threatened, for few are extremely narrow endemics and gypsum mining is a minor industry. Many gypsum endemics also respond favorably to minor disturbance of the soil surface, the breakage of the hard gypsum crust apparently facilitating the establishment of seedlings. Grazing does not seem to be a strong threat to most gypsum endemics although they may be palatable. Usually plants are very sparse, and they seem to be only occasionally "sampled" in passing by animals traversing the gypsum along well used trails to better pasture.

Erigeron rhizomatous, the Zuni Fleabane, is an example of an edaphic endemic that has had at least temporary reprieve from the threat of extinction. Known for about 50 years only from the clays of the Chinle Formation in the vicinity of Fort Wingate, McKinley County, it recently was discovered to occur on the same formation considerably to the south in the Datil Mountains. Perceived threats to the extinction of the species as a whole thus were alleviated. However, the greatest relief from pressure against this species has been depression in the uranium mining industry. The Zuni Fleabane occurs on or near a number of uranium mining claims. Should mining activity resume at a later date, considerable controversy may surround this plant, now listed as "threatened" and therefore protected by federal law (USFWS, 1987).

A species that has been surrounded by controversy is *Atriplex pleiantha,* the Mancos Saltbush, a small annual for several decades known only from a single collection made in northwestern New Mexico, a species so different that recently it has been placed in its own genus, *Proatriplex*. Its preservation obviously preserves a unique genetic combination. It was once very seriously listed for immediate legal protection, but recent discoveries of large populations away from mining concerns has alleviated pressure on the species. It is used here as an example of the kinds of information and arguments brought to bear when preservation conflicts with land use, and the difficulty of making well-informed, timely decisions when faced with relentless development. Information on which the survival of a species may depend is often hard to obtain, yet to stall large economic ventures until indecision is resolved is very expensive and undesirable. To list a species in substantial ignorance of critical biological attributes, though it may save it from rapidly moving threats, could undermine the credibility of those supporting the Endangered Species Act if the species is not actually in peril, eventually weaken the Act itself, and ultimately increase the jeopardy of extinction for all rare species.

Ironically, it was the environmental efforts of the coal mining industry that brought this species increasingly under scrutiny for formal listing. Environmental surveys of the peculiar clays of the Mancos or Fruitland formations showed that major populations of *Proatriplex* occurred near and within the boundaries of a single large coal mine. The estimation of degree of threat involved (1) the perpetual problem of absence of complete information regarding distribution of New Mexico plant species, (2) the seasonal fluctuation in levels of numbers of populations, and numbers of plants in populations, and (3) the biological significance of this plant's "annualness."

After the original discoveries of *Proatriplex pleiantha* within and near the boundaries of the coal mine and associated power plant, searches for the species were conducted elsewhere in the Four Corners Region by specialists in arid land plants and local botanists. Small populations were found in adjacent Colorado and Utah. Sympathizers with mining interest pointed out that the existence of other populations meant that the mining industry did not seriously threaten the species with decline; conservationally oriented interests argued that the only large, dense populations of the species known, with millions of individuals, occurred on the mine.

Plants of this spring annual must be sought between time of germination in early spring and their death by early summer. After that their skeletal stalks are difficult to distinguish from those of common annual *Atriplex* growing with it. As with many annuals of seasonally xeric habitats, *Proatriplex* is highly responsive to timing and amount of winter moisture. In some years thousands of plants may grow in a very limited area; in other years not one plant grows at that exact site. Thus, to effectively estimate presence of *Proatriplex,* searches for it must be done at the right time in years with sufficient winter moisture. Proponents for mining interests argued that those small populations found off mine may be small because the year when they were discovered was "bad"; they may be larger in "good" years, and other populations will be discovered, particularly in "good" years. This very well may be correct, but those deciding whether protective measures are needed proposed caution and repeated searches to estimate the relative proportions of the entire species that occurred on and off the mine.

The nature of an annual species, as applied to this *Proatriplex,* was also a point of argument. In many instances annuals are "weedy" and opportunistic, the first plants onto a disturbed site early in stages of ecological succession. However, particularly on the winter-wet, summer-dry, warm deserts to the west of the Four Corners, particularly in the extreme environmental situations provided by unusual soils, several annuals are highly specialized narrow endemics, surviving the drought period as a dormant seed, occupying a unique ecological situation during the wet period, tolerating little disturbance of their habitat. They are not at all weedy. This appears to characterize *Proatriplex*. Other species of annual *Atriplex* in the Four Corners area are rather weedy, doing well after disturbance, and inhabit a broader range of soil types. The mining interest, naturally, held these plants up as examples. The counter argument was that annuals may very well be early successional plants in many situations, but that annualness may also be an adaptation to a precise microenvironment that is present only irregularly, and then for only a brief period.

WATER DEVELOPMENT AND
MANAGEMENT
(Table 11.3, column 8)

New Mexico is an arid state, yet extensive areas are suitable for habitation and ranching, which means water is an especially valuable resource. Springs, seeps, stream banks, and river plains form more or less isolated habitat markedly different from the surrounding terrain, and the native plants that occupy them, if still existing, are sometimes rare and local. However, most situations offering water are heavily impacted. Springs and seeps are often developed, and the natural situation has been altered by diversion of flow. If open to access, they are often focal points of use by domestic and wild animals. Most river plains have been heavily utilized during the 400 years since the settlement of New Mexico, cleared for farming or grazing, or disturbed by cutting of firewood, posts, and vigas (ceiling timbers). In many places, little of riparian woodland remains. Where suitable, rivers are dammed to provide a more reliable supply of water throughout the year, and extensive bottoms of riparian habitat are inundated. Elsewhere streams and rivers are channeled and diked to control flooding, the flood plain usually then dedicated to agriculture or residential use.

The impact on the natural vegetation has been

so severe in some areas that rare plants may have disappeared from them before the first plant collectors visited the state in the mid-1800s. That three plants, *Brickellia chenopodina, Cleome multicaulis,* and *Helianthus praetermissus,* each have been found only once in the state (and then only many decades ago, even though intensive searches have been made for them), suggests riparian habitats have changed significantly. Single, old records from one habitat suggest that other species may have already been lost from similar sites prior to any botanical study, no matter how cursory. As mentioned elsewhere, the situation with two of these is not clear. *Brickellia chenopodina* may not represent a species except in the most typological sense, but rather a peculiar form of a common species that happens to have been collected and named in the early days of taxonomy in the state. *Helianthus praetermissus* may have actually come from eastern Arizona (Fletcher, 1979b), but even so, it has not been seen anywhere since. Although there are problems with interpretation of two of these plants, all three serve to illustrate the point that riparian habitats have been severely impacted. The third, *Cleome multicaulis,* is known from adjacent states in the United States and in Mexico, but, except from southern Colorado where it is locally common, most collections are several decades old, an indication that it is rare and declining in most areas as increasing pressure is put on wet areas.

Other species of plants are tightly restricted to local water sources and are therefore vulnerable to the drying of their habitat, but do not now seem particularly threatened. They occur in areas of rugged terrain with little potential for development. Such is the case for *Oenothera organensis,* a beautiful, large-flowered, very distinct Evening Primrose whose closest relatives are 300 mi. (500 km) to the south, in the mountains of Mexico. It is endemic to the springs of the Organ Mountains alone, in the south-central part of New Mexico. These mountains are under pressure for recreation by the citizens of Las Cruces and El

Paso, but much of the range is part of a military withdrawal and therefore receives protection by its removal from unrestricted public use. Species such as these are periodically reviewed by botanists and government administrators to assess the threats that may impinge upon them.

The Pecos Sunflower, *Helianthus paradoxus,* is a small annual sunflower confined to marshes and moist open areas in southeastern New Mexico and western Texas. It was named in 1958 from Texas (Heiser, 1958) and for 25 years was believed to be a very local endemic. At the time of publication of the name of *H. paradoxus,* the author noted that the species may have been collected a century earlier, in 1851, by the Sitgreaves Expedition along the Rio Laguna, at about the same time as *H. praetermissus* was collected, and perhaps in the same vicinity. This situation remains confused, perhaps permanently so, for in a later, more comprehensive publication (Heiser, 1969) no mention is made of *H. paradoxus* from New Mexico and Fletcher (1979b), based on the chronology of the Sitgreaves journal, believes *H. praetermissus* actually to have been collected in eastern Arizona. The early specimens are fragmentary, making identification difficult, and one cannot tell from the localities cited by expedition collectors exactly where the sites are today.

In Texas, usage of marshy areas for livestock, combined with presumably normal annual fluctuations in population level, seemed to indicate that the species was very near extinction. However, in the early 1980s, the species was discovered in southeastern New Mexico by workers of the Soil Conservation Service. It grows on marshy ground along irrigation ditches in the Pecos Valley, in this instance the activities of man apparently providing habitat where there may previously have been none. Populations are small, though vigorous. Further discussion occurs in the last section of this chapter.

Eriogonum gypsophilum, the Gypsum Wild Buckwheat, is listed as federally threatened. It is another example of a species strongly limited by edaphic features, for, as its name implies, it is restricted to almost pure gypsum substrate. Even though there are hundreds of square miles of gypsum habitat in southeastern New Mexico, for 70 years after its discovery the species was known from only one small outcrop, the total extent of the population occupying about $1/2$ sq. mi. (0.66 km^2) (Limerick, 1984). Recently, a botanist from the Bureau of Land Management discovered two more small populations about 30 mi. (48 km) south of the previously known one. Plants are numerous and vigorous in all three, and thus the threat of extinction is considerably alleviated.

The first known population of Gypsum Wild Buckwheat occurs very near the construction site of the new Brantley Dam on the Pecos River north of Carlsbad, a project that now provides better flood control for that downstream city by replacing the old McMillan Dam and nearly full sedimented lake it forms. Near the start of the project, officials from the Bureau of Reclamation, the planning and administering agency, became aware of the potential problem presented by this threated species. Detailed surveys were performed that precisely located virtually all plants of *Eriogonum gypsophilum* in the known population, and a search was made to locate other populations (none were found) on apparently suitable habitat from the same geological formation nearby (Spellenberg, 1977). Recommendations were presented that minimized impacts of construction, highway relocation, development for recreation, and access to water by livestock. The project and the attention paid to the potential for impacting a threatened species serves as an example of one of the thousands of mitigations that progressed well in the past decade since the passage of the Endangered Species Act.

RECREATIONAL USE OF LAND
(Table 11.3, column 9)

Outdoor recreation is popular and important throughout the United States. In the Southwest, the canyons of the lower mountains are extensively visited for their scenery and isolation. The higher mountains provide cool, green relief from summer heat of the deserts and plains. Covered with snow in winter, they are the southernmost ski areas in the country. People seeking recreation hike, ski, hunt, picnic, camp, build cabins, and drive their vehicles across open areas in search of secluded places. Easily accessible areas, particularly meadows and canyon bottoms with running water, are especially utilized. To service this demand, construction of the homes and cabins, roads, water supplies, and support businesses all add considerably to the destruction of the natural environment. This is especially true of the Sacramento and White mountains in south-central New Mexico and the southeastern flank of the Rocky Mountains in northern New Mexico, the mountain habitats most easily accessible to Texas residents, and therefore shared as recreation areas with residents of New Mexico.

Springs and streams are always attractive as picnic areas and campsites, because of their pleasant atmosphere and the availability of water for drinking and camp chores. Frequent visitation and heavy use results in trampling of the margins of these sites, and plants that require these habitats disappear quickly. If streamside plants have attractive flowers that would make a pretty bouquet on the picnic table or a novel addition to the garden, as would the handsome, yellow-flowered Chapline's Columbine (*Aquilegia chaplinei*), flowers may be picked or whole plants dug up. This direct and regular pressure can result in reduced reproductive potential, the population not replacing itself as plant senesce and die, or the destruction of the population through loss of individuals to attempts at transplantation.

In addition to Chapline's Columbine, there are several other species of special concern in the state that are impacted by this type of human use of habitat. *Cirsium vinaceum*, the Mescalero Thistle (mentioned under the sections on logging and grazing), grows in canyon bottoms in the Sacramento Mountains, some of its few populations near picnic and camp sites. *Oenothera organensis*, the Organ Mountains Evening Primrose, is an endemic restricted to springs and seeps, some visited by hikers. *Senecio quarens*, the Gila Groundsel (mentioned in the section on logging), is found at a few wet sites in the popular Mogollon Mountains.

Other montane species are not particularly representative of streamside or seep vegetation, but are nevertheless impacted by increased mountain visitation. *Allium gooddingii*, Goodding's Onion (discussed in the sections on grazing and logging), found in several of the southern mountain ranges, often occurs near trails or habitation. *Astragalus altus*, the Tall Milkvetch, from the Sacramento Mountains, is common along roads. *Delphinium alpestre*, the Alpine Larkspur from the high meadows of northern New Mexico, suffers occasionally from foot traffic. *Hedeoma apiculatum*, the McKittrick Pennyroyal, a chasmophile from the scenic canyons of the Guadalupe Mountains, occurs along some heavily used trails and is occasionally picked. *Potentilla sierra-blancae*, Sierra Blanca Cinquefoil, a species from high areas of the White Mountains of south-central New Mexico, also occurs in some areas that are heavily trampled by day hikers.

Perhaps even more serious is the destruction resulting from the recreational use of off-road vehicles, such as conventional four-wheel drives, the purely recreational all terrain vehicles, off-road motorcycles, and snowmobiles. The effects of such recreation are common within a few miles of most maintained roads in New Mexico on public land where vehicular traffic is possible, and in many private areas where access is not denied. This abuse of the landscape is very difficult to control in a sparsely populated state, where there is little sympathy for its regulation. On all sites, vehicles mash and kill plants, or soil is compacted, altering the potential for seedling establishment. The tracks left behind concentrate water, so that gullying often occurs.

New Mexico plants most seriously impacted by off-road vehicles are those that grow on open, rolling badland terrain. On the gypseous hills of southeastern New Mexico, *Eriogonum gypsophilum,* Gypsum Wild Buckwheat (discussed under the section on water development), and *Astragalus gypsodes* are affected. In northwestern parts, *Erigeron rhizomatous,* Zuni Fleabane, from the clay badlands of west-central New Mexico (discussed under the section on mineral extraction), *Gilia formosa,* Beautiful Gilia, from the vicinity of Aztec, and *Sclerocactus mesaverde,* Mesa Verede Cactus from barren clay hills in the Four Corners Area (Heil, 1984), all suffer damage from vehicles. *Opuntia arenaria,* Sand Prickly Pear (discussed under the section on urban development), from the sand hills between El Paso and Las Cruces, in south-central New Mexico, is also impacted. Throughout much of the state *Toumeya papyracantha,* Grama Grass Cactus (discussed under urban development and grazing), a species of open, rolling, sometimes grassy areas, is damaged severely at some sites. In several areas, the Bureau of Land Management has allowed hilly terrain to be sacrificed as play areas for those seeking this type of recreation, a somewhat successful trade-off for closing other sites to off-road vehicular use.

COLLECTORS
(Table 11.3, column 10)

The collecting of plants by individuals is a highly focused activity that can cause serious decline of plant populations and reduction in survival potential for a species. Scientific collection for research and the deposition of specimens in public museums may have reduced a few populations of the rarest species, but collectors now are usually careful to minimize impacts. Scientific collecting also is often regulated by law.

Restrictions not only are imposed by various levels of government, but in reality also by the time and economics of procuring and processing specimens, as well as by the reduced scientific significance of new collections as floras become well known.

A far more serious threat is the collection of plants for personal enjoyment and financial gain. Cacti are most commonly sought in the Southwest, but populations of orchids and species of the genus *Penstemon* may also be reduced or obliterated. Because of their intrinsic interest, and particularly if they are showy, orchids are picked wherever they are found, or are dug for gardens, an eventual sure death for the plants. The showier species are also restricted to wet sites, and as described in the section above, the habitat of these sites is often altered by intensive use. No orchids are endemic to New Mexico, so the threat to the species as a whole is not as great as it would be with more narrowly distributed species. However, when the entire North American ranges are considered, some orchids are obviously in decline.

Conservation problems surrounding cacti are as severe as those involving the pitcher plants of the eastern United States. Both families are taken from the wild for personal enjoyment and for profit. As a group, cactus growers are highly organized. Catalogues are published offering species for sale. At one time many vendors collected plants from the wild, but among the most responsible growers there is serious concern for the preservation of cacti in the wild, and plants they sell are seed-grown and greenhouse propagated. Although the offering of wild-collected plants by cactus retailers is now much reduced, recently dug plants are still offered at Saturday flea markets in the Southwest. Regional and national meetings of cactus and succulent societies result in a concentration of individuals who may, if unscrupulous, seriously deplete populations along their route to, or in the vicinity of, the convention. Several semiscientific journals regularly have articles describing journeys where many species of cacti were collected, observed, or photographed; trends in more recent articles encourage the last two methods of "collection." Strict state, federal, and international regulations regarding collecting and importation of any cacti, and laws regulating the commerce in rare cacti are now in effect. Recently severe fines have been levied upon violators.

Regulation of illicit collecting activities in the wide, empty West is difficult in the case of unscrupulous commercial operations and virtually impossible in the case of individual poaching. The locations of most populations of rare cacti are very well known, and information regarding others is very easy to obtain from scientific literature, government environmental impact reports, and public hearings. Some individuals may enter an area and take, for personal use, only the number of plants estimated to have little effect on the population, but a similar approach repeated by each person poaching from the population will result in an attenuated decline. Thoroughly unscrupulous collectors furtively remove all individuals seen in an area and disappear before they are observed and reported.

The poaching and illicit trade of cacti is an international problem. Cactus fanciers live worldwide, and this New World family of plants is under attack throughout its range. Mexican and South American cacti are enthusiastically dug by the poor and sold to dealers for a pittance. In the United States, cactus poachers from foreign countries have detailed directions published by their local clubs and societies to populations of rare cacti in the United States. As tourists they dig plants, package them, and ship them home via first class airmail. The packages are infrequently opened and inspected by customs officials.

Two instances illustrate the problem. The wonderfully awkward Night-Blooming Cereus, *Cereus greggii* (discussed under effects of grazing), has an odd stick-like appearance that makes it difficult to see in the wild. The species is widespread among creosotebush and mesquite from western Texas to southern Arizona, southward into Mexico, but it is rarely common. For most of the year, it is not a particularly attractive plant in the garden. However, when it blooms with numerous large, white or pale pink, sweet-scented nocturnal flowers for a few days in the early summer, it is a real delight. For this reason, the plant is actively sought but difficult to find. Once discovered, word about population sites usually spreads quickly, and all too often the area is reduced to a series of holes in the ground. Although once present along the western base of the Organ Mountains, near Las Cruces, the species has not been seen there in decades. Similar depredations in the last few years have been inflicted upon *Sclerocactus mesa-verdae* population on Navajo Nation land in the Four Corners area (Heil, 1984). Pressure upon populations of cacti is likely to increase as migration from the northern states brings more and more individuals to the region, and as curtailed supplies of water increase the need for desert landscaping.

The second example is *Pediocactus knowltonii,* Knowlton's Cactus, and provides a dramatic example of the deleterious impact of cactus poachers, but the successful implementation of recovery practices. This species was discovered in the late 1950s and was described and named in a popular cactus journal a few years later. It is a very distinct, diminutive, attractive cactus. It occurs on only one hillside in northwestern New Mexico and nowhere else. Immediately after the publication of its discovery the population declined precipitously, from an estimated 100,000 highly local clustered stems in 1960 to 1,000 scattered individuals in 1980 (Heil, 1985). Collecting reduced the population of this coveted cactus to 1% of its original level in only 20 years.

Pediocactus knowltonii nevertheless promises to be an example of successful recovery of a species. Increased commercial availability of plants that are legitimately greenhouse-grown, in combination with the potential of severe fines for collecting the species in the wild, have apparently reduced depredation of the population. The estimate of individuals in the natural habitat is now in excess of 7,000 (Heil, 1985). Furthermore, workers in the New Mexico Department of Natural Resources, with the help of a privately owned cactus greenhouse, have successfully grown plants from cuttings. The cuttings have been reintroduced to the wild in a site that appears environmentally suitable but is of very difficult access. This is a very well structured experiment of reintroduction, monitored annually. Initial evidence is that imposed regulations on collecting, the commercial availability of legally grown plants, and the successful introduction at a new site have given the species a reprieve.

ROAD DEVELOPMENT AND
MANAGEMENT,
INCLUDING HERBICIDE SPRAYING
(Table 11.3, column 11).

Throughout the United States is a netlike road system that ranges from broad interstate highways to remote dirt tracks. New Mexico is no exception; roads are nearly ubiquitous, and therefore their development and maintenance pose direct threats to many species of special concern. The tally of their impact in Table 11.3 indicates only direct effects; it does not include a corollary impact of increased access by humans to areas that otherwise would rarely be visited. The effects of roads are so great and so clear that there is little need for discussion. The building of new roads, the scraping and mowing of road margins, and the control of right-of-way vegetation by herbicides all have direct, negative impacts on all but the most weedy vegetation.

There is, however, a less obvious benefit from roads that has the potential to counteract their destructive effects. Roads are narrow transects across many vegetation types. Their right-of-way fences exclude large animals from the roadway and thereby provide a haven for shoulder vege-

tation from the damaging grazing practices that all too often occur on adjoining land. Road rights-of-way across extensive areas of intensive agriculture are often the only glimpses of the original vegetation for miles. Although a road right-of-way is a band so narrow that it is not very suitable as a preserve for pristine vegetation types, small populations of rare species that appear early or midway in ecological succession may persist or even flourish. This is particularly evident on the older two-lane highways in the state, where roadside maintenance is minimal, where roadbanks have stabilized, or where streams that cross under the road have well-developed streamside vegetation. The New Mexico Highway Department, county road departments, and the federal land-management agencies, through the activity of local botanists, are now beginning to recognize some of the botanical heritage that roads fortuitously preserve.

ENCROACHMENT BY EXOTICS
(Table 11.3, column 12)

Encroachment by exotics affects very few species of special concern in New Mexico, and it is questionable that any would be affected were it not for habitat disturbance in the first place (see Chapter 2). Alteration of habitat and change in species composition have produced disturbed areas of considerable extent along water courses and around springs at lower elevations throughout the state. The two most noticeable exotics are Saltcedar, *Tamarix ramosissima* (synonym *T. pentandra* as used in the Southwest), and Russian Olive, *Elaeagnus angustifolia*. Saltcedar forms almost impenetrable thickets in wet areas at the lower elevations; if the habitat is periodically disturbed, virtually all other species are excluded. Mostly in the northern part of the state, Russian Olive, has become so common as to be a nuisance. Species of special concern, if ever present, probably could not survive in the new habitats produced by these weedy trees.

The Mescalero Thistle, *Cirsium vinaceum*, discussed several places elsewhere in this chapter, serves to illustrate the complexity involved in estimating the impact of exotics. As Table 11.3 shows, more threats impinge upon this species than upon any other.

The Mescalero Thistle is restricted to springs and seeps in the Sacramento Mountains, isolated in the south-central part of the state. Another thistle, *Carduus nutans*, the musk thistle, may have been introduced into the western United States for its ornamental value; it is a handsome accent plant in the garden. A second thistle-like species, *Dipsacus sylvestris* (Teasel), also has been introduced in wet areas in these mountains. Both are now unwelcome, aggressive, obnoxious range weeds throughout much of the western United States. Both grow very well around springs and seeps and in marshy valley bottoms, crowding into populations of *Cirsium vinaceum*. The three species plants are of similar height and require similar living space; the presence of the other two obviously means less room for the *Cirsium*. Moreover, continual disturbance to *Cirsium* sites, in this region commonly by grazing, probably favors the introduced weeds. Even though a significant population of the Mescalero Thistle recently has been fenced, it is still too early to determine if protection from grazing in its habitat will result in a decline of the *Carduus* or *Dipsacus*.

An equally complex, and potentially more harmful, impact to the Mescalero Thistle might have occurred if certain efforts to control the musk thistle had been carried out by agriculturists without consultation with a conservation biologist. Several introduced European weeds that economically damage rangelands in the western United States were brought under control by introducing specific insect pests from their home range. A weevil that parasitizes the seed of *Carduus nutans* had been introduced in the Intermountain West and seemed to be of promise in the control of this thistle. Introduction of the weevil in the Sacramento Mountains was proposed, but it was

pointed out that *Carduus* and *Cirsium* are very closely related. There was no assurance that the weevil might be as (or more) effective in eliminating the rare *Cirsium*, which might have no natural defenses against the insect, as it would the noxious *Carduus*. Studies of the effect of the weevil on the *Cirsium* are now under way.

Although not specifically addressing the impact of exotic species on a rare one, the Pecos Sunflower, *Helianthus paradoxus* (discussed earlier under the section on water development and management), illustrates another potential biological problem for plant species when two close relatives are in proximity. *Helianthus petiolaris*, a native annual in the Southwest prefers dry, disturbed areas. In the time before settlement and disturbance of the landscape, the two species probably did not come in contact. Although native, in one sense *H. petiolaris* may be considered an exotic when, along roadsides and tops of ditch banks, it approaches the habitat of *H. paradoxus* in disturbed areas in the Pecos River Valley. As judged from observations on morphology (P. Knight, pers. comm.), where the two come into contact introgression may be occurring. That is, the two species are hybridizing, and across the hybrid "bridge," genes from one species may enter the other. If this is the case, the genotype of *H. paradoxus* might become diluted with genes from the more common *H. petiolaris*, particularly around the margins of wet areas, where hybrid genotypes might be more adaptive. In prime habitat for *H. paradoxus*, the interplay of several factors could result in either the survival or extinction of the pure *H. paradoxus* genetic line. Such factors are the degree of disturbance, frequency of visitation by insects bringing in foreign versus same-species pollen, relative rates of pollen tube growth when simultaneously pollinated with *H. petiolaris* and *H. paradoxus* pollen, and the intensity of selection in wet habitats for only those genotypes that are pure *H. paradoxus*. This natural experiment will be interesting to follow over the years.

ACKNOWLEDGMENTS

Several people have provided information and critical reviews during the preparation of this chapter. The assistance of Reggie Fletcher (U.S. Forest Service), Paul Knight (N.M. Department of Natural Resources), and Charles MacDonald (U.S. Fish and Wildlife Service) is gratefully acknowledged.

Table 11.1. Distribution of rare, endemic, and threatened or endangered plants in New Mexico, by broad geographic region and by county within regions. See Fig. 11.1 for map of six sectors of New Mexico. (Data extracted primarily from NMNPPAC 1984.) E = endangered, T = threatened, and O = other species of special concern.

Region	Number of species			Region	Number of species		
	E	T	O		E	T	O
NORTHWEST				**SOUTH-CENTRAL**			
Cibola	0	0	9	Doña Ana	1	0	27
McKinley	0	1	6	Lincoln	0	0	24
San Juan	1	1	9	Otero	3	1	27
Region total	1	2	26	Region total	2	1	62
NORTH-CENTRAL				**SOUTHEAST**			
Bernalillo	0	0	13	Chavez	1	0	5
Los Alamos	0	0	2	De Baca	0	0	1
Rio Arriba	0	0	9	Eddy	0	3	18
Sandoval	0	0	10	Lea	0	0	1
Santa Fe	0	0	11	Roosevelt	0	0	1
Taos	0	0	10				
Valencia	0	0	7	Region total	1	3	27
Region total	0	0	34	**SOUTHWEST**			
				Catron	0	1	17
NORTHEAST				Grant	0	0	15
Colfax	0	0	3	Hidalgo	0	0	11
Curry	0	0	1	Luna	0	0	11
Guadalupe	0	0	2	Sierra	1	0	14
Harding	0	0	5	Socorro	0	0	21
Mora	0	0	1				
Quay	0	0	0	Region total	1	1	57
San Miguel	0	0	6				
Union	0	0	0				
Region total	0	0	11				

Table 11.2. Distribution of rare, endemic, and threatened or endangered plants in New Mexico by family, indicating number of named entities that are "good" taxa, the number that are doubtful (?) and the number of listed endangered (E) and threatened (T) taxa.

	Taxa		Total		
	Good	?	entities	E	T
Apiaceae	4	0	4	0	0
Apocynaceae	1	0	1	0	0
Asteraceae	26	6	32	0	2
Boraginaceae	2	0	2	0	0
Brassicaceae	4	0	4	0	0
Cactaceae	16	9	25	4	2
Capparidaceae	1	0	1	0	0
Caryophyllaceae	3	0	3	0	0
Chenopodiaceae	2	0	2	0	0
Commelinaceae	1	0	1	0	0
Crossomataceae	1	0	1	0	0
Cucurbitaceae	1	0	1	0	0
Euphorbiaceae	1	0	1	0	0
Fabaceae	23	4	25	0	1
Hydrophyllaceae	1	0	1	0	0
Lamiaceae	6	0	6	1	1
Liliaceae	1	0	1	0	0
Loasaceae	1	0	1	0	0
Malvaceae	1	2	3	0	0
Martyniaceae	1	0	1	0	0
Nyctaginaceae	1	0	1	0	0
Onagraceae	1	0	1	0	0
Papaveraceae	1	0	1	1	0
Poaceae	2	0	2	0	0
Polemoniaceae	3	0	3	0	0
Polygalaceae	2	0	2	0	0
Polygonaceae	3	2	5	0	1
Portulaceae	2	0	2	0	0
Ranunculaceae	2	0	2	0	0
Rosaceae	3	0	3	0	0
Saxifragaceae	2	0	2	0	0
Scrophulariaceae	11	1	12	0	0
Valerianaceae	1	0	1	0	0
Totals	131	24	153	5	7

Table 11.3. Plant species that are endemic, rare, or in decline in New Mexico, the approximate area of the state occupied by the range of the species, and the estimated sources influencing the species (primarily from NMNPPAC 1984).

Those that are federally listed are denoted with an asterisk (*), the category threatened or endangered indicated by (T) or (E) following the scientific name. The level that each species is ranked in the federal scheme of classification is given in parentheses following the scientific name; (–) means that the name does not appear on federal candidate list (*Federal Register* 9/27/85 and 2/21/90). See discussion in text regarding definition of federal categories.

The number of species a specific threat affects is indicated in parentheses following the listing of the threats below.

Column numbers signify: 1 = number of counties of occurrence; 2 = rough estimate of % of state's area occupied by species range. Possible or potential influence on species at present: 3 = urban development (9); 4 = tilled agriculture (3); 5 = grazing by wildlife or domestic livestock (19) (see Ch. 2, 7, 8); 6 = logging (7); 7 = mineral extraction (5); 8 = water development and management (11); 9 = recreational use of land (21); 10 = collectors (23); 11 = road development and management, including herbicide spraying (44); 12 = threat by other organisms (5) (see Ch. 2).

Scientific, common, and family names	1 #Co	2 %NM	3 Urb	4 Agr	5 Grz	6 Log	7 Min	8 Wat	9 Rec	10 Col	11 Rds	12 Org
Abronia bigelovii (3C) Tufted Sand Verbena (Nyctaginaceae)	3	5					*		*			
Agastache cana (–) Grayish-White Giant Hyssop (Lamiaceae)	5	20										
Agastache mearnsii (–) Mearn's Giant Hyssop (Lamiaceae)	2	5										
Aletes filifolius (3C) Threadleaf False Carrot (Apiaceae)	9	35										
Aletes sessiliflorus (–) Sessile-Flowered False Carrot (Apiaceae)	4	5										
Allium gooddingii (1) Goodding's Onion (Liliaceae)	2	5			*	*			*			
Amsonia fugatei Fugate's amsonia (Apocynaceae)	1	<1										
Apacheria chiricahuensis (3C) Cliff Brittlebush (Crossomataceae)	2	1–5										
Aquilegia chaplinei (3C) Chapline's Columbine (Ranunculaceae)	1	<1						*	*			
Argemone pleiacantha ssp. *pinnatisecta* (E) Sacramento Prickly Poppy (Papaveraceae)	1	1						*			*	

Table 11.3. (continued).

Scientific, common, and family names	1 #Co	2 %NM	3 Urb	4 Agr	5 Grz	6 Log	7 Min	8 Wat	9 Rec	10 Col	11 Rds	12 Org
Aster blepharophyllus (2) Playa aster (Asteraceae)	1	<1						*				
Aster horridus (–) Spiny Aster (Asteraceae)	3	1–5										
Aster laevis var. *quadalupensis* (–) Smooth aster (Asteraceae)	1	<1							*			
Aster neomexicanus (–) New Mexico Aster (Asteraceae)	1	1										
Astragalus accumbens (3C) Zuni Milk Vetch (Fabaceae)	1	1				*					*	
Astragalus altus (3C) Tall Milk Vetch (Fabaceae)	1	1				*			*		*	
Astragalus castetteri (3C) Castetter's Milk Vetch (Fabaceae)	2	1–5									*	
Astragalus chuskanus (–) Chuska Mtn. Milk Vetch (Fabaceae)	1	1									*	
Astragalus cyaneus (–) Cyanic Milk Vetch (Fabaceae)	3	1–5	*								*	
Astragalus feensis (–) Santa Fe Milk Vetch (Fabaceae)	3	1–5									*	
Astragalus gypsodes (2) Gypsum Milk Vetch (Fabaceae)	1	<1							*		*	
*Astragalus humillimus (E) Mancos Milk Vetch (Fabaceae)	1	1					*				*	
Astragalus kentrophyta var. *neomexicanus* (–) Spiny-leaf Milk Vetch (Fabaceae)	7	25							*		*	
Astragalus kerrii (–) Kerr Milk Vetch (Fabaceae)	1	1										
Astragalus knightii (–) Knight's Milk Vetch (Fabaceae)	1	<1										

Table 11.3. (continued).

Scientific, common, and family names	1 #Co	2 %NM	3 Urb	4 Agr	5 Grz	6 Log	7 Min	8 Wat	9 Rec	10 Col	11 Rds	12 Org
Astragalus micromerius (–) Chaco Milk Vetch (Fabaceae)	3	15										
Astragalus mollissimus var. *mathewsii* (–) Mathew's Woolly Milk Vetch (Fabaceae)	4	5									*	
Astragalus monumentalis vars. *cottamii* and *monumentalis* (3C) Cottam's and Monument Valley Milk Vetches (Fabaceae)	1	1									*	
Astragalus naturitensis (3C) Naturita Milk Vetch (Fabaceae)	1	<1										
Astragalus neomexicanus (–) New Mexico Milk Vetch (Fabaceae)	2	1–5									*	
Astragalus oocalycis (3C) Arborales Milk Vetch (Fabaceae)	2	1–5									*	
Astragalus puniceus var. *gertrudis* (3B) Taos Milk Vetch (Fabaceae)	2	1–5									*	
Astragalus ripleyi (2) Ripley's Milk Vetch (Fabaceae)	2	<1			*						*	
Astragalus siliceus (3C) Flint Mtns. Milk Vetch (Fabaceae)	1	1									*	
Astragalus wittmanii (3C) One-flowered Milk Vetch (Fabaceae)	2	1–5									*	
Atriplex griffithsii (3C) Griffith's Saltbush (Chenopodiaceae)	2	1		*	*						*	
Atriplex pleiantha (2) Mancos Saltbush (Chenopodiaceae)	1	1		*			*				*	
Besseya oblongifolia (–) Sierra Blanca Kittentails (Scrophulariaceae)	1	<1										
Brickellia chenopodina (–) Gila Brickelbush (Asteraceae)	1	<1						*				

Table 11.3. (continued).

Scientific, common, and family names	1 #Co	2 %NM	3 Urb	4 Agr	5 Grz	6 Log	7 Min	8 Wat	9 Rec	10 Col	11 Rds	12 Org
Castilleja organorum (–) Organ Mtn. Paintbrush (Scrophulariaceae)	4	5–10										
Castilleja wootoni (–) Wooton's Paintbrush (Scrophulariaceae)	2	1–5										
Cereus greggii (2) Night-blooming Cereus (Cactaceae)	4	5	*		*				*	*		
Chaetopappa elegans (2) Sierra Blanca Cliff Daisy (Asteraceae)	1	<1										
Chaetopappa hersheyi (2) Hershey's Cliff Daisy (Asteraceae)	1	<1										
Chrysothamnus spathulatus (–) Spoonleaf Rabbitbrush (Asteraceae)	4	20										
Cirsium gilense (–) Gila Thistle (Asteraceae)	1	<1										
Cirsium inornatum (–) Plain Thistle (Asteraceae)	5	30										
Cirsium vinaceum (T) Mescalero Thistle (Asteraceae)	1	<1			*	*		*	*		*	*
Cleome multicaulis (2) Slender Spiderflower (Capparaceae)	1	<1			*			*				*
Coryphantha duncanii (3C) Duncan's Pincushion Cactus (Cactaceae)	1	<1								*		
Coryphantha organensis (–) Organ Mtn. Pincushion Cactus (Cactaceae)	1	<1										
Coryphantha scheeri var. *scheeri* (–) Scheer's Pincushion Cactus (Cactaceae)	5	<1								*		
Coryphantha scheeri var. *valida* (–)	5	15								*		
Coryphantha sneedii var. *leei* (T) Lee's Pincushion Cactus (Cactaceae)	1	<1								*		

Table 11.3. (continued).

Scientific, common, and family names	1 #Co	2 %NM	3 Urb	4 Agr	5 Grz	6 Log	7 Min	8 Wat	9 Rec	10 Col	11 Rds	12 Org
Coryphantha sneedii var. *sneedii* (E) Sneed's Pincushion Cactus (Cactaceae)	3	1–5	*							*		
Crataegus wootoni (–) Wooton's Hawthorne (Rosaceae)	2	5										
Cryptantha paysonii (–) Payson's Hiddenflower (Boraginaceae)	5	20										
Dalea scariosa (3C) La Jolla Prairie Clover (Fabaceae)	3	1–5	*								*	
Delphinium alpestre (–) Alpine Larkspur (Ranunculaceae)	1	1							*			
Draba mogollonica (3C) Mogollon Whitlowgrass (Brassicaceae)	4	5										
Echinocereus fendleri var. *kuenzleri* (E) Kuenzler's Hedgehog Cactus (Cactaceae)	2	1			*					*		
Echinocereus lloydii (E) Lloyd's Hedgehog Cactus (Cactaceae)	3	1			*		*			*	*	
Erigeron acomanus (–) Acoma Fleabane (Asteraceae)	2	1–5										
Erigeron hessii (2) Hess's Fleabane (Asteraceae)	1	<1										
Erigeron rhizomatus (T) Zuni Fleabane (Asteraceae)	2	1					*		*			
Erigeron rybius (–) Sacramento Mtn. Fleabane (Asteraceae)	2	1–5									*	
Erigeron scopulinus (–) Rock Fleabane (Asteraceae)	3	5										
Erigeron sivinskii (–) Sivinski's Fleabane (Asteraceae)	1	<1										
Erigeron subglaber (–) Pecos Fleabane (Asteraceae)	1	1										

Table 11.3. (continued).

Scientific, common, and family names	1 #Co	2 %NM	3 Urb	4 Agr	5 Grz	6 Log	7 Min	8 Wat	9 Rec	10 Col	11 Rds	12 Org
Eriogonum aliquantum (–) Springer Wild Buckwheat (Polygonaceae)	1	<1										
Eriogonum densum (3B) Woolly Wild Buckwheat (Polygonaceae)	1	<1										
Eriogonum gypsophilum (T) Gypsum Wild Buckwheat (Polygonaceae)	1	1						*	*		*	
Eriogonum jamesii var. *wootoni* (–) Wooton's Wild Buckwheat (Polygonaceae)	2	1–5									*	
Escobaria orcuttii var. *koenigii* (–) Koenig's Pincushion Cactus (Cactaceae)	1	<1								*		
Escobaria orcuttii var. *macraxina* (–)	1	<1								*		
Big Hatchet Pincushion Cactus (Cactaceae)	1	<1										
Escobaria sandbergii (–) Sandberg's Pincushion Cactus (Cactaceae)	2	1								*		
Escobaria villardii (–) Villard's Pincushion Cactus (Cactaceae)	1	<1								*		
Euphorbia strictior (3C) Tall Plains Spurge (Euphorbiaceae)	3	1–5										
Ferocactus wislizenii (–) Southwestern Barrel Cactus (Cactaceae)	5	15	*							*	*	
Gilia formosa (2) Beautiful Gilia (Polemoniaceae)	1	1							*		*	
Haplopappus microcephalus (2) Small-headed Goldenweed (Asteraceae)	1	<1										
Hedeoma apiculatum (T) McKittrick Pennyroyal (Lamiaceae)	1	<1							*			
Hedeoma pulcherrimum (3C) Mescalero Pennyroyal (Lamiaceae)	2	1–5									*	

Table 11.3.   (continued).

Scientific, common, and family names	1 # Co	2 % NM	3 Urb	4 Agr	5 Grz	6 Log	7 Min	8 Wat	9 Rec	10 Col	11 Rds	12 Org
Hedeoma todsenii (E) Todsen's Pennyroyal (Lamiaceae)	1	<1			*							
Helianthus paradoxus (1) Pecos Sunflower (Asteraceae)	2	5						*				*
Helianthus praetermissus (3A) Lost Sunflower (extinct) (Asteraceae)	1	<1						*				*
Heuchera pulchella (−) Sandia Alumroot (Saxifragaceae)	3	5										
Heuchera wootonii (−) Wooton's Alumroot (Saxifragaceae)	2	5										
Hymenoxys olivacea (−) Olivaceous Bitterweed (Asteraceae)	1	<1										
Hymenoxys vaseyi (−) Vasey's Bitterweed (Asteraceae)	2	1–5										
Iliamna grandiflora (−) Wild Hollyhock (Malvaceae)	2	1–5										
Ipomopsis sancti-spiritus (2) Holy Ghost Gilia (Polemoniaceae)	1	<1										
Ipomopsis pinnatifida (−) Bent-flowered Gilia (Polemoniaceae)	1	<1										
Lepidospartum burgessii (2) Gypsum Scalebroom (Asteraceae)	1	<1									*	
Lesquerella aurea (3C) Golden Bladderpod (Brassicaceae)	1	<1					*				*	
Lesquerella gooddingii (3B) Goodding's Bladderpod (Brassicaceae)	1	1										
Lupinus sierra-blancae (−) Sierra Blanca Lupine (Fabaceae)	1	<1									*	
Mammilaria wrightii var. *wilcoxii* (−) Wilcox's Pincushion Cactus (Cactaceae)	1	1								*		

Table 11.3. (continued).

Scientific, common, and family names	1 #Co	2 %NM	3 Urb	4 Agr	5 Grz	6 Log	7 Min	8 Wat	9 Rec	10 Col	11 Rds	12 Org
Mammilaria wrightii var. *wrightii* (–) Wright's Pincushion Cactus (Cactaceae)	10	35			*					*		
Mammilaria viridiflora (–) Green-Flowered Pincushion Cactus (Cactaceae)	1	<1								*		
Machaeranthera amplifolia (–) Organ Mtn. Aster (Asteraceae)	1	<1										
Mentzelia perennis (–) Gypsum Blazing Star (Loasaceae)	4	10									*	
Mertensia viridis var. *caelestina* (–) Alpine Bluebell (Boraginaceae)	4	1–5										
Nama xylopodum (3C) Cliff Nama (Hydrophyllaceae)	2	1										
Oenothera organensis (2) Organ Mtn. Evening Primrose (Onagraceae)	1	<1							*			
Opuntia arenaria (2) Sand Prickly Pear (Cactaceae)	1	1	*	*					*	*	*	
Opuntia clavata (–) Dagger-thorn Cholla (Cactaceae)	13	30									*	
Opuntia viridiflora (2) Santa Fe Cholla (Cactaceae)	2	1	*						*	*	*	
*Pediocactus knowltonii (E) Knowlton's Cactus (Cactaceae)	1	<1								*	*	
Penstemon alamosensis (2) Alamo Beardtongue (Scrophulariaceae)	2	1–5		*						*		
Penstemon cardinalis ssp. *cardinalis* (–) White Mtn. Beardtongue (Scrophulariaceae)	2	1										
Penstemon cardinalis ssp. *regalis* (–) Guadalupe Beardtongue (Scrophulariaceae)	1	1										

Table 11.3. (continued).

Scientific, common, and family names	1 #Co	2 %NM	3 Urb	4 Agr	5 Grz	6 Log	7 Min	8 Wat	9 Rec	10 Col	11 Rds	12 Org
Penstemon dasyphyllus (–) Gila Beardtongue (Scrophulariaceae)	2	1–5										
Penstemon inflatus (–) Bluebell Beardtongue (Scrophulariaceae)	7	10										
Penstemon neomexicanus (–) New Mexico Beardtongue (Scrophulariaceae)	2	1–5										
Penstemon pseudoparvus (–) San Mateo Beardtongue (Scrophulariaceae)	1	1										
Perityle cernua (2) Nodding Cliff Daisy (Asteraceae)	1	<1										
Perityle quinquiflora (–) Five-flowered Rock Daisy (Asteraceae)	1	<1										
Perityle staurophylla (3C) Starleaf Rock Daisy (Asteraceae)	3	5										
Phlox caryophylla (3C) Pagosa Phlox (Polemoniaceae)	1	1									*	
Polygala rimulicola var. *mescalerorum* (2) Mescalero Milkwort (Polygalaceae)	1	<1										
Polygala rimulicola var. *rimulicola* (3C) Guadalupe Milkwort (Polygalaceae)	1	<1										
Potentilla sierra-blancae (3C) Sierra Blanca Cinquefoil (Rosaceae)	1	<1								*		
Proboscidea sabulosa (2) Dune Unicorn Plant (Martyniaceae)	3	5–10										
Pseudocymopterus *longiradiatus* (–) Desert Parsley (Apiaceae)	2	1–5										
Pteryxia davidsonii (–) Davidson's Cliff Carrot (Apiaceae)	3	1–5										

Table 11.3. (continued).

Scientific, common, and family names	1 #Co	2 %NM	3 Urb	4 Agr	5 Grz	6 Log	7 Min	8 Wat	9 Rec	10 Col	11 Rds	12 Org
Puccinellia parishii (2) Parish's Dropseed (Poaceae)	1	<1			*			*				
Rumex tomentellus (–) Mogollon Dock (Polygonaceae)	1	<1			*				*			
Salvia summa (–) Supreme Sage (Lamiaceae)	2	5										
*Sclerocactus mesa-verdae (T) Mesa Verde Cactus (Cactaceae)	1	1–5	*						*	*	*	
Sclerocactus whipplei vars. *heilii* and *reevesii* (2) (3C) Hardwall Cactus (Cactaceae)	1	<1								*		
Scrophularia laevis (–) Organ Mtn. Figwort (Scrophulariaceae)	1	<1										
Scrophularia macrantha (1) Mimbres Figwort (Scrophulariaceae)	2	1										*
Senecio cardamine (3C) Heartleaf Senecio (Asteraceae)	1	<1				*						
Senecio quaerens (2) Gila Groundsel (Asteraceae)	1	<1			*	*		*	*			
Senecio sacramentanus (–) Sacramento Groundsel (Asteraceae)	5	15										*
Senecio spellenbergii (–) Spellenberg's senecio (Asteraceae)	1	<1		*								
Sibara grisea (3C) Gray Sibara (Brassicaceae)	1	<1			*							
Sicyos glaber (–) Smooth Cucumber (Cucurbitaceae)	1	<1										
Silene plankii (3C) Plank's Catchfly (Caryophyllaceae)	4	10			*							
Silene wrightii (3C) Wright's Catchfly (Caryophyllaceae)	5	10										

Table 11.3. (continued).

Scientific, common, and family names	1 \# Co	2 % NM	3 Urb	4 Agr	5 Grz	6 Log	7 Min	8 Wat	9 Rec	10 Col	11 Rds	12 Org
Sophora gypsophila var. *guadalupensis* (2) Guadalupe Mtn. Mescal Bean (Fabaceae)	1	<1										
Sphaeralcea procera (2) Porter's Globemallow (Malvaceae)	1	<1										
Sphaelalcea wrightii (–) Wright's Globemallow (Malvaceae)	2	5									*	
Stipa curvifolia (3C) Curlleaf Needlegrass (Poaceae)	2	5			*							
Talinum humile (2) Pinos Altos Flame Flower (Portulacaceae)	1	<1										
Talinum longipes (–) Long-Stemmed Talinum (Portulacaceae)	1	<1										
Tetradymia filifolia (–) Threadleaf Horsebrush (Asteraceae)	5	15									*	
Toumeya papyracantha (2) Grama Grass Cactus (Cactaceae)	9	25	*		*				*	*	*	*
Tradescantia wrightii (3C) Wright's Spiderlily (Commelinaceae)	4	20										
Trifolium longipes var. *neurophyllum* (–) Mogollon Clover (Fabaceae)	1	1–5			*			*				
Valeriana texana (2) Texas Tobacco Root (Valerianaceae)	2	1–5										
Vauquelinia pauciflora (2) Few-flowered Rosewood (Rosaceae)	1	<1			*							

Table 11.4. Nature of endemism in New Mexico plant species whose geographic range is estimated to occupy 5% or less of the overall area of the state. 49 entities are listed; the remaining 99 are endemics where no single strong environmental factor is identifiable and are not listed here.

Edaphic endemics	On cliffs or rock faces, or at bases of cliffs	Dependent on very wet habitat
Abronia bigelovii	*Apacheria chiricahuensis*	*Aquilegia chaplinei*
Aletes sessiliflorus	*Chaetopappa elegans*	*Aster blepharophyllus*
Amsonia fugatei	*Chaetopappa hersheyi*	*Aster neomexicanus*
Astragalus accumbens	*Erigeron acomanus*	*Cleome multicaulis*
Astragalus chuskanus	*Erigeron scopulinus*	*Cirsium vinaceum*
Astragalus gypsodes	*Haplopappus microcephalus*	*Helianthus paradoxus*
Astragalus humillimus	*Hedeoma apiculatum*	*Helianthus praetermissus*
Astragalus knightii	*Nama xylopodum*	*Iliamna grandiflora*
Astragalus siliceus	*Perityle cernua*	*Oenothera organensis*
Astragalus wittmanii	*Perityle quinqueflora*	*Sporobolus parishii*
Atriplex griffithsii	*Perityle staurophylla*	
Atriplex pleiantha	*Polygala rimulicola*	
Erigeron rhizomatus	*Pteryxia davidsonii*	
Erigeron siuinskii	*Salvia summa*	
Eriogonum gypsophilum	*Sibara grisea*	
Euphorbia strictior	*Silene plankii*	
Gilia formosa	*Silene wrightii*	
Hedeoma todsenii		
Lepidospartum burgessii		
Opuntia arenaria		
Pediocactus knowltonii		
Proboscidea sabulosa		
Sclerocactus mesa-verdae		
Senecio spellenbergii		

CHAPTER 11 REFERENCES

SPECIES OF SPECIAL CONCERN

Ayensu, E. S. and R. A. DeFilipps, 1978. Endangered and threatened plants of the United States. Smithsonian Institution Press, Washington, D.C. 403 pp.

Callicot, J. B., 1986. On the intrinsic value of nonhuman species. Ch. 6, pp. 138–172. In: B. G. Norton (ed.). The preservation of species. The value of biological diversity. Princeton University Press, Princeton, NJ. 305 pp.

Cully, A. C., D. E. House, and P. J. Knight, 1985. Status report on *Atriplex pleiantha*. Prepared under contract for U.S. Fish and Wildlife Service, Region 2, Albuquerque. 32 pp. typescript.

Ehrlich, P. R., 1980. The strategy of conservation,

1980–2000. Ch. 19, pp. 329–344. In: M. E. Soule and B. A. Wilcox (eds.). Conservation Biology. An evolutionary-ecological perspective. Sinauer Assoc., Inc. Sunderland, MA. 395 pp.

Ehrlich, P. R. and A. Ehrlich, 1981. Extinction. The causes and consequences of the disappearance of species. Random House, New York, NY. 305 pp.

Ferguson, D. and S. Brack, 1986. Status report on *Cylindropuntia viridiflora*. Prepared under contract for the U.S. Fish and Wildlife Service, Region 2, Albuquerque, 19 pp. typescript.

Fletcher, R., 1978. Status report on *Cirsium vinaceum*. Prepared by U.S. Forest Service, Region 3, Albuquerque, for U.S. Fish and Wildlife Service, Region 2, Albuquerque.

————, 1979a. Provisional status report supplement, *Astragalus altus*. In-house document prepared by U.S. Forest Service, Region 3, Albuquerque, 10 pp. typescript.

————, 1979b. *Helianthus praetermissus* and *Helianthus paradoxus*. In-house review for U.S. Forest Service, Region 3, Albuquerque. 34 pp., including typescript and photocopies of literature.

————, 1980. Provisional status report; *Senecio quarens*. U.S. Forest Service, Region 3, Albuquerque. 5 pp. typescript.

————, 1984. *Allium gooddingii* status report supplement. Prepared by the U.S. Forest Service, Region 3, Albuquerque, for the U.S. Fish and Wildlife Service, Region 2, Albuquerque. 28 pp. typescript.

Heil, K. D., 1984. Mesa Verde cactus (*Sclerocactus mesae-verdae*) recovery plan. Prepared under contract for the U.S. Fish and Wildlife Service, Region 2, Albuquerque. 63 pp. typescript.

————, 1985. Recovery plan for the Knowlton cactus (*Pediocactus knowltonii*) L. Benson. U.S. Fish and Wildlife Service, Region 2, Albuquerque. 53 pp.

Heil, K. D. and S. Brack, 1986. Sneed and Lee pincushion cacti (*Coryphantha sneedii* var. *sneedii*, *Coryphantha sneedii* var. *leei*) recovery plan. Prepared under contract for the

U.S. Fish and Wildlife Service, Region 2, Albuquerque. 53 pp.

Heiser, C. B., Jr., 1958. Three new annual sunflowers (*Helianthus*) from the southwestern United States. Rhodora, 60: 272–283.

————, 1969. The North American sunflowers (*Helianthus*). Mem. Torr. Bot. Club, 22: 1–218.

Knight, P. J., 1985. Status report on *Sphaeralcea procera*. Prepared under contract for the U.S. Fish and Wildlife Service, Region 2, Albuquerque. 128 pp. typescript.

Koopowitz, H. and H. Kaye, 1983. Plant extinction: a global crisis. Stone Wall Press, Inc., Washington, D.C. 239 pp.

LaDuke, J. C., 1985. A new species of *Sphaeralcea* (Malvaceae). Southw. Naturalist, 30: 433–436.

Limerick, S., 1984. Recovery plan for gypsum wild buckwheat (*Eriogonum gypsophilum* Wooton and Standley). Prepared by U.S. Fish and Wildlife Service, Region 2, Albuquerque. 34 pp.

Lovejoy, T. E., 1979. The epoch of biotic impoverishment. pp. 5–10. In: Endangered Species: a symposium. Great Basin Naturalist Memoirs, No. 3.

————, 1980. Foreword, pp. ix–x. In: M. E. Soule and B. A. Wilcox (eds.). Conservation Biology. An evolutionary-ecological perspective. Sinauer Assoc., Inc. Sunderland, MA. 395 pp.

————, 1986. Species leave the ark one by one. Ch. 1, pp. 13–27. In: B. G. Norton (ed.). The preservation of species. The value of biological diversity. Princeton University Press, Princeton, NJ. 305 pp.

Martin, W. C. and C. R. Hutchins, 1980, 1981. A flora of New Mexico. Cramer, Vaduz. 2 vols. 2591 pp.

Miller, J. T., 1985. Living in the environment (4th ed.). Wadsworth Publishing Co., Belmont, CA. 460 pp. + suppl.

Moran, J. M., M. D. Morgan, and J. H. Wiersma, 1986. Introduction to environmental science (2nd ed.). W. H. Freeman & Co., NY. 709 pp.

NMNPPAC (New Mexico Native Plant Protection

Advisory Committee), 1984. A handbook of rare and endemic plants of New Mexico. University of New Mexico Press, Albuquerque, NM. 291 pp.

Reveal, J. L., 1981. The concept of rarity and population threats in plant communities. pp. 41–47. In: L. E. Morse and M. S. Henifin (eds.). Rare plant conservation: geographical data organization. NY Botanical Garden, Bronx, NY.

Slobodkin, L. B., 1986. On the susceptibility of different species to extinction: elementary instructions for owners of a world. Ch. 9, pp. 226–242. In: B. G. Norton (ed.). The preservation of species. The value of biological diversity. Princeton University Press, Princeton, NJ. 305 pp.

Smithsonian Institution, 1975. Report on endangered and threatened plant species of the United States. 94th Congress House Document No. 94-51. U.S. Govt. Printing Office, Washington, D.C. 200 pp.

Soreng, R. J., 1981. Status report on Lesquerella aurea. Prepared under contract for the U.S. Fish and Wildlife Service, Region 2, Albuquerque. 31 pp. typescript.

Soule, M. E. and B. A. Wilcox, 1980. Conservation biology: its scope and challenge. Ch. 1, pp. 1–8. In: M. E. Soule and B. A. Wilcox (eds.). Conservation Biology. An evolutionary-ecological perspective. Sinauer Assoc., Inc. Sunderland, MA. 395 pp.

Spellenberg, R. W., 1977. A report on the investigation of Eriogonum gypsophilum and Haplopappus spinulosus subspecies of laevis in the vicinity of the Brantley Reservoir, Eddy Co., N.M. Prepared under contract for the U.S. Bureau of Reclamation, Amarillo. 140 pp. typescript.

———, 1981. Status report on Penstemon alamosensis. Prepared under contract for the U.S. Fish and Wildlife Service, Region 2, Albuquerque. 21 pp. typescript.

———, 1982. Status report on Allium gooddingii. Prepared under contract for the U.S. Fish and Wildlife Service, Region 2, Albuquerque. 53 pp. typescript.

Spellenberg, R., R. Worthington, P. Knight, and R. Fletcher. 1986. Additions to the flora of New Mexico. Sida 11: 455–470.

Spellenberg, R. W., C. Leiva, and E. Lessa, 1988. An evaluation of Eriogonum densum (Polygonaceae). Southw. Naturalist, 33:71–80.

Terborgh, J. and B. Winter, 1980. Some causes of extinction. Ch. 7, pp. 119–132. In: M. E. Soule and B. A. Wilcox (eds.). Conservation Biology. An evolutionary-ecological perspective. Sinauer Assoc., Inc. Sunderland, MA. 395 pp.

USFWS (United States Fish and Wildlife Service), 1987. Endangered and threatened species of Arizona and New Mexico. U.S. Fish and Wildlife Service, Region 2, Albuquerque, NM. 124 pp.

Vermeij, G. J., 1986. The biology of human caused extinction. Ch. 2, pp. 28–49. In: B. G. Norton (ed.). The preservation of species. The value of biological diversity. Princeton University Press, Princeton, NJ. 305 pp.

Whitmore, T. C., 1980. The conservation of tropical rainforest. Ch. 17, pp. 303–318. In: M. E. Soule and B. A. Wilcox (eds.). Conservation Biology. An evolutionary-ecological perspective. Sinauer Assoc., Inc. Sunderland, MA. 395 pp.

Woodwell, G. M., 1977. The challenge of endangered species. pp. 5–10. In: G. T. Prance and T. S. Elias (eds.). Extinction is forever. NY Botanical Garden, Bronx, NY. 437 pp.

12

THE FUTURE OF
NEW MEXICO VEGETATION

NEED FOR PRESERVATION

All New Mexico citizens should want to protect examples of their diverse and often unique natural heritage. This natural heritage includes geological, zoological, botanical, and scenic features. Pride in our natural heritage and the desire to set aside examples for ourselves and future generations should be reason enough to ensure the safeguarding of these natural features. However, in today's complex society, portions of the state's flora, fauna, geology, and space have commercial value, and there is strenuous competition for their utilization. Aesthetic motivations, such as pride and living museum potential, are seldom sufficient to compete with commercial demands. The public needs to be convinced that protection of our natural heritage is of real value and will be of lasting importance to the state.

The need for some sort of monitoring and/or preservation of the country's natural heritage has been recognized by many for a long time. Over 70 years ago, "Man's destruction of the natural landscape appeared so widespread and pervasive that in 1917 the newly organized Ecological Society of America appointed Shelford the chairman of a committee to find out what remained of wild, natural America and to promote the idea of a system of natural preserves" (Moir, 1972). In 1959, H. H. Iltis proposed a system of "scenic areas" for the state of Wisconsin: "First of all, it is here that we can preserve permanently the tremendous complexity of undisturbed natural biotic communities, so that we further the understanding of nature for the benefit of man by basic research into the ecology, taxonomy, and economic uses of its organisms, and to transmit such an understanding to students of this and later generations." Jenkins and Bedford (1973) proposed a preservation program for the United States: "In order that wise decisions can be made in environmental management, an understanding is needed of ecosystem functioning and reaction to change. To obtain this information we must have continuing knowledge of the undisturbed ecosystem as a base line against which to measure the effects of modifications. It is proposed that relatively undisturbed natural areas form the basic research tool for the establishment of such baselines." In 1972, the Illinois Nature Preserves Commission and the Illinois Department of Conservation published a document that included a list and explanations of the values of a system of "nature preserves" for the state:

> Pursuit of Knowledge
> Maintaining a Healthful Environment
> Guidance in Land Use
> Reserves of Breeding Stock
> Unknown Uses of Wild Creatures
> Natural Beauty
> Living Museums
> Outdoor Classrooms
> Sanctuaries

Due to their economic value, many of the natural features of New Mexico have been extensively used or modified by man during the past 200 years. Some of these natural features can be considered "renewable," but we must learn how to manage them if they are to be renewable in fact. Natural vegetation is one of these potentially renewable resources, one that has been of great economic importance; the lumber and livestock industries played a major role in the settling and development of the state. Ironically, some of the most extensive and most important portions of the state's vegetation have few if any parcels of sufficient size set aside to maintain their integrity.

There are virtually no large areas of prime grama grass or prime stands of subalpine and montane forest kept in a potential, natural or bench-mark status. It is puzzling that federal resource management agencies apparently have never followed a basic tenet of experimental science—setting up experimental controls! Had this been done, New Mexico would today contain adequate parcels of major vegetation types in stable and relatively pristine condition.

CONCEPT OF NATURAL AREA

One of the best ways to protect natural features is through "natural areas." This concept has been developed and used nationally, largely through the efforts of Dr. Robert Jenkins of The Nature Conservancy (TNC), a private organization with headquarters in Washington, D.C. A national Heritage Program was initiated in 1974 by the TNC to locate, identify, and inventory the nation's natural features. Once an inventory was completed, the relative density or scarcity of features could be used to determine priorities in securing protection. Plant and animal species and geological landmarks were features considered by the TNC to be building blocks of natural units. Some plant and animal species, called "narrow endemics" by scientists, have extremely small ranges. The nature of narrow endemics makes it likely that they are or will become threatened or endangered by human utilization of the nation's natural resources. With a dependable inventory, some research, and careful planning, it should be possible to locate sites for protection that are of sufficient size to permit the perpetuation of one or more critical species, provide stability for a number of plant and animal communities, and possibly even preserve a geological feature. Such sites would have discrete boundaries and would constitute natural areas. A natural area can be defined as any area where the integrity of its

naturalness is being safeguarded, independent of ownership or custodial agency. One or many natural features might be included in a natural area. Vegetation is automatically protected with the creation of natural areas, and is often the major ecosystem parameter that ultimately dictates the size and boundaries of a natural area. Portions of the landscape in New Mexico have been protected by various organizations and agencies; if the protection permits the perpetuation of the included natural features, such sites can be considered natural areas. The New Mexico office of The Nature Conservancy has obtained a number of qualifying areas, which it either maintains itself or turns over to a state or federal organization, with the stipulation that their natural-area status be maintained. Portions of some national parks and monuments contain such sites. A number of federal agencies utilize a use category called Research Natural Area (RNA), where research of a nondestructive nature is the primary activity. Normally there is little human disturbance on the RNAs, which allows the perpetuation of natural conditions and processes.

If natural areas are to be identified and set aside, there must be administrative units to act as custodians for the areas, in order to insure their preservation. In addition, the organizations identifying or creating natural areas must use a technique to identify, evaluate, and prioritize them. Many techniques have been proposed, such as that of Tans (1974), "Priority Ranking of Biotic Natural Areas" and of Gehlbach (1975), "A scheme for the evaluation of each area utilises the weighted values of climax condition, educational suitability, species significance, community representation, and human impact through multiplicative scoring to give a natural area score. With the weighted values-multiplicative scoring scheme, natural areas are clearly distinguished in priority for acquisition." The Forest Service has also developed detailed establishment procedures for research natural areas (RNA Task Group, 1984):

a. Natural areas are part of a comprehensive system; members of the system are samples of typical environments, ecosystems in all their variety from all geographic zones and landscapes.
b. Minimum human disturbance is an essential feature.
c. Natural areas are ecological in scope and criteria.
d. RNAs are given permanency and preservation to the greatest possible degree, in order to evaluate natural processes of ecosystem dynamics.
e. These withdrawals of ecosystems in relatively unmodified conditions are for the purposes of "research, study, observations, monitoring, and educational activities that are nondestructive, nonmanipulative, and that maintain unmodified conditions . . ."
f. Natural areas harbor genetic stock of future value to society in such areas as agriculture, silviculture, medicine, horticulture, range management, landscape restoration, aesthetics, and wildlife management.

Not all established Forest Service RNAs fit these characteristics, and many RNAs of other federal agencies also fall short. Natural areas administered by other organizations often have vague characteristics and are loose policies. There are many problems in obtaining protected status for natural areas, and their continued administration is costly. Constant threats to their protected status come both from within the custodial agencies and from the general public.

Franklin (1984) outlined some likely future problems for Forest Service RNAs and natural areas in general, in a keynote address for a natural areas symposium. Under the heading "Use It Or Lose It," he remarked that,

establishing a research natural area or reserve does not insure its existence inperpetuity—regulation, law, or ownership, not withstanding. Federal research natural areas are going to be reviewed periodically by the responsible agency . . . Many questions will be posed at each review. The most critical question may be, "Has anybody used this natural area?" However much we may argue (and believe) that reserves have value even without any use, managers and the public are going to find such arguments unconvincing. Managers already complain constantly of the real or imagined lack of scientific use of existing research natural areas. Each cycle of land-use planning—of reassessment— will be a moment of truth in which concrete evidence of use by the scientific community will be essential . . . The importance of using natural areas is not confined to Forest Service or Bureau of Land Management research natural areas. It will almost certainly come to apply to all lands exempted from normal social uses for scientific purposes. The Nature Conservancy and other private reserves are commonly granted tax exemptions based on scientific and other benefits to the public. We can be sure that this contribution will be periodically examined.

Laissez faire management of natural areas is the third danger area. Simply the absence of management plans for most of the Federal research natural areas suggests that we have a serious problem . . . The Nature Conservancy is far ahead of the Federal agencies; stewardship plans have been developed for the majority of its preserves, and intensive management to achieve specific preservation objectives is characteristic of many of their properties. We argue that the natural areas are invaluable, yet the management attention they are receiving is not consistent with those purported values. Management plans are a first step and can help clarify our objectives, as well as define management needs.

As with management planning, The Nature Conservancy is ahead of the Federal agencies in volunteer involvement with management and use of natural areas. Many Nature Conservancy perserves have management committees composed of interested scientists and laypersons. These committees sometimes develop and implement the management plans, although many State and regional offices of The Nature Conservancy have professional stewardship positions, and larger preserves have full-time directors and management staffs. Sometimes universities have assumed responsibility for management and protection of The Nature Conservancy reserves.

NATURAL AREAS OF NEW MEXICO

Table 12.1 is a list of natural and partial natural areas in New Mexico. Included in the table are size of area, area designation used by the custodial organization, and the custodial organization. The areas listed can be considered natural areas because they are relatively undisturbed by human activities and are of sufficient size to maintain their naturalness. Table 12.1 also includes a list of partial natural areas in New Mexico, areas that have some naturalness, but are also considerably impacted by one or more of the following human activities: grazing by domestic livestock, manipulations of native plant and/or animal populations, heavy recreational use, and developments such as ditches, channels, cultivation, and construction of recreational facilities. Notice that some RNAs have been classified as partial natural areas and some wilderness areas are included in the natural category. The delegation of areas to these two categories is based upon current conditions. Should there be a change in philosophy of the management agency or a change of agency, an area could warrent delegation to the other category. From the standpoint of safeguarding examples of New Mexico's vegetation, it is to be hoped that all shifts would be from partial natural to natural status.

The types of vegetation found within the state's natural areas are lised in Table 12.2. Vegetation information for some areas is general, such as for Montane Coniferous Forest, found in the Humphries, Neblett, Sandia, and Urraca areas. The information for Forest Service RNAs is more precise, such as Subalpine Fir on Buck Mountain. Wetland vegetation is found in the Ink Pot, Lake St. Francis, Rattlesnake Springs, and Rio Grande Marsh areas. Detailed inventories are needed in order to identify gaps and to establish priorities for the acquisition of areas containing vegetation types not yet protected.

PROJECTIONS AND PROPOSALS

The succession from plains-mesa grassland to juniper savanna will probably continue in many areas of the state. At the lower (drier) boundaries of plains-mesa grassland, many acres of grama grassland will become desert grassland, and much of the present desert grassland will become Chihuahuan or Great Basin desert scrubland. On many sites, these successional trends, which range users consider deterioration of grassland, were set in motion early in this century; subsequent "range management" efforts are unlikely to halt, let alone reverse the trend. Some natural forest vegetation may be found in the future, if the "let burn" efforts of the Forst Service continue, but protection from fire could cause the loss of species that were members of the natural vegetation of the state's forests.

Examples of New Mexico's natural vegetation could be safeguarded under a sound natural areas system. A thorough inventory of all existing natural areas needs to be made in order to identify gaps and to prioritize other areas for inclusion within the system, either through "establishment" (Forest Service RNAs) or acquisition (TNC). The Forest Service has already itemized some needs and gaps in their RNA system (RNA Task Group, 1984). The inventory, identification of gaps, and prioritization suggestions could best be accomplished through a strong state effort, because the existing and potential natural areas are under federal, state, and private administration or ownership. Many states have such systems, some for a considerable time. Wisconsin established its system in 1951, and Illinois established its nature preserves system in 1963. The need for a state system for New Mexico has been recognized by researchers and managers for many years. The following was proposed to a past Secretary of Natural Resources by a citizens' Natural Areas Committee and illustrates one possible approach:

SAFEGUARDING NEW MEXICO'S NATURAL HERITAGE

1. An inventory of the state's "natural features" should be maintained. This inventory should include the number, size, and location of the various "features." A natural feature can be physical such as a caldera, lava flow, mountain cirque, Cambrian outcrop, or badlands. A feature can be biotic such as a plant or animal species, vegetation type (community, association, etc.), or an ecosystem such as a marsh or pond.

2. An inventory of the state's Natural Areas should be maintained. This inventory should include size, location, and natural features included for each Natural Area. A natural area is any area where the integrity of its naturalness is being safeguarded, independent of its ownership or custodial agency. A natural area has legal boundaries and is under the stewardship of some public or private administrative entity. One to many natural features might be included in a natural area. Some existing administered areas which at first might seem to be natural areas, would not be included in the system. These could be considered Partial Natural Areas. Examples would be most Wilderness areas where livestock grazing is permitted; parks which are heavily impacted by the public and/or have public facilities; or state and federal game refuges on which manipulations or controls of plant or animal populations are being practiced.

3. If the Governor were to decree the establishment of a Natural Areas "system," the existing natural areas would constitute the foundation for the state system. We would then be in a position to recommend additions to the system. The natural features and Natural Areas inventories would aid in identifying gaps in the system and indicate how they could be filled. This would give valuable direction to The Nature Conservancy, Forest Service, etc. The data contained in these inventories would supply substance to arguments proposing protective use categories on public lands or to fund raising efforts for the acquisition of sites.

Some of the most interesting and diverse vegetation patterns found in the United States are native to New Mexico. Exmaples of this diverse heritage can still be saved for use and enjoyment by future generations, but setting aside and monitoring small samples of each major type will require a persistent and concerted effort. We are fortunate in New Mexico to have much of our natural vegetation under the custody of various federal agencies, all of which have mechanisms that allow nonuse or at least relatively nondisruptive use of parcels for which they are responsible. These agencies are making use of their protective measures to a greater or lesser degree. The New Mexico Department of Game and Fish is inadvertently preserving some types of vegetation, as they acquire "wildlife habitats," and The Nature Conservancy is continuing to do an excellent job of securing parcels of the state's natural vegetation for preservation. As valuable as these various programs are, there exists much unevenness in type and degree of protection and redundancy in some of the holdings. New Mexico needs to have some mechanism for maintaining an inventory of protected areas and for monitoring protective efforts. Eventually, a state-sponsored operation could supply coordination and provide supportive data for the preservation activities of the various public and private organizations.

Table 12.1. Natural and Partial Natural Areas of New Mexico.

Designation Symbols: RNA = Research Natural Area; RA = Research Area; P = Preserve; W = Wilderness Area; ACEC = Area of Critical Ecological Concern; WA = Wildlife Area; WFA = Waterfowl Area; FWA = Fish and Wildlife Area; LTER = Long Term Ecological Reserve; WR = Wildlife Refuge; Rec = Recreational Area; NM = National Monument; BA = Botanical Area.

   * = establishment not completed.

Administrative Unit Symbols: BLM = Bureau of Land Management; USFS = U.S. Forest Service; FWS = U.S. Fish and Wildlife Service; NPS = National Park Service; TNC = The Nature Conservancy; NMGF = New Mexico Department of Game and Fish; UNM = University of New Mexico.

	Size			Administrative
	hectares	(acres)	Designation	Unit
NATURAL AREAS				
Albuquerque Volcanos	176	(440)	P	City of Albuquerque
Arellano Canyon	256	(641)	*RNA	USFS
Barker	2193	(5415)	WA	NMGF
Bernalillo Watershed	454	(1120)	*RNA	USFS
Bitter Lake	120	(300)	*RNA	FWS
Cañada Bonito	122	(300)	*RNA	USFS
Carlsbad Caverns	12,235	(30,210)	W	NPS
Clayton Pass	122	(300)	*RNA	USFS
Chupadera	2183	(5289)	RNA	FWS
Corrales Bosque	178	(400)	P	TNC
Edward Sargent	8262	(20,400)	FWA	NMGF
Gila Riparian	92	(230)	P	TNC
Gila River	161	(402)	RNA	USFS
Gray Ranch	128,681	(321,703)	P	TNC
Haynes Canyon	294	(610)	*RNA	USFS
Heart Bar	323	(797)	WA	NMGF
Humphries	4402	(10,868)	WA	NMGF
Ink Pot	.8	(2)	RNA	FWS
Jornada del Muerto	4455	(11,000)	RNA	FWS
Lake Lucero	2333	(5760)	RNA	NPS
Lake St. Francis	284	(700)	RNA	FWS
Largo Mesa	122	(300)	*RNA	USFS
Little Costilla Peak	263	(650)	*RNA	USFS

Table 12.1. (continued).

| | Size | | | Administrative |
	hectares	(acres)	Designation	Unit
Little Water Canyon	385	(950)	*RNA	USFS
Maxwell	32	(80)	RNA	FWS
McCrystal Meadow	230	(590)	*RNA	USFS
Mesita de los Ladrones	200	(500)	*RNA	USFS
Monument Canyon	259	(640)	RNA	USFS
Mount Taylor	271	(670)	RA	UNM
Neblett	13,412	(33,116)	WA	NMGF
Organ Mountain	1940	(4851)	ACEC	BLM
Otero Mesa (4 separate units)	1480	(3700)	ACEC	BLM–Ft. Bliss
Owl Canyon	847	(2090)	*ACEC	BLM
Prairie Chicken	8100	(20,000)	WA	NMGF
Rabbit Trap	122	(300)	*RNA	USFS
Rattlesnake Springs	6	(14)	P	TNC
Rio de los Pinos	34	(850)	FWA	NMGF
Rio Grande Marsh	39	(97)	RNA	FWS
Salt Creek	3897	(9621)	W	FWS
San Andres	23,173	(57,218)	WR	FWS
Sandia	9255	(22,851)	W	USFS
San Pascual	1296	(3200)	RNA	FWS
Sevilleta	89,819	(221,775)	LTER	FWS–UNM
Tres Piedras	1320	(3260)	WA	NMGF
Turkey Creek	540	(1335)	*RNA	USFS
Upper McKittrick Canyon	335	(827)	*RNA	USFS
Urraca	5617	(13,870)	WA	NMGF
Vegosa	218	(537)	RNA	FWS
White Sands	59,347	(146,535)	NM	NPS
William G. Telfer	294	(727)	*RNA	USFS
PARTIAL NATURAL AREAS				
Alamo Hueco Mountains	1856	(4640)	*ACEC	BLM
Aldo Leopold	81,817	(202,016)	W	USFS
Aden Lava Flow	1620	(4000)	RNA	BLM
Antelope Pass	3384	(8460)	*ACEC	BLM
Apache Box	1056	(2640)	*ACEC	BLM
Apache Camp	89	(220)	RNA	FWS
Apache Kid	17,860	(44,650)	W	USFS
Bandelier	8550	(21,110)	W	NPS
Bernardo	637	(1573)	WFA	NMGF
Big Hatchet Mountains	1664	(4160)	*ACEC	BLM
Bishop's Cap	768	(1920)	*ACEC	BLM
Bisti	1607	(3968)	W	BLM
Bitter Lake	5150	(12,716)	WR	FWS
Black Mesa	2184	(5460)	ACEC	BLM

Table 12.1. (continued).

| | Size | | | Administrative |
	hectares	(acres)	Designation	Unit
Black River	1600	(4000)	*ACEC	BLM
Blue Range	11,722	(29,304)	W	USFS
Blue Water Canyon	256	(640)	ACEC	BLM
Bosque del Apache	13,060	(32,247)	WR	FWS
Cabezon Peak	2000	(5000)	ACEC	BLM
Cañon del Ojo	24,076	(60,190)	ACEC	BLM
Cañon del Norte/Ladron	416	(1040)	ACEC	BLM
Capitan Mountains	14,329	(35,822)	W	USFS
Capulin	314	(775)	NM	NPS
Carrizozo Malpais	6840	(17,000)	*ACEC	BLM
Central Peloncillo Mountains	5120	(12,800)	ACEC	BLM
Chaco Canyon	8711	(21,509)	NM	NPS
Chama River Canyon	20,372	(50,300)	W	USFS
Chosa Draw Caves	880	(2200)	ACEC	BLM
Clancy	877	(2166)	FWA	NMGF
Colorado Farms	162	(400)	WFA	NMGF
Cooke's Range	1576	(3940)	*ACEC	BLM
Cruces Basin	7200	(18,000)	W	USFS
Dark Canyon	593	(1480)	ACEC	BLM
De-Na-Zin	9668	(23,872)	W	BLM
Dome	2106	(5200)	W	USFS
Dripping Springs/Ice Canyon	160	(400)	*ACEC	BLM
Elk Springs	4000	(10,000)	ACEC	BLM
El Morro	518	(1278)	NM	NPS
Filmore Canyon	48	(120)	*ACEC	BLM
Fort Stanton Cave	432	(1080)	*ACEC	BLM
Franklin Mountains	1608	(4020)	*ACEC	BLM
Gallinas	156	(385)	RNA	FWS
Gila	220,320	(544,000)	W	USFS
Gila Middle Box	288	(720)	ACEC	BLM
Gila Lower Box	1052	(2631)	ACEC	BLM
Gila River	163	(402)	RNA	USFS
Gilia formosa	2280	(5700)	ACEC	BLM
Granite Gap	672	(1680)	*ACEC	BLM
Guadalupe Canyon	1476	(3691)	*ACEC	BLM
Hogback	3792	(9480)	ACEC	BLM
Indian Hollow	256	(640)	*ACEC	BLM
Indian Well	2081	(5138)	W	FWS
Jackson Lake	340	(840)	WFA	NMGF
Kiowa Short Grass	120	(300)	*RNA	USFS
La Joya	1438	(3550)	WFA	NMGF
Las Conchas	24	(60)	BA	USFS
Las Vegas	2519	(6219)	WR	FWS

Table 12.1. (continued).

	Size		Designation	Administrative Unit
	hectares	(acres)		
Latir Peak	8100	(20,000)	W	USFS
Little McKittrick Draw	4	(10)	ACEC	BLM
Lonesome Ridge	916	(2290)	ACEC	BLM
Manzano	14,788	(36,970)	W	USFS
Marquez	6075	(15,000)	WA	NMGF
Mathers	98	(242)	RNA	FWS
Maxwell	1224	(3021)	WR	FWS
Mescalero Sands	2517	(6293)	*ACEC	BLM
Mimbres River	5	(13)	P	NMGF
North Ladron	512	(1280)	ACEC	BLM
Ojito	6000	(15,000)	ACEC	BLM
Overflow Wetlands	1728	(4320)	*ACEC	BLM
Pecos	89,333	(223,333)	W	USFS
Pecos River/Canyons	2076	(5190)	ACEC	BLM
Reese Canyon	880	(2200)	ACEC	BLM
Rio Chama	5265	(13,000)	FWA	NMGF
Rio Nutria	240	(600)	P	TNC–NMGF
Sabo	10	(25)	P	TNC
Sacramento Escarpment	1456	(3640)	ACEC	BLM
Sandia Mountain	14,893	(37,232)	W	USFS
San Luis Mesa	3204	(8010)	ACEC	BLM
San Pedro Breaks	480	(1200)	ACEC	BLM
San Pedro Parks	16,658	(41,132)	W	USFS
Sawtooth Mountains	48	(120)	ACEC	BLM
Seven River Hills	216	(540)	ACEC	BLM
South Texas Hill Canyon	544	(1360)	ACEC	BLM
Springs Riparian	64	(160)	ACEC	BLM
Tent Rocks	4000	(10,000)	ACEC	BLM
Torgac Cave	192	(480)	*ACEC	BLM
Tres Piedras	1320	(3260)	WA	NMGF
Uvas Valley	896	(2240)	*ACEC	BLM
Water Canyon	1150	(2840)	WA	NMGF
Wheeler Peak	19,549	(48,873)	W	USFS
White Mountain	12,624	(31,171)	W	USFS
White Sands			NM	NPS
Withington	7643	(18,870)	W	USFS
Yeso Hills	256	(640)	ACEC	BLM

Table 12.2. Vegetation Found in Some Natural and Partial Natural Areas of New Mexico.

TUNDRA VEGETATION
   Alpine Tundra   Little Costilla Peak
FOREST VEGETATION
   Subalpine Coniferous Forest   Urraca
      Bristlecone Pine   Clayton Pass
      Subalpine Fir   Buck Mountain
      Engelmann Spruce–Subalpine Fir   Arellano Canyon, Sargent
   Montane Coniferous Forest   Humphries, Neblett, Sandia, Urraca
      White Fir   Haynes Canyon
      Douglas Fir   Arellano Canyon, Monument Canyon
      Douglas Fir–White Fir   Mount Taylor, Sargent
      Ponderosa Pine   Clancy, Heart Bar, Monument Canyon
   Successional-Disturbance
      Aspen   Sargent
WOODLAND-SAVANNA
   Coniferous Woodland   Central Peloncillo, Heart Bar, Humphries, Largo Mesa, Rio de los Pinos, San Andres, Sandia, Urraca, Vegosa
   Mixed Woodland   Central Peloncillo, Gallinas, Owl Canyon
   Juniper Savanna   Mesita de los Ladrones, Rabbit Trap, Sevilleta
GRASSLAND
   Subalpine-Montane Grassland   Barker, Buck Mountain, Canada Bonito, Humphries, Mount Taylor, Sargent
   Plains-Mesa Grassland   Bernalillo Watershed, Carlsbad Caverns, Chupadera, Gallinas, Mathers, Maxwell, Otero Mesa, Prairie Chicken, Rabbit Trap, San Andres, Sevilleta, Tres Piedras, Vegosa
   Desert Grassland (*ecotone*)   Albuquerque Volcanos, Carlsbad Caverns, Central Peloncillo, Chupadera, Chuchillo Trap, Gila River, Lake St. Francis, Otero Mesa, Owl Canyon, Rabbit Trap, San Pascual, Sevilleta, Tres Piedras
SCRUBLAND VEGETATION
   Montane Scrubland
      Mountain Mahogany–Mixed Scrub   Guadalupe Mountain, Monument Canyon, Rio de los Pinos, San Andres
      Mixed Evergreen Sclerophyll Scrub   Central Peloncillo, Owl Canyon
   Plains-Mesa Sand Scrubland
      Plains Sand Scrub   Mathers, Prairie Chicken
      Mesa Sand Scrub   Rio Grande Marsh, San Pascual, White Sands
   Great Basin Desert Scrubland   Urraca
   Chihuahuan Desert Scrubland   Chupadera, Cuchillo Trap, Otero Mesa
RIPARIAN VEGETATION
   Alpine Riparian   McCrystal Meadow
   Montane Riparian   Barker, Clancy, Gila Riparian, Gila River, Heart Bar, Neblett, Rattlesnake Springs, Rio de los Pinos, San Andres, Sargent, Turkey Creek
      Blue Spruce   Arellano Canyon, Little Water Canyon
   Floodplain-Plains Riparian   Corrales Bosque
   Arroyo Riparian   San Pascual
   Closed Basin–Playa–Swale–
   Alkali Sink Riparian   Bitter Lake, Lake Lucero, Maxwell, Otero Mesa, Sevilleta, White Sands

CHAPTER 12 REFERENCES

THE FUTURE OF
NEW MEXICO VEGETATION

Cain, S. A., 1947. Characteristics of natural areas. Ecological Monographs, 17: 187–200.

Carother, S. W., 1977. Importance, preservation, and management of riparian habitat: an overview. pp. 2–4. In: R. R. Johnson and D. A. Jones (eds.). Importance, preservation, and management of riparian habitat: a symposium. U.S. Department of Agriculture, Forest Service General Technical Report RM-43. Ft. Collins, Colorado.

Conway, M., 1981. Wildlife for everyone. New Mexico Wildlife Magazine, 26, Number 5: 1–45. New Mexico Department of Game and Fish, Santa Fe, New Mexico.

Franklin, J. F., 1984. Keynote comments: prophylaxes for our research natural areas. pp. 1–4. In: J. L. Johnson, J. F. Franklin, and R. G. Krebill (coordinators). Research natural areas: baseline monitoring and management. Proceedings of a symposium, Missoula, Montana. U.S. Department of Agriculture, Intermountain Forest and Range Experiment Station, General Technical Report Int-173. Ogden, Utah.

Franklin, J. F. and J. M. Trappe, 1968. Natural areas: needs, concept, and criteria. Journal of Forestry, 66: 456–461.

Gehlbach, F. R., 1975. Investigation, evaluation, and priority ranking of natural areas. Biological Conservation, 8: 79–88.

Hubbard, J. P., 1977. Importance of riparian ecosystems: Biotic considerations. In: R. R. Johnson and D. A. Jones (eds.). Importance, preservation, and management of riparian habitat: a symposium. U.S. Department of Agriculture, Forest Service General Technical Report RM-43. Ft. Collins, Colorado.

Illinois Nature Preserves Commission, 1972. Comprehensive plan for the Illinois nature preserves system: part 1: guidelines. Illinois Nature Preserves Commission, Rockford, Illinois.

Iltis, H. H., 1959. We need more scientific areas. Wisconsin Conservation Bulletin, 24: 3–8.

Jenkins, R. E. and W. B. Bedford, 1973. The use of natural areas to establish environmental baselines. Biological Conservation, 5: 168–174.

Johnson, G. M. and P. M. Emerson (eds.), 1984. Public lands and the United States economy: balancing conservation and development. Westview Press, Boulder, Colorado.

Johnson, R. R. and D. A. Jones (eds.), 1977. Importance, preservation, and management of riparian habitat: a symposium. U.S. Department of Agriculture, Forest Service General Technical Report RM-43. Ft. Collins, Colorado.

Martin, S. C., 1972. Semidesert ecosystems—who will use them. How will we manage them? Journal of Range Management, 25: 16.

Moir, W. H., 1972. Natural Areas. Science, 177: 396–400.

New Mexico Field Office, 1984. Across our great state—the preserves The Nature Conservancy, Albuquerque, New Mexico.

Peterson, R. S. and E. Rasmussen, 1986. Research natural areas in New Mexico. U.S. Department of Agriculture, Rocky Mountain Forest and Range Experiment Station, Albuquerque, New Mexico.

RNA Task Group, 1984. Research natural areas: progress report. U.S. Department of Agriculture, Forest Service, Region 3, Albuquerque, New Mexico.

Scientific Areas Preservation Council, 1973. Wisconsin scientific areas. Department of Natural Resources, Madison, Wisconsin.

Tans, W., 1974. Priority ranking of biotic natural areas. The Michigan Botanist, 13: 31–39.

SUBJECT INDEX

alien (exotic) species, 20, 130, 151, 154, 192, 207

alkali sink, 154, 167

altitude and latitude, 4, 27, 88, 89, 101, 103, 127, 148

animal habitat, 9, 47, 59, 128, 147, 167

Animas Mountains, 31, 66, 91, 127

Animas Valley, 104

apomictic species, 193

aspect dominance, 109, 110

Aspen Basin, 50, 53

Associations of species, 35, 39–41

available moisture and moisture availability, 27–31, 88, 89, 123, 127, 148, 168

bajadas, 29, 104, 108, 109, 152

Big Lue Mountain, 69

Bitter Lake, 154

Bitter Lake National Wildlife Refuge, 167

blackjack pines, 69

Black Range, 57, 66

bosque, 152

botanical gardens, 183

Brazos Canyon, 127, 147

broadleafed evergreens, 66, 67, 68, 69, 85, 90

Burro Mountains, 90

canyon effect, 31

Capitan Mountains, 57, 61, 89

Carrizozo Malpais, 168, 169, 173

catchments (water), 29, 93, 105, 127, 128

Cebolleta Mountains, 127

cedar posts, 89

Chama Valley, 19

chasmophiles, 192

Chihuahuan Desert Scrub, 107, 108, 109, 131

Chuska Mountains, 16, 21

cienega, 166

Cimarron Range, 57, 71

cliff vegetation, 192, 222

climatic shifts, 131

climax vegetation, 33, 52 n 2, 55, 56, 59, 62, 64, 66, 69, 70, 102, 104, 107, 109, 123

closed basins, 1, 38, 123, 153, 170

community types, 35

coppice dunes, 129

critical habitat, 187

cryptogams, 55

cushion plants, 48, 49, 75

Datil Mountains, 89, 200

deep sand vegetation, 128

desert, 130

diatoms, 166

diseases and insects, 50, 54, 56, 60, 61, 63, 70, 71, 107, 208

disturbance and succession, 36, 50–72 passim, 87, 91, 104, 108, 109, 124, 129, 199, 201, 207, 208

diversity, 182, 183

dune vegetation, 128, 129

ecotones (transitional zones), 37, 47, 87, 91, 106, 108, 153

edaphic endemics, 189, 192, 193, 200, 222

emergent vegetation, 165, 167, 171

encinal, 90, 91, 128; Encinal Series, 40, 94

endangered species, 181, 185–86, 192

Endangered Species Act, 181, 184, 185–87, 190, 193, 201

endemic plants, 51, 54, 55, 57, 169, 189, 191, 192, 193, 200, 222, 226

environments, special, 54, 59, 61, 63

erosion, 1, 3, 48, 92, 93, 105

escarpments, 1, 88, 147

evergreens, broadleafed (nonconiferous), 66–69, 85, 90

evergreens, faculative, 66; riparian, 145

exotic (alien) species, 20, 130, 151, 154, 192, 207

exposed rock vegetation, 169, 174

exposure effects, 27–29

extinction, 180–82

faculative evergreens, 66; riparian, 145

fellfield vegetation, 48; Fellfield Series, 39, 73. *See also* cushion plants

fire, adaption to, 69; behavior of, 63; cool and hot, 55; prescribed, 67; research on, 32. *See also* disturbance and succession
floating-leaf species, 165, 170
flooding, 148, 151
floristic regions, 187
forbs (herbs), listed, aquatic, 171; bajada, 109; forest, 53–56, 60–62, 64, 66, 68, 69, 78–80; grassland, 103–9, 112, 115, 118; malpais, 173; riparian, 148, 153, 154, 157, 158; scrubland, 129, 132, 135, 138, 139, 141; tundra, 75; wetland, 169; woodland and savanna, 96
forests, 16, 18, 37, 50–72; alternation of generations in, 72; disturbance and scrub, 124–25, and woodland, 87; gallery forests, 148, 151, 152; grass in, 108; old growth, 56, 63, 192, 198; Series lists for, 39, 40, 73–74; tundra/timberline, 47; woodland different from, 85
fossils, 9, 14, 15
Four Corners Region, 195, 201
Franklin Mountains, 194

gallery forests, 148, 151, 152
gene pool, 182
geofloras, 13–17, 85, 101
Gila River, 190
Grants Malpais, 167, 168, 169, 173
grass, 15, 16, 18, 19, 29, 37, 40, 101–11, 112, 114, 117; aquatic, 171; bajada, 108; forest, 53, 55, 59, 62–70 passim, 77, 108; malpais, 173; riparian, 147, 148, 150, 153, 154, 156, 157; savanna, 95; scrubland, 128, 129, 132, 135, 137, 139, 140; shift to scrub from, 131; shift to woodland from, 88; woodland grasses, 89, 90, 92, 95; tundra, 49, 75
gravelly scree, 49, 61
grazing and livestock, 19, 64, 70, 71, 103–9, 130, 131, 151, 192, 195
growing season, 4
Guadalupe escarpment, 147
Guadalupe Mountains, 16, 89, 90, 127, 129, 189, 204
gypsophiles, 169, 200
gypsum, 129, 153, 169, 175

habitat types, 35, 52 n 2
Hachita Mountains, 127
halophytes, 147, 154
herbs. *See* forbs
High Plains, 188
human induced disturbance, 20, 91, 151, 192. *See also* disturbance and succession
hybrids, 90, 150, 193, 208

impoundment of waters, 36, 151

insects and diseases, 50, 54, 56, 60, 61, 63, 70, 71, 107, 208
International Nature Conservancy, 184
International Union for the Conservation of Nature and Natural Resources (IUCN), 180, 184
introduced (alien/exotic), 20, 130, 151, 154, 192, 207
inventory of natural areas, 229

Jemez Mountains, 29, 31, 57, 62, 71, 90, 101, 127
Jicarilla Mountains, 87
Jornada Experimental Range, 108, 129

krummholz, 47, 48, 52

lakes, 15, 154, 165, 166, 167
latitude, 4, 14, 27
lava flow vegetation, 168, 173; sink hole ponds, 167

Magdalena Mountains, 57, 62
malpais vegetation, 168, 169, 173
Manzano Mountains, 127
marshes, 165, 166, 167
McCrystal Meadow, 147
meadows, 71, 101–2, 104, 148
Mexico, 13, 16, 36, 57, 88, 90, 105, 131
middens, 9, 10
Mimbres River, 151
mixed forest, 58–66; mixed woodland, 87, 90
Mogollon Mountains, 53, 54, 55, 57, 61, 62, 66, 68, 189, 204
moisture availability and available moisture, 27–31, 88, 89, 123, 127, 148, 168
montane grassland, 64; scrub, 88, 92; understory, 127
Mount Taylor, 57, 71

Nacimiento Mountains, 90
narrow endemics, 191, 192, 226
National Environmental Policy Act (NEPA), 184
National Environmental Protection Act, 190
natural areas, 226–34
Nature Conservancy, The (TNC), 226, 229, 230
nature reserves, 183
Navajo Reservoir Basin, 166
New Mexico Department of Game and Fish, 229
nonseed (seedless) species, 166, 170, 174

obligate species, 127, 145, 146, 148, 169
old growth forest, 56, 63, 192, 198
ooze communities, 166
Organ Mountains, 91, 127, 147, 169, 202, 204

parasite, 67

Pecos River, 128, 131, 151
Pecos Valley, 203, 208
Pecos Wilderness, 52, 53, 71, 72, 103
Peloncillo Mountains, 69, 89, 107, 127
photography, 10, 92
phreatophytes, 146, 151
phytoplankton, 165
plant associations, 52 n 2
playa vegetation, 153
pollen, 9, 10
population growth, 179–81
precipitation, 5; desert, 129; shifts in, 16
primeval vegetation, 10
pseudoriparian vegetation, 127, 146
pygmy forest, 169

ramets, 71
rare species, 180, 190, 193, 194, 200
relicts, 57, 89, 106, 132, 145
reproduction in plants, 151, 193, 195
Research Natural Area (RNA), 226–29
restricted species, 145
ridgetop vegetation, 57, 63, 64, 67
Rio Chama, 128
Rio Grande, 128, 131, 151, 152, 195
riparian species, 38, 40, 123, 145–60, 169, 202
river systems, 1, 88. *See also rivers by name*
rivulet vegetation, 49
rock crusts, 166
rock field vegetation, 49; rocky surface vegetation, 168
Rocky Mountains, 13, 16, 54, 189
rosette shrubs, 69, 123, 127
rubbleland, 49

Sacramento Mountains, 48, 54, 57, 61, 62, 66, 68, 87, 89, 104, 167, 196, 197, 199, 204, 208
salinity, 154
salt marsh, 167
San Agustin Plains, 9, 16, 21, 87, 104
San Andres Mountains, 127, 197
San Francisco Mountains, 62, 66, 90
San Juan Mountains, 61, 71
San Juan River, 128
San Mateo Mountains, 57, 62, 69
San Pedro Parks, 127
sand scrub, 170
Sandia Mountains, 89, 92, 127, 194
Sangre de Cristo Range, 15, 48, 52, 53, 54, 104
Santa Fe, 193
savanna, 19, 37, 40, 87, 88, 91, 94, 109, 128
scrubland, 16, 17, 19, 38, 40, 93, 123–41
seeps, 167

semiriparian species, 127, 145, 146, 152
seral vegetation, 31, 54, 55, 56, 61, 62, 66, 68, 70, 87, 108
Series category, the, 36, 37, 39–41, 73, 74, 93, 110, 159, 160
shinnery vegetation, 38, 128, 129; Shinoak Series, 40
shrubs, listed, bajada, 108; forest, 53, 54, 55, 57, 59, 60–69, 76, 77; grassland, 104, 106–9, 113, 116; malpais, 169, 173; riparian, 150, 151, 152, 154; rosette, 123, 127; savanna, 91, 92, 95; scrubland, 123, 127–30, 132, 134, 136, 137, 139, 140; wetland, 169; woodland, 89, 90, 91, 92, 95
Sierra Blanca Peak (Region), 48, 72, 102, 103
Smithsonian Institution, 185
snowbank vegetation, 49
snow fence effect, 48
soils, 6; alkaline, 168; Blue Spruce, 59; closed basin, 153; endemics and, 192, 193; factors for, 29, 48 n 1; fellfield, 48; fines as, 169; grass and, 108, 109, 110; meadow, 52; riparian, 146, 148, 153; seep, 53; turf, 49; volcanic, 63
solar radiation, 54, 57
South America, 17
species, defined, 181, 185, 190; annual, 201; dynamic equilibrium of, 32; federal categories for, 180–87; recovery and reintroduction of, 206, 207; threats to, 191–93, 211–21
springs, 167
streamside forests, 59
submersed species, 166, 167, 170
succession, forest, 31–32, 53, 54, 56, 59–60, 62–64, 65, 67, 69, 69–70, 70–72, 87; grass, 19, 64, 104, 105, 108, 109, 131, 132, 228; riparian, 152; rocky habitat, 169; savanna, 228; scrubland, 19, 125, 131, 132; wetland, 167; woodland, 19, 92
Successional-Disturbance category, 36, 40, 41
survey records of vegetation, 10, 18, 22, 132
swale, 105

taiga, 52
temperature, 4, 27, 48 n 1
thickets, 48, 49
threatened species, 181, 185–86, 192–93
Thurber Fescue meadows, 102
Tobosa Swale, 105
transitional zones (ecotones), 37, 47, 87, 91, 106, 108, 153–54
trees, accidental-associated, 62; associated, 53, 54, 58–69, 173; climax, 55, 56, 59, 69, 70; coclimax, 62, 64, 66; diagnostic, 51; dominant, 48, 60, 66, and codominant, 52, 53, 90; forest list, 76; on malpais, 173; riparian, 148, 150, 152, 154, 155; seral, 54, 55,

56, 61, 62, 66, 68, 70; woodland and savanna, 94
Tularosa Basin, 128
Tularosa Mountains, 90
turf, 48, 54

understories, 32, 53, 55, 57, 58, 59, 61, 62, 70, 90, 92, 150, 152, 169
U.S. Fish and Wildlife Service, 145, 184, 185
U.S. Forest Service, 35, 36, 50, 57, 65, 196, 227

vertical layering, 35
vines, 156
volcanic activity, 1; soils, 63

water catchments, 29, 93, 105, 127, 128

water, code system for bodies of, 165; endemic species and, 192; water as hostile, 195, and flooding, 148, 151
wetlands, 145, 192, 222
Wheeler Peak, 148
White Mountains, 89, 204
White Sands National Monument, 128, 129, 153
woodland, 16, 17, 19, 37, 40, 87, 88, 92, 127, 128; coniferous woodland, 85, 87
World Conservation Strategy, 184

xeric sites, 168

yellow pine, 59

Zuni Mountains (region), 68, 90, 127

INDEX OF
SELECTED KEY PLANTS

Abies: Corkbark Fir, 51, 57; in Series, 39, 53–54, 54–55, 73; White Fir, in habitat types, 61, 62, 63; in Series, 39, 54–55, 60, 62–63, 73, 74

Acacia, 132; in Series, 111, 133

Acer: Bigtoothed Maple, 127; in habitat type, 61; Boxelder, 148; in Series, 41, 159; Rocky Mountain Maple, 52, 57, 62; in habitat types, 55, 57, 61; in Series, 159

Agave, 66, 69, 108, 109, 123, 127, 128, 132; in Series, 111

Agropyron (wheatgrass), 15, 64, 101, 104; in Series, 110, 111

algae, 105, 110, 165–67, 170, 174; in Series, 167

Alisma, 165, 171; in Series, 167

Alnus (Alder), 59, 150; in Association, 148; in Series, 41, 159

Alpine Avens, 49, 53; in Series, 39, 73

Amelanchier, 127; in Series, 132

Andropogon (bluestem), 68, 90, 101, 128; in Series, 40, 110, 133

Apache Plume, 127, 169; in Series, 41, 160

Arctostaphylos spp., 67, 69, 90, 91, 127; in habitat types, 57, 59, 63, 67; in Series, 133

Aristada, 69, 105, 128; in Series, 133

Artemisia spp., 15, 19, 29, 38, 49, 61, 62, 69, 70, 87–91, 106, 109, 128–30; in habitat type, 68; in Series, 40, 111, 133

Ash, 148, 150; in Series, 41, 159

Aspen, 54–72 passim, 124, 145; in Association, 150; in Series, 40, 41, 74, 159

Atriplex spp., 38, 91, 109, 128, 130, 153, 154, 195; in Series, 111, 133, 160

Azolla, 165, 170; in Series, 167

Baccharis, 109, 152; in Series, 159

Berberis spp., 59, 62, 65, 108, 127, 169; in habitat types, 63, 65

Blepharoneuron (dropseed), 67, 68, 104, 128; in Series, 40, 110

Bluebells, 64, 104; in habitat types, 53, 54

Blue Spruce, 58, 149; in Association, 148; in habitat type, 59, 150; in Series, 39, 41, 58–59, 73, 159

Border Pinyon, 88; in Series, 93

Bouteloua (grama grasses), 15, 32, 40, 89, 101–9, 128; in Associations, 148, 150; in Series, 110, 133; Black Grama, 107, 109; Blue Grama, 104; in habitat types, 68, 69; Sideoats Grama, 105, 109

Boxelder, 148; in Series, 41, 159

Brickellia spp., 150, 151, 152, 169, 190; in Series, 41, 159, 160

Bristlecone Pine, in habitat types, 52, 64, 87; in Series, 39, 40, 52, 52–53, 64, 73, 93

Broadleaf Cottonwood Series, 159, 160

Bromus (brome grass), 53–55, 59, 61, 64, 65; in habitat type, 62

Bulrush Series, 167

Burrobush, 152; in Series, 41, 160

Cactaceae, 109, 132, 189, 192–97, 205, 210

Campanula spp., 64, 104; in habitat types, 53, 54

Carex spp., 49, 55, 61, 62, 66, 69, 147, 148, 150; in habitat type, 59; in Series, 73, 110, 167

Cattail Series, 165, 166, 167, 171

Ceanothus, 66, 69, 127, 128; in Series, 133

Celtis (hackberry), 148, 169; in Association, 150; in Series, 41, 159

Ceratoides, 106, 130; in Series, 133

Ceratophyllum, 166, 170; in Series, 167

Cercocarpus spp., 65, 66, 90, 169; in Association, 104, 127; in Series, 40, 132, 133

Chara, 166, 167, 170; in Series, 167

Chihuahua Pine, 66, 69; in Series, 40, 69, 74

Chrysothamnus, 29, 106, 129, 130, 152; in Series, 111, 159, 160

Colorado Pinyon, 87; in habitat type, 89; in Series, 93

Corkbark Fir. *See Abies:* Corkbark Fir

Cornus, 59, 148; in Series, 159

Cottonwoods, 59, 145, 149; in Association, 150; in Series, 41, 159, 160

Cowania, 127; in Series, 132

Creeping Mahonia, in habitat type, 63, 65; in Series, 73
Creosotebush, 17, 20, 131–32; in Association, 109; in Series, 40, 133
currants (Ribes), 52, 53, 54; habitat type, 87

Dasylirion (sotol), 123, 127; in Series, 111
Deschampsia, 49, 103, 147; in Series, 110, 159
Distichlis (saltgrass), 105, 168; in Series, 111, 160
Douglas-fir, 58, 145, 167, 169; in Series, 39, 40, 56, 60, 61, 64, 65, 65–66, 74, 167; Douglas-fir Belt, 169
Drummond Rush Community, 49
Duckweed, 166; in Series, 167

Eleagnus, 20, 151, 152; in Series, 160
Eleocharis, 167, 171; in Series, 167
emergent species Series, 167
encinal Series, 94
Engelmann Spruce, 48, 51, 62; in Series, 39, 52–53, 53–54, 54–55, 56, 57, 73
Ericameria (turpentinebush), 108; in Series, 133
Erigeron spp., 49, 53–55, 57, 59–62, 64, 66, 103
Eriogonum, 129, 191, 193; in Series, 133
Erioneuron (Fluff Grass), 109, 132; in Series, 133
Evergreen (Mixed) Series, 133

Fallugia, 127, 152, 169; in Series, 160
ferns, 170, 171, 174
Festuca (fescues), 49, 52, 55, 61, 62, 102–4; in Association, 104; in habitat types, 59, 63, 64, 65, 67; in Series, 40, 110
fleabanes, 59, 60; in habitat types, 55, 57, 61
Flourensia, 108, 131–32, 153; in Series, 133
Fluff Grass, 109, 132; in Series, 133
Foresteria, 152, 169; in Series, 159
Fourwing Saltbush, 38, 128, 130; in Series, 41, 160
Fraxinus (Ash), 148; in Association, 150; in Series, 41, 159
Fremont Cottonwood, 152; in Association, 150; in Series, 160

Gambel Oak, 65, 125, 150; in habitat type, 62; in Series, 65, 66, 68, 133
Garrya, 66, 68, 90, 169; in Series, 133
Geum, 49, 53; Alpine Avens Series, 39, 73
grama grasses. *See Bouteloua*
Gray Oak, 91; in Series, 68–69, 133
greasewood, 130, 152, 154; in Series, 41, 160
groundsels, 49, 61; in habitat types, 54, 55
Gutierrezia, 29, 106, 108, 128, 129; in Series, 111, 133

hackberry (*Celtis*), 148; in Association, 150; in Series, 41, 159

Hilaria, 15, 104, 105; in Series, 40, 110
Holodiscus, 61, 62, 64, 65, 167
Huckleberry, 54, 55, 60, 61; in habitat types, 53, 57

Indian Ricegrass, 55, 59, 104, 105; in habitat type, 68; in Series, 40, 105, 110, 133
Iodineweed, 154, 167; in Series, 41, 160

Jacob's Ladder, in habitat type, 53
Juglans spp. (walnut), 150, 152, 169; in Series, 41, 159, 160
Juncus (rush), 49, 103, 148; in Series, 167
Juniperus spp., 16, 20, 29, 48, 55, 57, 65–69, 86–92, 94, 109, 127, 169; in Association, 109; in habitat type, 59; in Series, 40, 93, 111, 133

Kentucky Bluegrass, associations, 103, 104; in habitat type, 150; in Series, 110, 159
Kinnikinnick, 63, 69; in habitat types, 57, 59, 63, 67
Kobresia turf association, 48–49
Koeleria, 64, 65, 101, 103

Larrea (creosotebush), 17, 20, 108, 109, 131–32; in Series, 40, 111, 133
Lathyrus (peavines), 59; in habitat types, 54, 55
Lemna (duckweed), 166, 170; in Series, 167
lichens, 49, 55, 105, 174
Ligusticum (Osha), 54
Limber Pine, 64; in Series, 57, 64
Linnaea (Twinflower), 59; in habitat type, 55
Luzula (woodrush), 53, 77
Lycium (wolfberry), 153; in Series, 160

Maples. *See Acer*
mesquite, 20, 106, 108, 129, 132, 146, 152, 153, 195; in Association, 109; in Series, 41, 160
milfoil, 166, 170; in Series, 167
moss, 54, 174; in habitat types, 55, 57
Mountain Mahogany, 65, 66, 90, 169; in Associations, 104, 127; in Series, 132, 133
Muhlenbergia, 59, 62–69, 89, 103, 107, 109, 129, 167, 171; in Association, 104; in habitat types, 63, 64, 65, 67; in Series, 110, 133

Narrowleaf Cottonwood, 59; in Association, 150; in Series, 159
Netleaf Oak, 66; in habitat type, 68
needlegrass (Stipa), 15, 87, 101, 104
New Mexico Locust, 67; in habitat type, 63; in Series, 133
Ninebark, in habitat type, 62

Nolina (sacahuista, beargrass), 68, 69, 109, 123, 127, 128; in Series, 111, 133

oaks, 38, 62, 65–69, 85, 89–92, 94, 125, 127, 128; in Association, 150; in Series, 40, 94, 133
Oceanspray, in habitat types, 61, 65
One-seed Juniper, 87, 88, 91; in habitat types, 89; in Series, 93, 94
Oryzopsis (ricegrass), 55, 59, 104, 105; in habitat type, 68; in Series, 40, 110, 133
Osha (*Lingusticum*), 54

peavines, 59; in habitat types, 54, 55
Philadelphus, 128; in Series, 132
Picea: Blue Spruce, 58, 149; in Association, 148; in habitat type, 150; in Series, 39, 41, 58–59, 73, 159; Engelmann Spruce, 48, 51, 62; in Series, 39, 52–53, 53–54, 54–55, 56, 57, 73
Pine Dropseed, 104; in Series, 110
Pinus spp., 16, 20, 52, 64–69, 86–89, 128; in Series, 39, 40
pinyons, 65, 66, 87, 88; in habitat type, 89; in Series, 68, 68–69, 93, 133
Platanus (sycamore), 146, 148, 150; in Series, 41, 159
Poa (bluegrass, Muttongrass), 15, 53, 59, 62–69, 101, 103; in habitat type, 150; in Series, 40, 110, 159
Ponderosa Pine, 89, 145, 169; in Series, 61–62, 62–64, 64, 66, 67, 68, 68–69, 74; Ponderosa Pine Belt, 169
pondweed (*Potamogeton*), 165, 171; in Series, 167
Populus spp., 53–72 passim, 58, 59, 62, 70–72, 124, 145, 149; in Association, 150; in Series, 40, 41, 74, 159
Potentilla (cinquefoil), 48, 49, 64, 103, 104
Prosopis spp., 20, 106, 108, 109, 129, 132, 146, 152, 153, 195; in Series, 41, 111, 133, 160
Prunus (chokecherry), 127, 150
Pseudotsuga (Douglas-fir), 16, 58, 145, 167, 169; in Series, 39, 40, 56, 60, 61, 64, 65, 65–66, 74, 167; Douglas-fir Belt, 169
Pteridium (brachen), 103

Quercus spp. (oaks), 38, 62, 65–69, 85, 89–92, 94, 125, 127, 128; as dominant, 150; in Series, 40, 66, 94, 133

Rabbitbrush Series, 41, 160. *See also Chrysothamnus*
Red Ozier Dogwood, 59, 148; in Series, 159
Rhus spp., 65, 66, 68, 89, 90, 108, 127, 152, 169; in Series, 132, 133, 159
Ribes spp., 48, 53, 54, 64, 65, 127; in habitat type, 52, 87
Robinia (New Mexico Locust), 62, 66, 67, 125, 145; in habitat type, 63; in Series, 133

Rocky Mountain Maple, 52, 57, 62; in habitat types, 55, 57, 61
Rosette Scrub Series, 127
rush, 103, 148, 167, 171; bulrush, 165, 167, 171; Rush Community, 49; in Series, 167
Russian Olive (*Eleagnus*), 20, 151, 152; in Series, 41, 160

sacaton, 105, 154; in Series, 40, 110
sagebrush and sage (*Artemisia* spp.), 15, 19, 29, 38, 49, 61, 62, 69, 70, 87–91, 106, 109, 128–30; in habitat type, 68; in Series, 40, 111, 133
Sagittaria (arrowhead), 165, 167, 172; in Series, 167
Salix spp., 48, 49, 57, 60, 61, 75, 77, 148, 150; in habitat type, 59; in Series, 39, 40, 41, 73
Salsola (Russian Thistle), 20, 129
saltbush (*Atriplex* spp.), 38, 91, 109, 128, 130, 153, 154, 195; in Series, 40, 41, 133, 160
saltcedar (*Tamarix*), 20, 151, 152, 154; in Series, 41, 160
Sapindus (soapberry), 128, 150, 152; in Series, 41, 160
Sarcobatus (greasewood), 130, 152, 154; in Series, 41, 133, 160
Schizachyrium (bluestem), 68, 69, 105; in Series, 110, 133
Scirpus (bulrush), 165, 167, 171; in Series, 167
Screwleaf Muhly, in habitat types, 63, 65, 67
sedges, 49, 55, 61, 62, 66, 67, 69, 75, 77, 103, 112, 147, 148, 171; in Community, 49; in habitat types, 59, 150; in Series, 39, 40, 73, 110, 159, 167
Senecio spp., 49, 53, 61, 66, 198; in habitat types, 54, 55
Sideoats Grama, 105, 109; in Series, 110
Shinnery Series, 109; Shinoak Series, 40, 133
Silverleaf Oak, 66, 67; in Series, 65–66
Skunkbush (*Rhus*), 65, 66, 68, 89, 90, 108, 127, 152, 169
Snowberry, 61–68, 127; in habitat type, 63
Soapberry, 128, 150, 152; in Series, 41, 160
Southwestern White Pine Series, 61
Sporobolus (dropseed), 15, 67, 101, 105, 130; in Series, 40, 110, 111, 133, 160
spruce. *See Picea*
Stipa (needlegrass), 15, 87, 101, 104; in Series, 133
Subalpine Fir, 48
submersed species in Series, 167
sycamore, 146, 148; in Association, 150; in Series, 41, 159
Symphoricarpus (snowberry), 61–68, 127; in habitat type, 63

Tamarix, 20, 151, 152, 154; in Series, 41, 160

tarbush, 108, 131–32, 153
Thimbleberry (*Rubus*), 59; in habitat type, 55
Thurber Fescue, 52, 102, 104; meadows, 52, 71, 102
Tobosa, 105; Series, 40, 110; Tobosa Swale, 105
Tufted Hairgrass Community, 49
Twinflower (*Linnaea*), 55, 59; in habitat type, 55
Typha, 165, 166, 171; in Series, 167

Utah Juniper, 91, 94

Vaccinium, 48, 54, 55, 60; in habitat types, 53, 57
Velvet Ash. *See* Ash
Veronica, in Series, 167

Walnut, 150, 152; in Series, 41, 159, 160

Water Fern Series, 167
Wavyleaf Oak, 66, 90, 91; in habitat types, 65, 66, 68;
    in Series, 133
wheatgrass, 15, 64, 101, 104; in Series, 110, 111
White Fir, in habitat types, 61, 62, 63; in Series, 39, 54–
    55, 60, 62, 62–63, 73, 74
White Oak, 90, 91; in habitat type, 69; in Series, 94
willows, 48, 49, 57, 60, 61, 75, 77, 145, 148, 150, 152;
    in habitat type, 59; in Series, 39, 40, 41, 73, 159, 160

Yerba de Pasmo (*Baccharis*), 109, 152; in Series, 159
Yerba Mansa (*Anemopsis*), 168, 171
Yucca spp., 68, 69, 105, 106, 108, 109, 123, 127, 128,
    129, 131, 153; in Series, 111, 133. *See also Agave*